DEVELOPMENTS IN APPLIED SPECTROSCOPY

Volume 7B

DEVELOPMENTS IN
APPLIED SPECTROSCOPY

Selected papers from the Annual Mid-America Spectroscopy Symposia

A Publication of the Chicago Section of the Society for Applied Spectroscopy

DEVELOPMENTS IN APPLIED SPECTROSCOPY

Volume 7B

edited by

E. L. Grove
Illinois Institute of Technology
Research Laboratories
Chicago, Illinois

Alfred J. Perkins
University of Illinois
College of Pharmacy
Chicago, Illinois

Selected papers from the Seventh National Meeting of the Society for Applied Spectroscopy (Nineteenth Annual Mid-America Spectroscopy Symposium) Held in Chicago, Illinois, May 13-17, 1968

PLENUM PRESS · NEW YORK–LONDON · 1970

Library of Congress Catalog Card No. 61-17720
SBN 306-38372-1

Plenum Press, New York
A Division of Plenum Publishing Corporation
227 West 17 Street, New York, N.Y. 10011

United Kingdom edition published by Plenum Press, London
A Division of Plenum Publishing Company, Ltd.
Donington House, 30 Norfolk Street, London W.C. 2, England

Printed in the United States of America

Preface

Volume 7 of *Developments in Applied Spectroscopy* is a collection of forty-two papers selected from those that were presented at the 7th National Meeting of the Society of Applied Spectroscopy, held (in place of the 19th Mid-America Symposium on Spectroscopy) in Chicago, May 13-17, 1968. These papers, selected by the editors and reviewed by persons knowledgeable in the field, are those of the symposium type and not those pertaining to specific research topics that one would expect to be submitted to a journal. It is the opinion of the committee that this type of publication has an important place in the literature.

The relatively large number of papers would result in quite a sizable volume if bound in one set of covers. For this reason, and to present the material in areas of more specific interest, Volume 7 was divided into two parts, Part A, Physical-Inorganic, and Part B, Physical-Organic Developments.

The 7th National Meeting was sponsored by the Chicago Section as host in cooperation with the St. Louis, New England, Penn York, Niagara-Frontier, Cincinnati, Ohio Valley, New York, Baltimore–Washington, North Texas, Rocky Mountain, and Southeastern Sections of the Society for Applied Spectroscopy and the Chicago Gas Chromatography Group.

The editors wish to express their appreciation to the authors and to those who helped with the reviewing. The latter include Dr. Elma Lanterman, Mr. John E. Forrette, Dr. Carl Moore, Dr. B. Jaselskis, Mr. H. G. Zelinski, Mr. D. J. Rokop, Mr. N. R. Stalica, Dr. Charles Reagan, Dr. Morris A. Wahlgren, and Dr. David Edgington.

Thanks should also be extended to the exhibitors for their part in the Symposium and the exhibitor seminars.

E. L. Grove
A. J. Perkins

Contents of Volume 7B

INFRARED AND RAMAN SPECTROSCOPY

INTERNAL-REFLECTION SPECTROSCOPY

NUCLEAR MAGNETIC RESONANCE SPECTROSCOPY

SPECTROCHEMICAL APPLICATIONS TO TEXTILES AND FIBERS

Contents of Volume 7A

X-RAY SPECTROSCOPY

EMISSION, FLAME, AND ATOMIC ABSORPTION SPECTROSCOPY

MASS SPECTROMETRY

NUCLEAR PARTICLE SPECTROSCOPY

INSTRUMENTAL PARAMETERS

Infrared and Raman Spectroscopy

Spectral Properties of Carbonate in Carbonate-Containing Apatites

Racquel Z. LeGeros, John P. LeGeros,
Otto R. Trautz, and Edward Klein

New York University
New York, N. Y.

A study of the infrared absorption spectra of carbonate – containing synthetic and biological apatites is reviewed. The implications of the results for the nature of the incorporation of the CO_3^{2-} ion into biological apatites is discussed.

INTRODUCTION

The inorganic phase of calcified tissues (i.e., teeth and bone) has been identified by x-ray diffraction as belonging to the apatite family, exemplified by hydroxyapatite, $Ca_{10}(PO_4)_6(OH)_2$. The apatite structure is a hospitable one, allowing varied substitutions to take place without a significant change in its symmetry. Substitutions in the apatite structure, however, are reflected by the lattice parameters of the apatite, i.e., F^- substituting for the OH^- ions causes a unit-cell contraction, while Sr^{+2} substituting for Ca^{2+} ions causes a unit-cell expansion.

The carbonate ion, CO_3^{2-}, is the chief impurity in the biological apatites, and the nature of its incorporation into the apatite structure has been the subject of long and intensive research in our and in many other laboratories. The proposed nature of carbonate incorporation in biological apatites is as follows: (1) it is present as an admixed phase of either calcite, $CaCO_3$, or magnesite, $MgCO_3$[1,2]; (2) it partially substitutes for the OH;

groups[3]; and (3) it substitutes for PO_4^{3-} groups.[5,8] The first two proposals were made on the basis of infrared absorption spectra of biological apatites, while the third one was based on the observations of *a*-axis contraction in the lattice of carbonate-containing mineral apatites (dahllite and francolite) as compared with the carbonate-free apatites (OH- apatite and F-apatite, respectively). Because carbonate-containing apatite crystals, whether mineral or synthetic, are too small for direct x-ray diffraction structural analysis, other methods, such as powder x-ray diffractometry and infrared absorption spectroscopy, have been used in an attempt to resolve this problem.

In our x-ray diffraction studies on synthetic apatites we have demonstrated that, without detecting a separate phase, a significant correlation exists between the carbonate content and the shortening of the *a* axis.[7-9] Further experimental evidence for the CO_3^{2-} for PO_4^{3-} substitution obtained in our infrared absorption studies are presented in this report.

EXPERIMENTAL

All infrared absorption spectra were made on Perkin-Elmer IR Grating Instrument 337. The samples were mixed with KBr (in concentrations of about 0.5–1.2 mg sample/300–320 mg KBr) and made into pellets. The range covered was from 4000 to 400 cm^{-1}. Among the biological apatites used were human enamel, dentine, and bone. Besides our synthetic apatites precipitated at 100°C,[8] we also examined some high-temperature apatites which had been prepared by Elliott (University of London) by passing dry CO_2 over OH-apatite at 1000°C.[3]

RESULTS AND DISCUSSION

On the Adsorption Theory

The planar CO_3^{2-} ion has four fundamental modes of vibration in its free state: the IR inactive stretching mode ν_1; the in-and-out-of-plane bending mode ν_2, the doubly-degenerate stretching mode ν_3; and the doubly-degenerate bending mode ν_4. However, the vibrations of the CO_3^{2-} ions in a crystalline field are considerably affected by its environment, and these effects are manifested in the IR spectra of several carbonate compounds (Fig. 1). We observe in these spectra that the normally IR-inactive vibration-mode ν_1 is absent in calcite, but present in aragonite and vaterite. The loss of degeneracy of the ν_4 mode is also observed. The ν_3 mode is seen to have lost its degeneracy

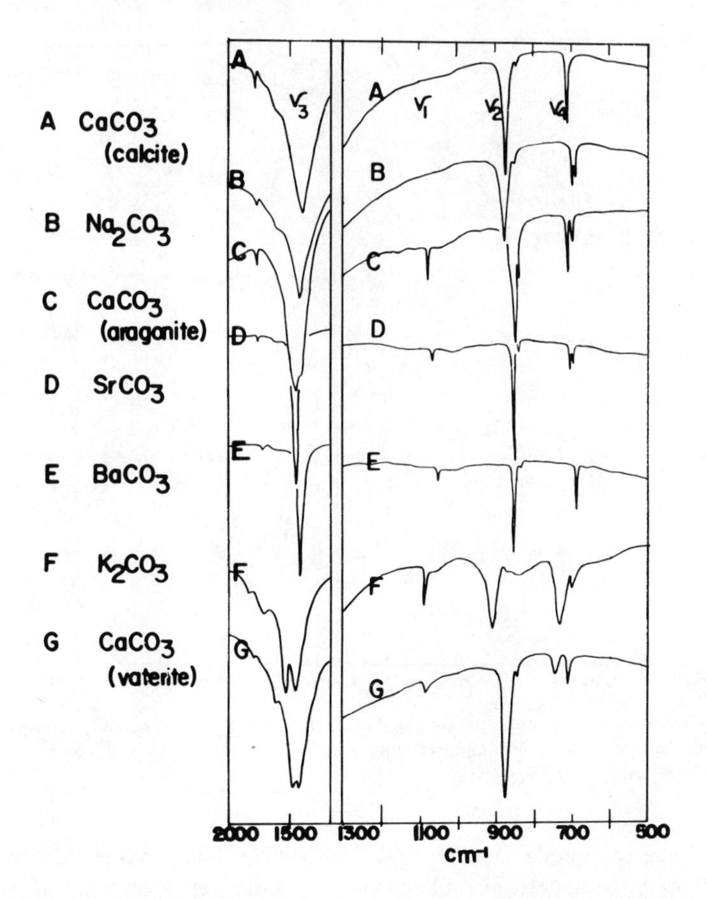

Fig. 1. Infrared absorption spectra of some carbonate compounds: calcite-type (*A*,*B*); aragonite-type (*C, D, E*); and vaterite-type (*F, G*). Calcite (*A*), aragonite (*C*), and vaterite (*G*) are three forms of calcium carbonate, $CaCO_3$.

in the spectra of K_2CO_3 and of vaterite, but not in the other carbonate compounds. Furthermore, depending on the weight of the cations, the bands are shifted to higher or lower frequencies. Although calcite, aragonite, and vaterite have the same chemical composition, i.e., $CaCO_3$, the spectra are different, reflecting their structural differences and therefore the differences in the environment of the $CO_3{}^{2-}$ groups.

The carbonate bands in carbonate-containing apatites as compared to the carbonate bands in calcite and aragonite are shown in Fig. 2. The absence of the ν_4 vibration bands in the carbonate-containing apatite is a characteristic

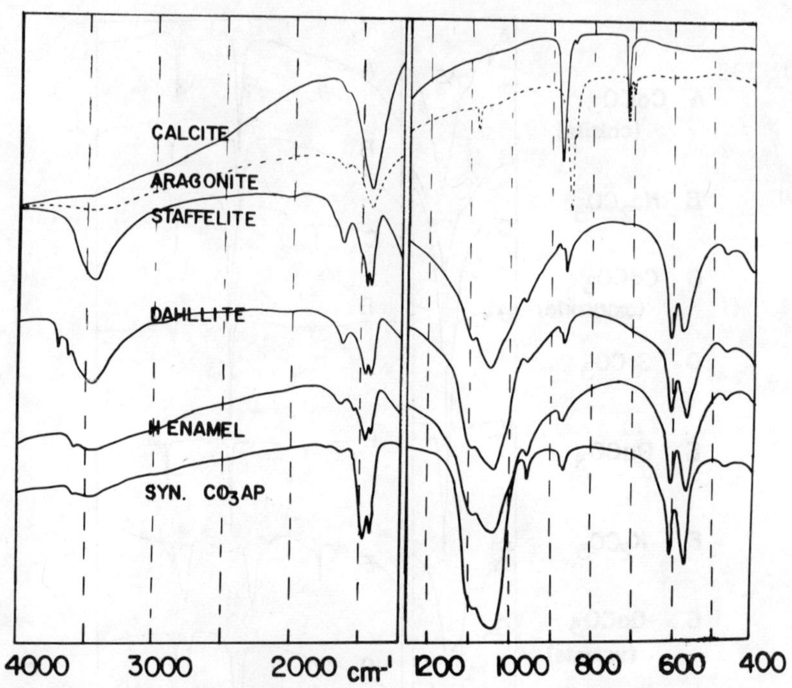

Fig. 2. Infrared absorption spectra of two forms of $CaCO_3$ (calcite and aragonite) compared to carbonate-containing mineral (staffelite and dahllite), biological (human enamel), and synthetic (precipitated) apatites.

feature. The ν_3 mode of the $CO_3{}^{2-}$ is split in the apatite spectra. The presence of ν_1 (symmetrical stretching mode) could not be ascertained, since it appears in the region of a strong $PO_4{}^{3-}$ band. However, because of the loss of degeneracy of ν_3, it can be assumed that ν_1 is also active. A doublet in the region of the nondegenerate ν_2 is observed in the spectra of biological apatites (Figs. 2 and 3), in dahllite, and in precipitated apatite. This anomalous feature will be discussed in the next section.

Since as little as 0.1% admixed carbonate can be detected in the spectrum of apatite, the absence of the $CO_3{}^{2-}$ ν_4 bands in all of the carbonate-containing apatites obviates the possibility of the carbonate being present as an admixed phase.

$CO_3{}^{2-}$ for OH^- or $CO_3{}^{2-}$ for $PO_4{}^{3-}$ Substitution in the Apatite Structure

Two types of synthetic carbonate-containing apatites are studied. Materials prepared at 900–1000°C have shown an expansion in the *a* axis

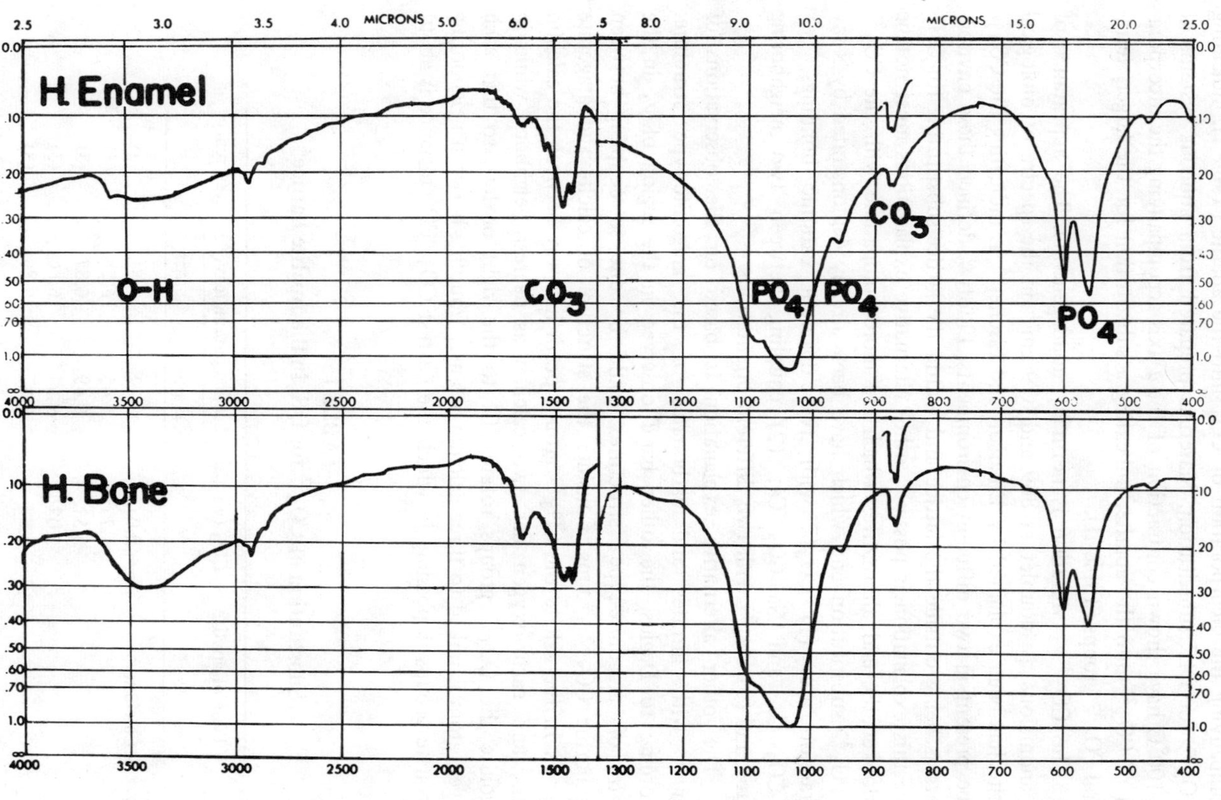

Fig. 3. Infrared absorption spectra of biological apatites (human enamel and human bone). The "anomalous" CO_3 ν_2 doublet is observed at 869 and 879 cm^{-1}.

accompanying the incorporation of CO_3 into the apatite (Table I), indicating a CO_3^{2-} for OH^- substitution. Materials prepared from aqueous systems at 95–100°C have shown contraction of the *a* axis accompanying the incorporation of CO_3^{2-} into the apatite, indicating a substitution for the larger, tetrahedral PO_4^{3-} groups (Table II).

The $CO_3^{2-}\nu_2$ vibration is normally nondegenerate. The appearance of the anomalous ν_2 doublet at 869 and 879 cm^{-1} in the spectra of biological apatite has been explained by Emerson and Fischer [4] as indicating the CO_3^{2-} to be present in two different environments. Elliott [3] defined these environments as being outside the lattice and within the lattice substituting for OH^-. While this explanation is possible, other alternative explanations must not be overlooked. Several alternatives which are more compatible with the CO_3^{2-} for PO_4^{3-} substitution and which have been already demonstrated by x-ray diffraction are: (1) CO_3^{2-} with two different cationic bondings, i.e., Ca–CO_3–Ca and Na–CO_3–Ca; (2) coupling between two neighboring groups; and (3) CO_3^{2-} with two different orientations.

The other alternative explanation is based on the observations of Decius[2] with nitrates and carbonates of the aragonite-type structure. According to Decius, the other band observed in the region of ν_2 in the spectra of aragonite-type structures could be due to coupling between neighboring CO_3^{2-} groups. Consider the structure of calcite and aragonite (Fig. 4). In the calcite the CO_3^{2-} groups occur half-way between the calcium layers, and each oxygen has two calcium as nearest neighbors, while in aragonite the CO_3^{2-} groups rotate 30° to the right or left, so that each oxygen atom is linked to three calcium atoms.[1] Coupling between neighboring carbonate groups is possible in which the nearest CO_3–CO_3 distance is about

TABLE I
Substitution of CO_3^{-2} for OH^- In the Apatite Lattice*

Type of apatite	CO_3 (wt. %)	Lattice parameters (Å)		References
		a axis	*c* axis	
Calcium	0.0	9.42	6.88	(3)
	1.77	9.45	6.88	(3)
	3.51	9.49	6.87	(3)
	5.03	9.54	6.86	(3)
	5.83	9.57	6.86	(3)
Strontium	0.0	9.76	7.28	(11)
	5.13	9.88	7.24	(11)

*Apatites prepared with dry CO_2 at 900 – 1000°C.

TABLE II
Substitutions of CO_3^{2-} for PO_4^{3-} In the Apatite Lattice*

Type of apatite	CO_3 (wt. %)	Lattice parameters (Å)		References
		a axis	*c* axis	
Calcium	0.6	9.43_6	6.87_9	(7–9)
	5.5	9.40_0	6.89_4	(7–9)
	10.5	9.37_7	6.90_8	(7–9)
	20.5	9.31_1	6.92_1	(7–9)
	22.2	9.30_4	6.92_3	(7–9)
Strontium	0.5	9.76_8	7.07_4	(7)
	5.7	9.73_9	7.26_1	(7)
Calcium–barium	0.4	9.48_9	6.89_5	(7)
(Ca_9Ba_1)	10.5	9.42_7	6.91_3	(7)
	16.1	9.38	6.92	(7)

* Apatites prepared from aqueous systems, 95–100°C.

2.86 Å (as in aragonite) to about 3.5 Å (as in KNO_3), but was not observed in calcite, in which the nearest carbonate groups are 4.98 Å apart. [2] In the apatite structure if CO_3^{2-} locates itself in place of a PO_4^{3-} group, the neighboring CO_3 groups would be about 3.4 Å apart; this distance might encourage coupling between the groups, thus causing an apparent doublet.

The third alternate explanation is a difference in orientation of the planar CO_3^{2-} groups in the structure. For example, in vaterite the CO_3 planes are standing vertical, parallel to the *c* axis of the crystal,[6] while in aragonite and calcite the CO_3 planes are horizontal, perpendicular to the *c* axis.[1] Because this difference in orientation is accompanied by a difference in the environment of the CO_3^{2-} groups, a splitting of the ν_3 in the spectra of vaterite is observed (Fig. 2).

McConnell[10] proposed that CO_3^{2-} groups in the apatite might sit in two different orientations – tilted perpendicular to the *c* axis and others parallel to the *c* axis. The slow acceptance of this proposal is due to the negative birefringence observed with carbonate-containing mineral apatites. The negative birefringence observed is incompatible with the vertical positioning of the CO_3 planes[13]. However, the observed negative birefringence does not exclude the possibility that while the carbonate groups are positioned predominantly horizontally, some groups may also be vertical, with the net birefringence still negative. In such a case the difference in orientation of the carbonate groups might cause a difference in their environments, thus causing the anomalous double of ν_2 and the loss of the degeneracy of ν_3.

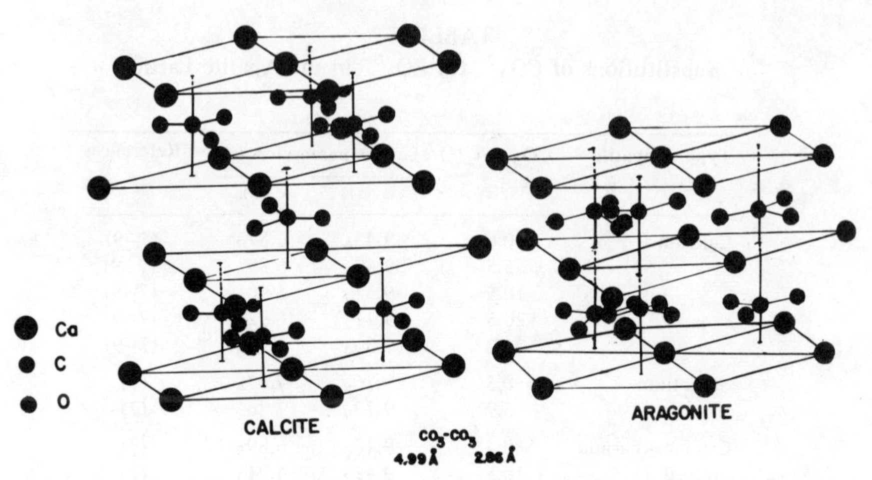

Fig. 4. Two forms of calcium carbonate, $CaCO_3$: calcite and aragonite. The closest CO_3 neighbors in aragonite are 2.86 Å and in calcite 4.99 Å apart. From Bragg[1].

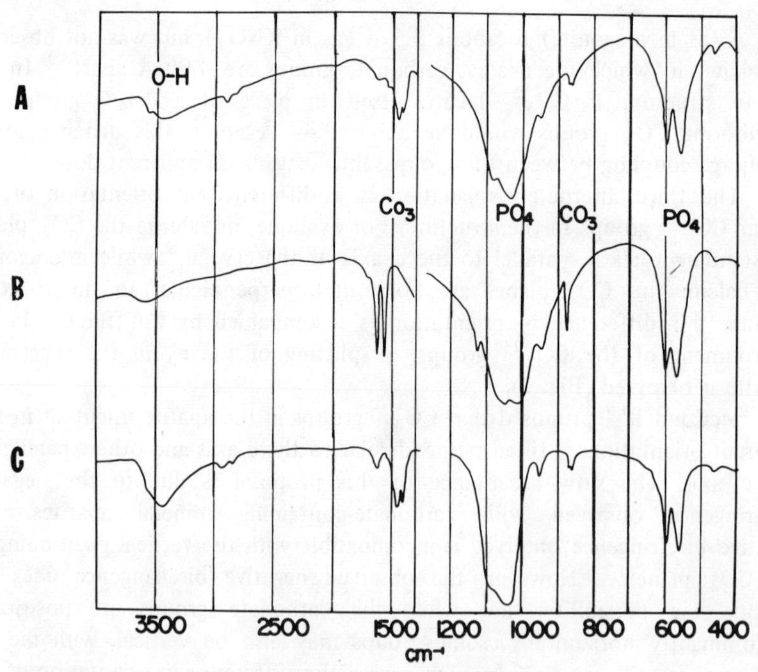

Fig. 5. Infrared absorption spectra of (*A*) enamel apatite (*A*) and two types of carbonate-containing synthetic apatites: (*B*) prepared at 1000°C by passing dry CO_2,[3] and (*C*) prepared at 100°C from aqueous systems.[7–9]

Figure 5 shows the spectra of human enamel apatite compared with those of two types of carbonate-containing synthetic apatites: one prepared at high temperature (which favors CO_3^{2-} for OH^- substitution) and one precipitated at 100°C (which favors CO_3^{2-} for PO_4^{3-} substitution). The biological apatites more closely resemble the spectra of the precipitated apatites, suggesting that the carbonate groups in both apatites are in similar environment.

SUMMARY

In view of the experimental evidence obtained both from x-ray diffraction and infrared absorption studies, we conclude that the carbonate in precipitated apatites (synthetic and biological) is: (1) *not* admixed as a separate phase of calcite or magnesite; (2) *not* substituting for OH^- groups; and (3) substituting for PO_4^{3-} groups.

ACKNOWLEDGMENT

This work was supported by USPHS Research Grant No. DE-00159. We gratefully acknowledge the technical assistance of K. Rajkowski and M. Klein in the earlier phase of the work and the valuable technical assistance of C. Strom and G. Domingo.

REFERENCES

1. W. L. Bragg, *Atomic Structure of Minerals*, Cornell University Press, New York (1937).
2. J. C. Decius, Coupling of the Out-of-Plane Bending Mode in Nitrates and Carbonates of the Aragonite Structure, *J. Chem. Phys.* **23**, 1290–1294 (1955).
3. J. C. Elliott, The Crystallographic Structure of Dental Enamel and Related Apatites, *Thesis*, University of London (1964).
4. W. H. Emerson and E. E. Fischer, The Infrared Spectra of Carbonate in Calcified Tissues, *Arch. Oral Biol.* **7**, 671–683 (1962).
5. J. W. Gruner and D. McConnell, The Problem of Carbonate Apatites, *Z. Krist.* **97**, 208–215 (1937).
6. S. R. Kamhi, On the Structure of Vaterite, $CaCO_3$, *Acta Cryst.* **16**, 770–772 (1963).
7. R. Z. LeGeros, Crystallographic Studies on the Carbonate Substitution in the Apatite Structure, *Thesis*, New York University (1967).
8. R. Z. LeGeros, Effect of Carbonate on the Lattice Parameters of Apatite, *Nature* **206**, 403–405 (1965).
9. R. Z. LeGeros, J. P. LeGeros, O. R. Trautz, E. Klein, and W. P. Shirra, Apatite Crystallites: Effect of Carbonate on Morphology, *Science* **155**, 1409–1411 (1967).

10. D. McConnell, The Crystal Chemistry of Carbonate Apatites and Their Relationship to the Composition of Calcified Tissues, *J. Dent. Res.* **31**, 53–63 (1952).
11. S. Mohseni-Koutchesfehani, Contribution a l'Etude des Apatites Barytiques, *Thesis*, University of Paris (1961).
12. A. S. Posner and G. Duyckaerts, Infrared Study of the Carbonate in Bone, Teeth and Francolite, *Experientia* **10**, 424–425 (1954).
13. O. R. Trautz, Crystallographic Studies of Calcium Carbonate Phosphates, *Ann. N. Y. Acad. Sci.* **85**, 145–160 (1960).

Polarized Infrared Reflectance of Single Crystals of Apatites

Edward Klein, John P. LeGeros,
Otto R. Trautz, and Racquel Z. LeGeros

New York University
New York, N. Y.

Experiments are described using the polarized specular-reflectance technique to determine the dipole directions and relative intensities of the PO_4 vibrations in F-apatite and hydroxyapatite single crystals. Use of this information to make assignments of the frequency bands is discussed.

INTRODUCTION

The apatites, typified here by hydroxyapatite, $Ca_{10}(PO_4)_6(OH)_2$, are basic calcium phosphates in which the $PO_4{}^{3-}$ groups serve as the main building blocks and are held together by the Ca^{2+} ions. The structure of the apatites is of interest to our research on the mineralization of tissues, since the apatites constitute the inorganic phases of the dental and skeletal hard tissues and their chemical reactivity and dissolution rates are affected by their structure. These biological apatites are not pure hydroxyapatite, but contain other constituents which affect their structure and chemical behavior. The phosphate groups in the apatite structure, as established by x-ray diffraction, are not perfectly tetrahedral in form. The deformation or distortion of the phosphate tetrahedra is also evident from infrared absorption spectra. It is this sensitivity of infrared spectroscopy to slight changes in the environment of the molecular groups, like the PO_4 groups, which makes infrared a useful tool in our studies of the apatite structure.

The usual infrared absorption spectroscopic methods use a powdered sample of the test material embedded with random orientation in a KBr disk.

Additional and more specific information about the PO_4 groups could be obtained by placing an oriented section of a single crystal of apatite in a beam of polarized infrared. However, because of the very strong absorption bands in the infrared region, the sections must be very thin (20 μ or less) to give sufficient details in the absorption spectra. The preparation of such thin sections by grinding is extremely difficult because of the apatite's brittleness, and the preparation by time-extrapolating etching methods is made unreliable by the irregularity of the surface (pits and grooves) due to the presence of a preferential etching direction in the apatite, especially in carbonate-containing apatites.

The polarized specular-reflectance technique allows the use of larger single crystals, the orientation of which can easily be varied in the course of an experimental series. It is the purpose of this report to show how we have obtained the dipole directions and the relative intensities of the PO_4 vibrations in the apatite, and from this information have made a reasonable assignment of the frequency bands.

EXPERIMENTAL

The single crystals of apatite used for these studies were F-apatite, $Ca_{10}(PO_4)_6F_2$, from Durango, Mexico, and OH-apatite, $Ca_{10}(PO_4)_6(OH)_2$, from Holly Springs, Georgia. Analyses of these materials have been previously reported.[5]

A specular-reflectance attachment was used with the Perkin-Elmer 337 Infrared Grating Spectrophotometer. Some materials were also analyzed using the Perkin-Elmer IR 621 instrument.* The IR 621 contained a wire-grid polarizer and gave complete polarization in the entire range between 4000 and 200 cm^{-1}, while the P-E 337 gave fair polarization only in the region from 700 to 400 cm^{-1}.

A schematic representation of the specular-reflectance technique is shown in Fig. 1. The IR sample beam from the glow-bar source is reflected by the first surface-silvered mirror to the surface of the crystal and from there to the second mirror, then to the slit, to the monochromator grating, and to the detector. The reference beam passes from the source through a comb attenuator directly to the monochromator and detector. In the IR 621 the wire-grid polarizer is placed in the common path before the grating, while in the experiments with the IR 337 the grating itself provided a certain polarizing effect, especially in the lower frequency range. Though the IR beams are polarized

*We appreciate the courtesy of Dr. R. Gore, Perkin-Elmer, Norwalk, Conn., for letting us use a P-E 621 instrument for these experiments.

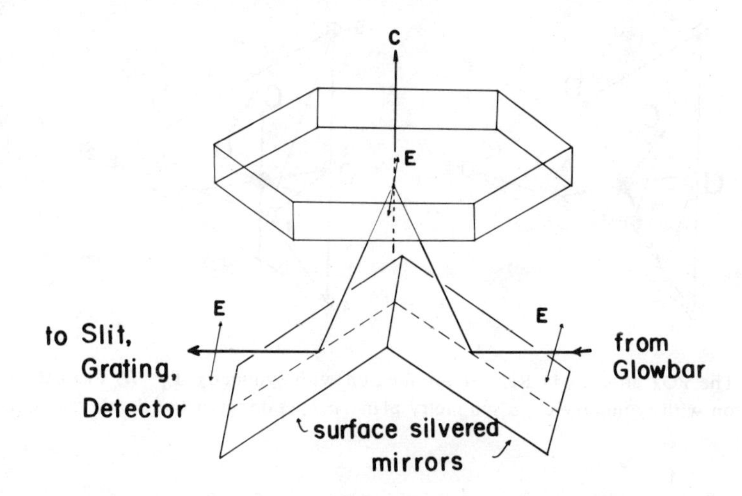

Fig. 1. A schematic representation of the specular-reflectance technique. EV indicates the electric vector of the light beam.

only at the detection end of their path, we can with certain adjustments consider the incident beam as being polarized with its electric vector E lying in the reflection plane as illustrated in Fig. 1. Any natural face, e.g., prism (1010), dipyramid face (1011), basal plane (0001), or any other plane, may be ground flat and polished and mounted in the instrument so that it coincides with the reflecting plane of the instrument. Of importance then is the direction of the crystal axes with respect to E as the orientation of the crystal (without displacing its reflecting surface) may still be varied by rotating it about the normal to the reflecting surface under examination. The spectra obtained with the crystal placed in such positions that its c axis is oriented (1) parallel, (2) about 45°, or (3) 90° to E of the instrument differed considerably (Fig. 6).

RESULTS AND DISCUSSION

The free phosphate group, PO_4, with site symmetry T_d, is a perfect tetrahedron and has the following vibrational modes: (1) the symmetrical P–O stretch ν_1 is normally IR-inactive; (2) the doubly-degenerate symmetrical P–O bend ν_2; (3) the triply-degenerate antisymmetrical stretch ν_3; and (4) the triply-degenerate antisymmetrical P–O bend ν_4. In a perfect tetrahedron all the P–O bonds would be equivalent ($P–O_1$ = $P–O_2$ = $P–O_3$ = $P–O_4$) and the three fundamental vibration frequencies of ν_3 and of ν_4 would be equal,

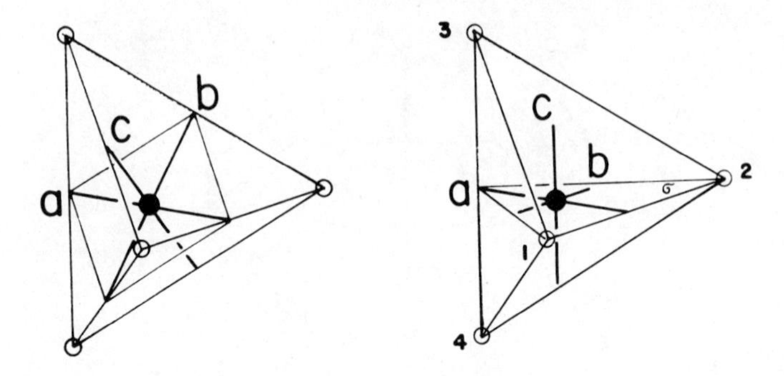

Fig. 2. The PO_4 group. (*A*) Regular tetrahedron with symmetry T_d. (*B*) distorted tetrahedron with symmetry C_s. σ symmetry plane; a, b, c: directions of dipole moments.

TABLE I

Frequency (cm^{-1}) Assignment for the PO_4 Vibrations in F-apatite from Reflectance and Absorption Spectra

	ν_3			ν_1	ν_4			ν_2	
	a	*b*	*c*		*a*	*b*	*c*	*a*	*c*
Reflectance	1100	1071	1058	959	604	572	570	336	305
Absorption	1090	1040		960	601	575		(*)	(*)

*Not observable in the range of P-E 337. Additional weak bands appear at 468 and 269 cm^{-1}.

while their three dipole moments (*a, b, c* in Fig. 2*A*) would also be equal and orthonormal to each other. In the PO_4 tetrahedron of OH-apatite the P–O distances are unequal: $P{-}O_1$ = 1.538 Å; $P{-}O_2$ = 1.547 Å; $P{-}O_3$ = $P{-}O_4$ = 1.529 Å;[3] and the symmetry is reduced from T_d to C_s; the symmetry plane σ passes through the horizontal edge, while the opposite vertical edge is parallel to the *c*-axis of the apatite (Fig. 2*B*). The directions of the dipole moments *a* and *b* lie in the symmetry plane, while *c* is perpendicular to it. The vector *a* represents the directions of the dipole moments for the following vibrations: ν_1, ν_{2a}, ν_{3a}, ν_{4a}; while *b* represents the dipole vectors of ν_{3b} and ν_{4b}; and *c* represents the dipole vectors of ν_{2c}, ν_{3c}, and ν_{4c} (Fig. 3). Thus six of the expected nine fundamental vibrations are considered to have their dipole vectors in the symmetry plane, while three vibrations have vectors antisymmetrical or perpendicular to the C_s plane. By checking the dipole directions of the nine vibrations in the reflectance spectra obtained in our experiments, we can now make definite assignments of the frequencies to the nine permitted fundamental vibration modes of the PO_4 groups in apatite, as tabulated in Table I.

The group symmetry of apatite is $P6_3/m$. The unit cell (Fig. 4) contains six PO_4 groups, three situated on one symmetry plane and three on the other symmetry plane half a unit cell below and rotated 60° about the *c*-axis.[1,4] A PO_4 group in the crystal responds maximally to one of the irradiating c frequencies when its *c*-dipole vector is parallel to *E* of the instrument. The

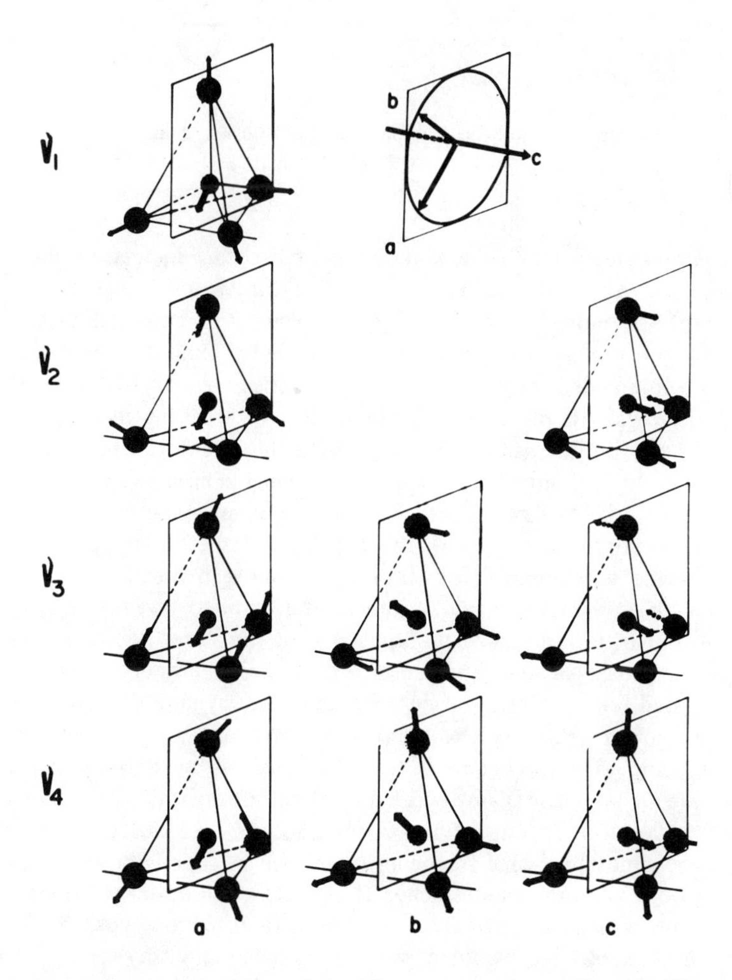

Fig. 3. Dipole moment directions of the PO_4 groups in the apatites. The PO_4 group has site symmetry C_s. From Herzberg[2].

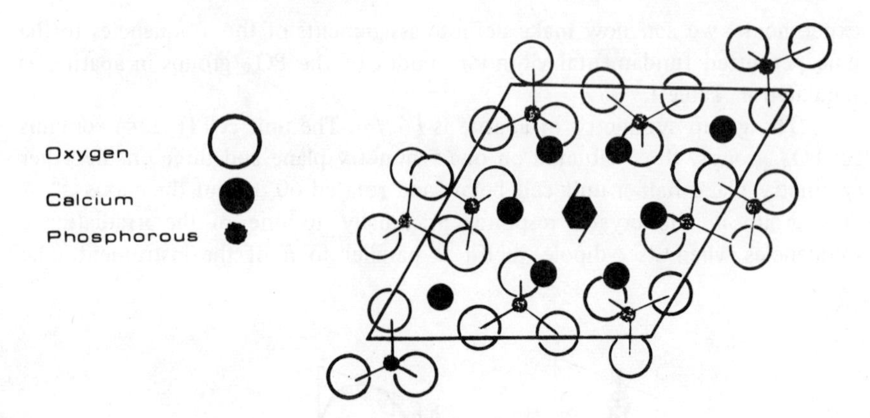

Fig. 4. F-apatite structure projected upon the basal plane. This shows the group symmetry of $P6_3/m$ for apatite.

response diminishes from the maximum with the cosine squared of the angle between the *c* vector and *E*. The response of the crystal is the sum of the responses of the individual groups (Fig. 5). Thus it is expected that the *c* components of the ν_2, ν_3 and ν_4 vibration modes exhibit their maximum intensity in the reflectance spectra from the prism plane (1010) when the *c*-axis of the crystal is placed parallel to *E* (Fig. 6*B*). Tilting the *c*-axis away from *E* reduces the intensities proportional to the cosine squared of the tilt angle until at 90° the intensities of the *c* components have become zero (Fig. 6*A*). While the *c* vectors of all the six PO_4 groups are parallel to each other (and to the *c*-axis of the crystal), the *a* vectors of the six groups lie in the symmetry plane with intervals of 60° rotation between them.

When the symmetry plane of the crystal is tilted away from parallel to *E* the intensities of the *a* components of the four vibration modes (and simultaneously of the *b* components) diminish proportionally to the cosine squared of the tilt angle between the symmetry plane and *E* from a maximum (Fig. 6*A*) to zero (Fig. 6*B*). At a tilt angle of the symmetry plane of 45° (Fig, 6*C*) the intensities of the *a* components (as well as of the *b* components) are expected to be half of their respective maxima.

The reflectance spectra of F-apatite (Figs. 6*A* and 6*B*) show that ν_{4b} and ν_{4c} have almost the same frequencies, and will be superimposed in spectra containing both vibration modes. Thus at a crystal position in which both the *c*-axis and the symmetry plane are at 45° to *E* (approximate position for the spectrum in Fig. 6*C*) we would expect a combined intensity of $\nu_{4b} + \nu_{4c}$ equal to half the sum of the maximal intensities of ν_{4b} and ν_{4c}. A rotation of the crystal about the normal to the reflecting surface (without displacing the surface) will increase the angle between the c-axis and E and consequently lower the intensity of ν_{4c} and increase the intensities of ν_{4a} and ν_{4b}. Thereby, the ν_{4b} band seems to separate from ν_{4c} and to appear as a small peak on the shoulder of ν_{4c}, or in the saddle between ν_{4c} and ν_{4a}, its frequency gradually shifting from 573 to 583 cm^{-1}. This is the first time that the ν_4 band has been split into a distinct triplet, as postulated by group theory for a tetrahedal C_s molecule.[2] The apparent frequency shift is explained by the convolution of the three frequency bands.

If we take the height of the reflectance bands as an approximate measure of their intensities, then we obtain from Figs. 6*A* and 6*B* an

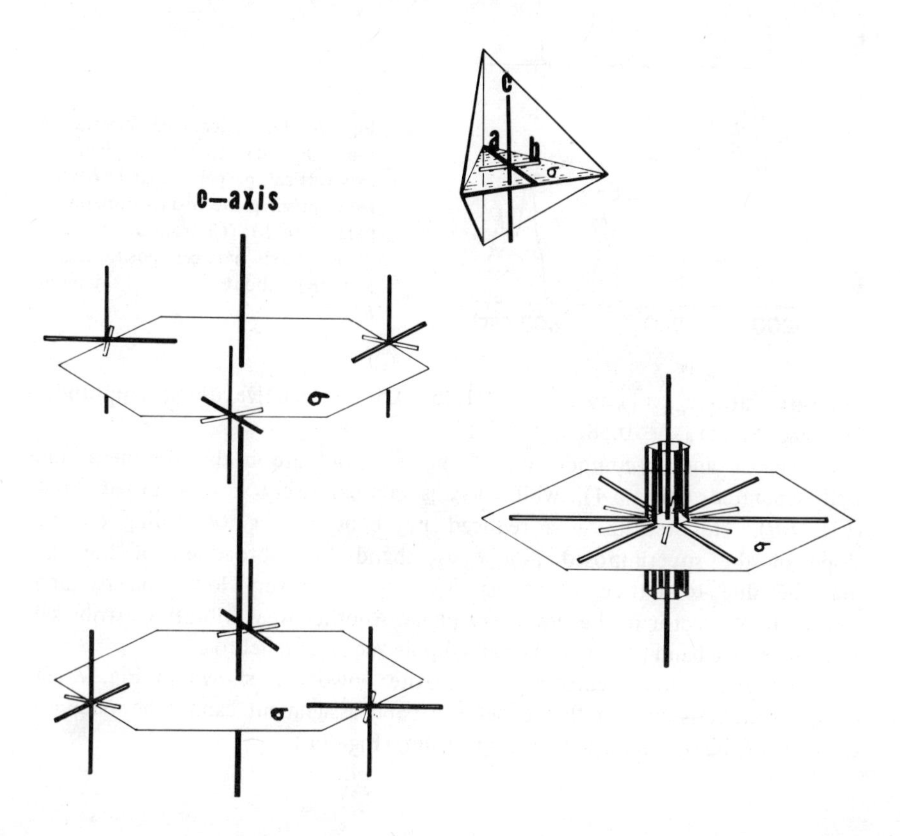

Fig. 5. The arrangement of the dipole vectors of the six PO_4 groups in the apatite Crystal with regard to the *c*-axis and the symmetry planes *σ*. In each group the *a* and *b* vectors are at right angles, both in the symmetry plane, while the *c* vector is perpendicular to the plane.

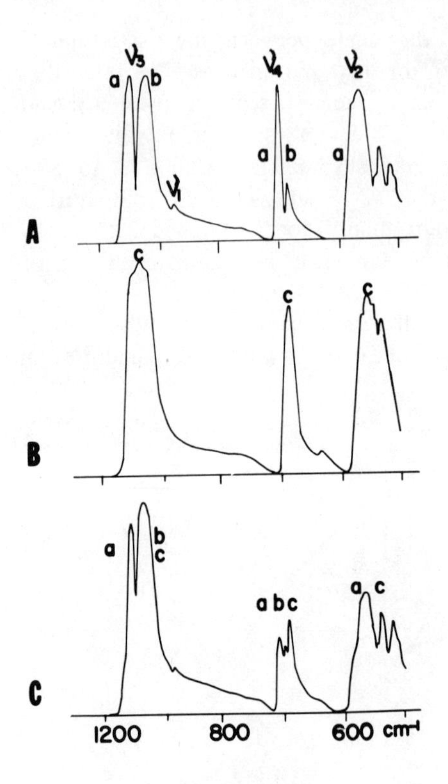

Fig. 6. IR reflectance spectra of F-apatite. (*A*) from basal plane (*c*-axis vertical, perpendicular to *E*); (*B*) from prism plane (*c*-axis horizontal, parallel to *E*); (*C*) from dipyramidal plane [*c*-axis between positions (*A*) and (*B*), about 45° to horizontal *E*].

intensity ratio $\nu_{4a}:\nu_{4b}:\nu_{4c}$ = 2.38:1.06:2.05; the relative dipole amplitudes are: ν_{4a} =1.3; ν_{4b} = 0.58; ν_{4c} = 1.12.

The *a* and *b* components of the ν_3 band are both prominent and well separated (Fig. 6*A*), while ν_{3c} is comparatively a very broad band (Fig. 6*B*). This results in a reduced ν_{3a} band in Fig. 6*C* riding on the slope of the superimposed ν_{3b} + ν_{3c} bands. The broadness of the ν_{4a} may be due to lattice vibrations. The ν_1 has a very low intensity and has its dipole vector in the symmetry plane. Another weak vibration attributed to a difference band (468 cm^{-1}) has its dipole vector parallel to *c*.

An absorption spectrum of F-apatite powder is shown in Fig. 7 for comparison. It is evident that a definite band assignment cannot be made on the basis of the powder absorption spectrum. (Fig. 7a.)

SUMMARY

The polarized, infrared, single-crystal, external-reflectance technique is a

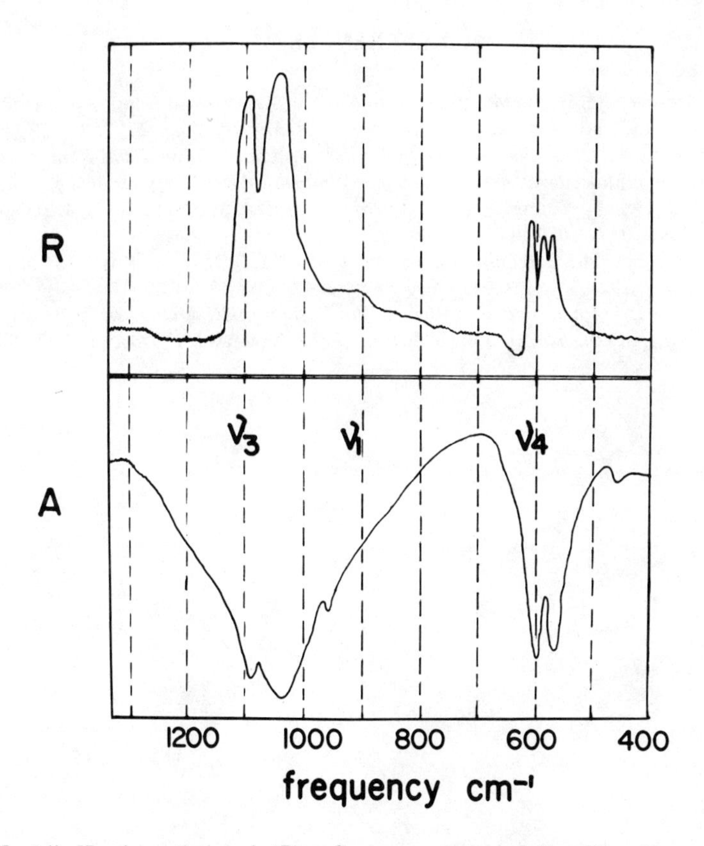

Fig. 7. (*A*) IR absorption and (*R*) reflectance spectra of F-apatite. The reflectance spectrum was from the basal plane.

more powerful method for the analysis of the vibration spectra of crystalline structures than the powder absorption techniques when structural information is desired. Definite band assignments can be made, and dipole directions for the fundamental vibrations of the molecular groups can often be obtained. This information is necessary for consideration of bond directions and strengths in the structure of the crystal and of the influence exerted upon them by substitutions in the structure.

ACKNOWLEDGEMENT

We acknowledge the support of the USPHS Grant No. DE-00159 and the assistance of G. Domingo in the photographic phase of this work.

REFERENCES

1. C. A. Beevers and D. B. McIntyre, The Atomic Structure of Fluor-Apatite and Its Relation to That of Tooth and Bone Materials, *Mineral. Mag.* **27**, 254–259 (1947).
2. G. Herzberg, *Molecular Spectra and Molecular Structure. Infrared and Raman Spectra of Polyatomic Molecules*, D. Van Nostrand, Princeton, New Jersey (1946), p. 100.
3. M. I. Kay, R. A. Young, and A. S. Posner, Crystal Structure of Hydroxyapatite, *Nature* **204**, 1051–1052 (1964).
4. S. Naray-Szabo, The Structure of Apatite (CaF) $Ca_4(PO_4)_3$, *Z. Krist.* **75**, 387–398 (1930).
5. C. Palache, H. Berman, and C. Frondel, *The System of Mineralogy of J. D. Dana and E. S. Dana*, 7th ed., Vol. II, John Wiley and Sons, New York (1951), pp. 877–887.

Analytical Infrared Spectra of Particulate Alpha-Aluminas

Conrad M. Phillippi

Air Force Materials Laboratory
Wright-Patterson Air Force Base, Ohio

A class study was performed of the infrared spectra of various powdered alpha-aluminas dispersed in KBr pellets, from which a typical infrared spectrum is established for qualitative analytical purposes. By reference to the polarized reflection spectra of single-crystal corundum and ruby it is demonstrated that reflection losses at the surfaces of the particles dominate the overall form of the powder transmission spectrum, and that in the case of this material the spectral features cannot be regarded as simple absorption bands. It also is shown that the fundamental vibrational frequencies of the corundum crystal lattice do not necessarily coincide with the minima in the transmission spectrum of its powder. A possible explanation for some of the minor spectral features is advanced.

INTRODUCTION

The term "analytical" appears in the title of this paper for two different reasons: It reflects the applied objective of this study, but also, and of greater fundamental significance, it is used to call attention to the difference between the intrinsic infrared spectrum of the alpha-alumina crystal lattice and a spectrum which is strongly dependent on the analytical technique used.

A solid inorganic material is conveniently analyzed by reducing it to a fine powder, dispersing it in a transparent medium such as potassium bromide, and measuring the infrared transmission spectrum of the resulting pellet. By extension from similar infrared techniques for organic analysis such a transmission spectrum is tacitly assumed to bear an inverse relationship to the absorption spectrum of the material, and the absorption maxima in turn are assumed to coincide with vibrational frequencies of the crystal lattice. Extensive application of this potassium bromide pellet technique to particulate materials has shown it to be extremely valuable from a practical standpoint, but it is subject to a wide variety of anomalies and inconsistencies which make data obtained by it open to question from a fundamental standpoint. Among the many processes known to influence the powder spectrum of a material, optical effects may alter the shape and position of the transmission maxima and minima found in a potassium bromide pellet. It is frequently the case that these optical effects will override the intrinsic spectrum of the material under analysis. It is possible to interpret the spectrum of a powdered material by referring to its corresponding single-crystal spectra and to theoretical predictions from factor group analysis. By this means some of the spectral perturbations can be isolated from the intrinsic vibrational spectrum of the crystal lattice. Alpha-alumina is the subject of this study (1) because of its widespread practical significance, (2) because it has already been the subject of many theoretical and single-crystal studies, and (3) because the wide selection of powdered aluminas available allows a class study to be made of these spectral variations. Alpha-alumina is also known as corundum and sapphire, and, with the addition of a trace of chromium oxide, ruby.

EXPERIMENTAL

Figure 1 compares the powder spectra of ten different samples of alpha-alumina. Each was verified to be the alpha form by x-ray diffraction. Number 1 also contains some beta form, and No. 4 contains a trace of theta form, but the spectral variations of these two do not exceed those of the alpha form. The small band at 695 cm^{-1} indicated by the asterisks comes from contamination by the plastic vial used in vibratory blending of the potassium bromide and sample powders.[1] These samples represent a wide variety of sources and preparation methods. All but two, Nos. 3 and 6, were incorporated into the pellet as-received without further grinding. Those two were ground dry in a boron carbide mortar and pestle until no further graininess was felt through the pestle. Aluminas falling in the reagent category are Nos. 1, 2, 5, 7, and 9. Those falling into the category of abrasives are Nos. 4, 8, and 10. Number 3 is the natural mineral, while No. 6 is the powder of a crushed synthetic crystal of the type used for infrared optical components. Particle-size distributions were measured on four of these powders, Nos. 1, 7, and 9 as-received and No. 6 as it was pulverized for incorporation into the pellet. The results of these counts may be presented in several ways. One way is to say that the arithmetic mean "diameters" of all four samples fall between 0.16 and 0.30μ. Another way is to weight these distributions by the effective area of all

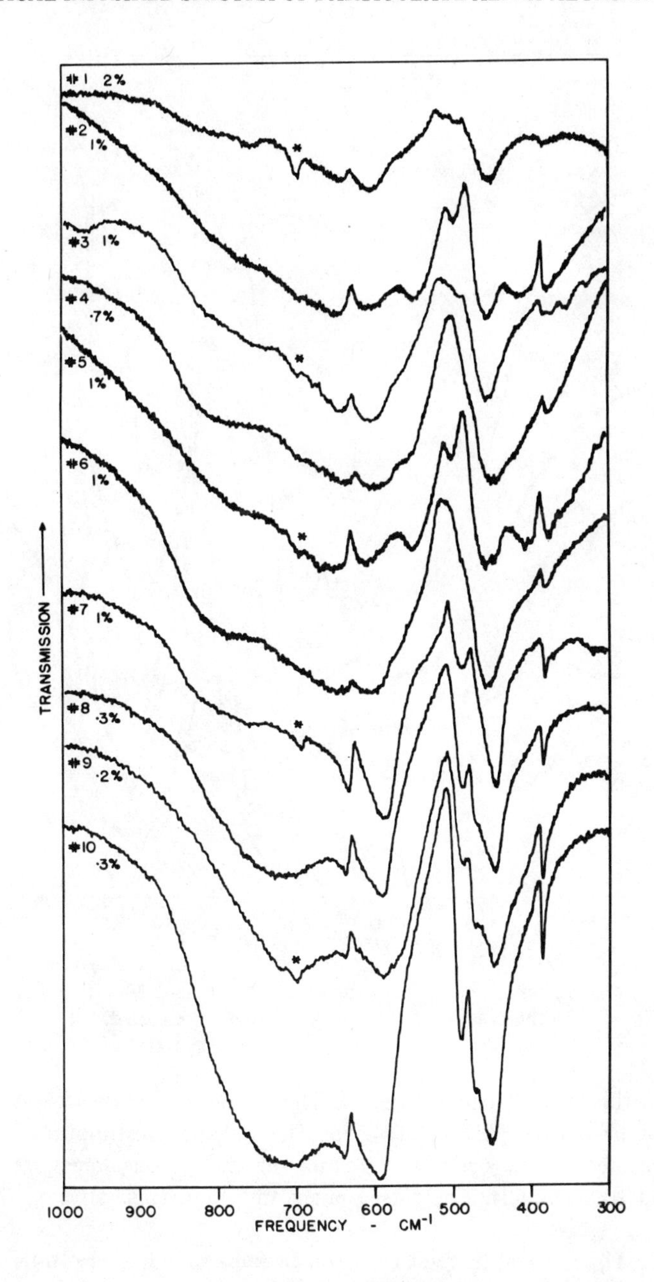

Fig. 1. Infrared transmission spectra of ten particulate alpha-aluminas in KBr pellets, representing a variety of particle-size distributions and modes of preparation.

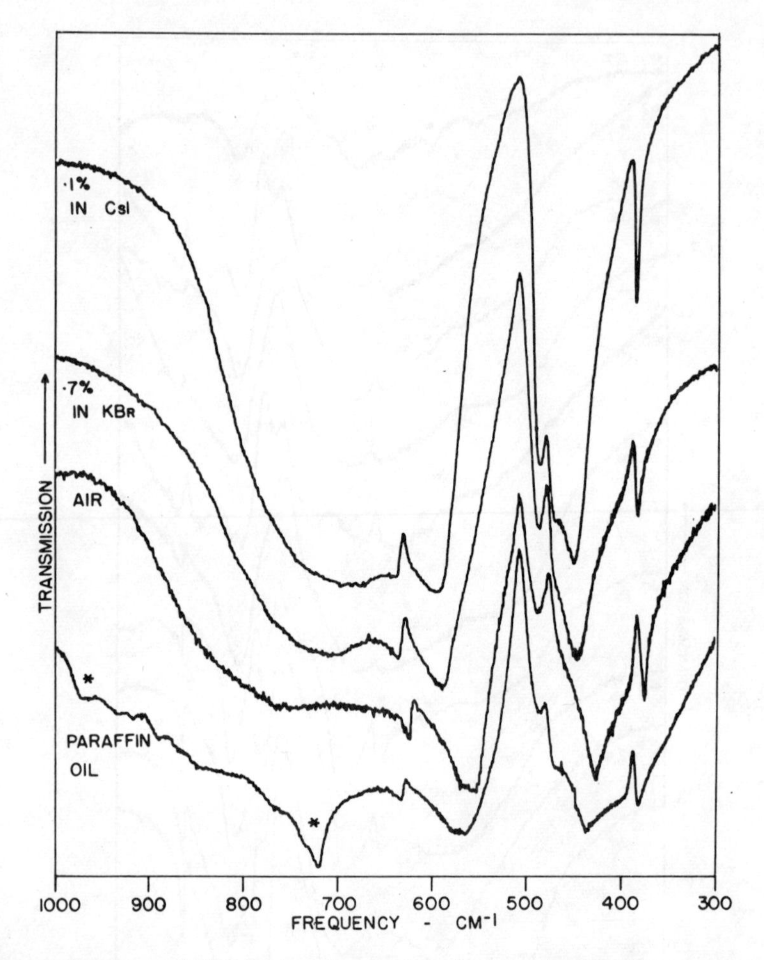

Fig. 2. Infrared transmission spectra of particulate alpha-alumina No. 10 in various media. In the two lower spectra CsI substrates were used.

particles within each diameter interval. The maxima of these weighted distributions all lie between 0.24 and 0.55μ. The weighted distribution of sample No. 1 indicates that a significant fraction of the particles are large. Sample Nos. 8 and 10 are reported by their sources to have particle sizes on the order of 1μ.

The spectral features common to these aluminas are: intense transmission minima in the 650- and 450-cm^{-1} regions, a prominent transmission maximum in the 550-cm^{-1} region which contains variable structure, and sharp, weak features at 630 and 380 cm^{-1}. The feature at 630 cm^{-1} always appears as a transmission maximum, while the other one can appear as either a maximum or minimum. Other broad, weak structure is variously superimposed on these common features. These features are not called absorption bands for reasons to be presented shortly.

To examine the effect of the medium on the spectrum, one alumina, No. 10, was blended in four different media without further grinding, as shown in Fig. 2. In the spectrum marked AIR the particles were simply supported on a cesium iodide substrate. Paraffin oil introduces additional bands between 1000 and 700 cm^{-1}. As a general statement, the medium used

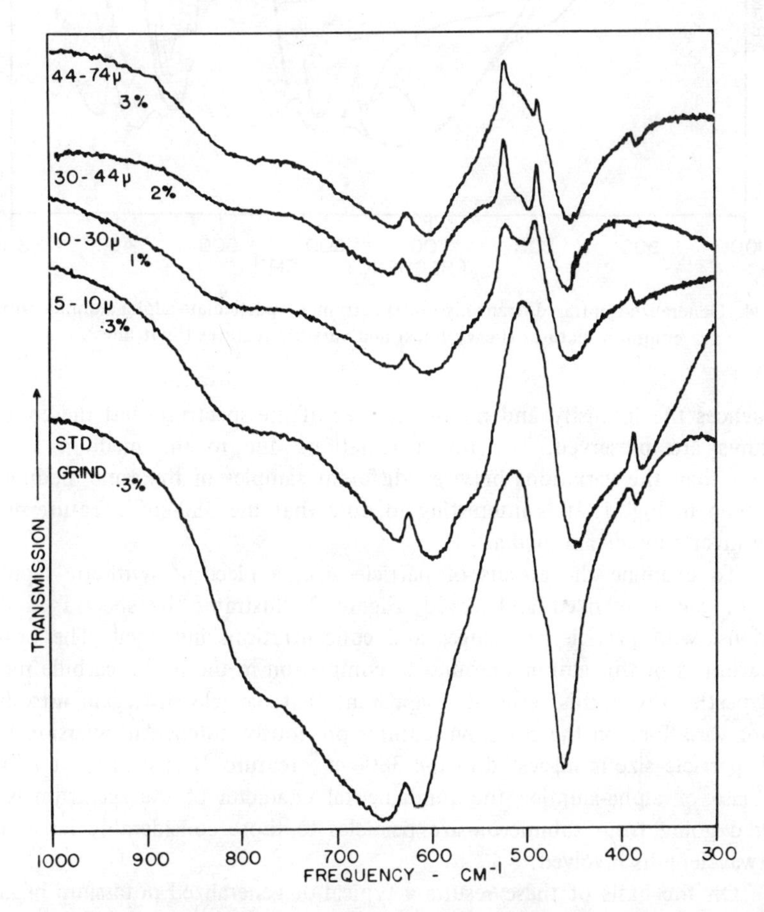

Fig. 3. Infrared transmission spectra of particulate alpha-alumina No. 6 sieved into various particle-size ranges.

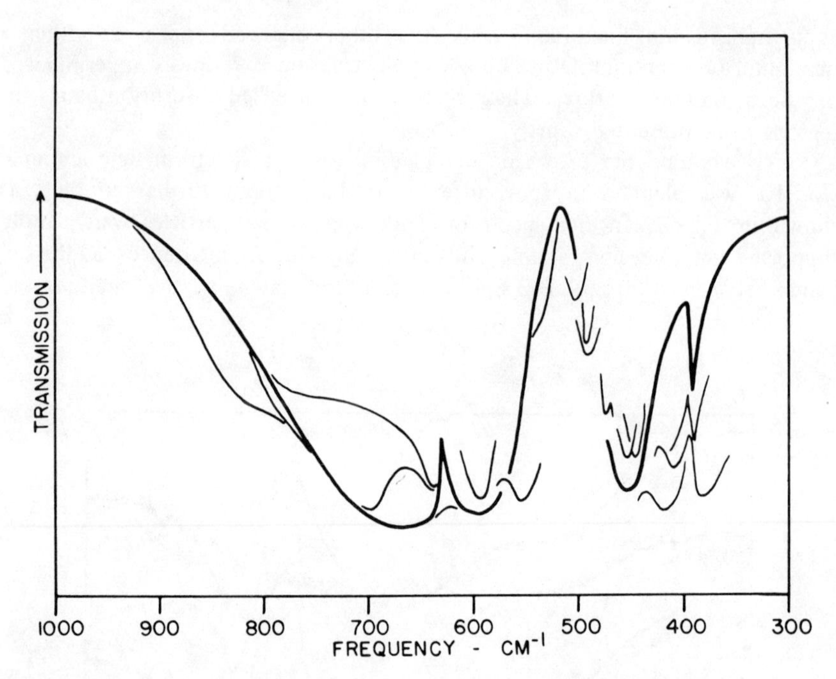

Fig. 4. Generalized infrared transmission spectrum of particulate alpha-alumina showing common features (heavy lines) and variable features (light lines).

influences the intensity and minor features of the spectrum, but the common features are preserved. The minor variations due to the medium are not greater than the variations between different samples in the same medium, as was seen in Fig. 1. It is interesting to note that the 380-cm^{-1} feature practically inverts in certain media.

To examine the effects of particle size, a piece of synthetic sapphire, No. 6, was pulverized and sieved. Figure 3 illustrates the spectra of these sections, with particle-size ranges and concentrations indicated. The bottom spectrum is of this alumina ground to completion in the boron carbide mortar and pestle. From this series it is apparent that particle size again introduces minor variations on the common features previously noted. An inversion trend with particle size is suggested in the 380-cm^{-1} feature. It is significant that for the case of alpha-alumina the fundamental character of the spectrum is not altered going from submicron-size particles to those considerably larger than the wavelengths involved.

On the basis of these results a typical or generalized potassium bromide spectrum of particulate alumina can be established for qualitative analytical purposes. This is shown in Fig. 4. The common features are indicated as heavy lines and the variable features which have been identified in this study are shown as light lines. Particle-size effects and the influence of the medium are responsible for some of these variations.

DISCUSSION

A strong correspondence is found between the transmission spectrum of powdered alumina and the specular reflection spectrum of single-crystal sapphire. This is roughly illustrated in Fig. 5 from a crystal of synthetic sapphire and a reflectance attachment which were available, but which were not ideal for this purpose. Two different orientations of the crystal surface produced differences in the reflection spectra indicated by the divergent solid lines. Well-isolated ordinary- and extraordinary-ray spectra may be found in the literature[2,3], but this figure does serve to call attention to this correspondence. The two intense transmission minima in the 650- and 450-cm^{-1}

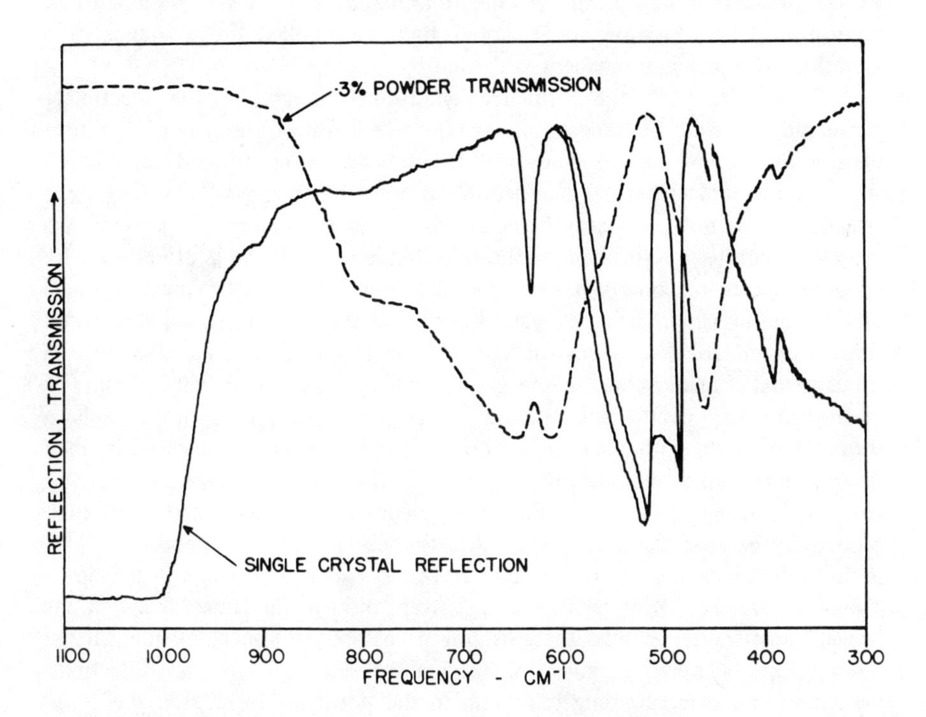

Fig. 5. Correspondence between the spectra of specular reflection and powder transmission of a synthetic sapphire sample as a single crystal and as pulverized.

regions correspond to the two strong reststrahlen bands. The sharp, weak reflection features at 630 and 380 cm^{-1} which are shown in the literature to be associated only with the ordinary ray, have their counterparts in the powder transmission spectra. The differences between the polarized reflection components may be associated with some of the minor spectral variations among the powder samples.

The reflectionlike character of the transmission spectra of particulate aluminas dispersed in potassium bromide shows that reflection processes at the surfaces of the crystallites, rather than absorption losses in their interiors, *dominate* the form of the powder spectra. Even with particle dimensions far smaller than the wavelengths involved radiation is selectively scattered out of the spectrometer beam by the randomly-oriented crystallite surfaces, and the resultant powder spectrum is more nearly an inverted reflection spectrum than an inverted absorption spectrum. This is not to say that absorption effects are not present in the spectrum, but that, since reflectivity is as great as 80 or 90% in the reststrahlen regions, only a small fraction of the radiation incident on a crystallite could enter the interior to be absorbed. Thus it is misleading to interpret a minimum in the powder transmission spectrum of alumina as an absorption band, because there would then be implied the existence of a vibrational frequency coincident with that band.

This subject of the influence of optical effects on the spectra of powdered inorganic materials is under intensive investigation at our laboratory because of its impact on analytical results employing infrared techniques. Also, in materials where this reflection process is strong and the vibrational frequencies do not necessarily coincide with transmission minima, research into crystal structure by infrared methods is hindered. It is well established by classical dispersion theory that a strong reflection band lies near a single vibrational fundamental of a crystal lattice, but there is not a unique relationship between the frequency of the reflection maximum and that of the fundamental. Furthermore, in the case of adjacent strong oscillators a sort of interaction may occur, and the net reflectivity spectrum is not simply a summation of the individual reflection spectra. Indeed, the very opposite may occur, for as Spitzer and Kleinman[4] point out "when a sharp weak resonance lies close to a strong resonance at longer wavelength so as to fall within the high reflectivity band of the latter, it produces a *minimum* in the reflectivity." This is precisely the case with the 635-cm^{-1} resonance of alumina, as will be shown subsequently. To reiterate this point, even though fundamental vibrations "cause" reststrahlen bands, there is not necessarily a direct and identifiable relationship between a given vibrational frequency and its reflectivity maximum and corresponding minimum in the powder transmission spectrum.

Interpretation of the powdered alumina spectra is aided by referring to theoretical considerations and spectral studies of single-crystal samples. A factor group analysis of the corundum lattice by Bhagavantum and Venkatarayudu[5] predicts six infrared-active frequencies, of which two are the extraordinary ray and four are the ordinary ray. Single-crystal alumina has been the subject of many experimental infrared and Raman studies. Among these, Barker[2] reports the polarized reflection spectra at normal incidence. He treats these according to a classical dispersion analysis to obtain the classical oscillator parameters. Haefele[3] arrives at the same objective by a considerably different route employing ruby crystals, which may be considered to be alumina with an 0.04% trace of chromium oxide. These two different methods give two sets of fundamental frequencies which are essentially in agreement, except that Haefele finds two more strong frequencies than theory predicts to be infrared-active, and Barker finds additional frequencies which he identifies as forbidden modes of vibration. In Fig. 6 the results of these two methods

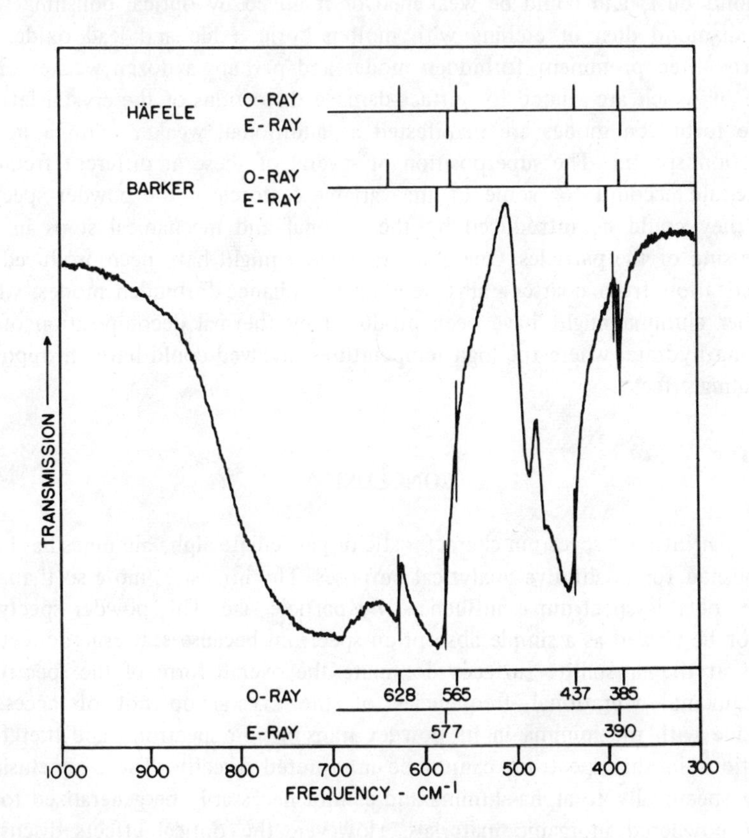

Fig. 6. Comparison of the fundamental frequencies of the alpha-alumina lattice with its powder transmission spectrum.

are compared, and the averaged values are projected onto a typical spectrum of powdered alumina. This illustrates the point that these powder transmission minima cannot be regarded as absorption bands in the original sense of the term.

Superimposed on the common spectrum of powdered alumina is a variety of weaker bands, some of which could be attributed to differences between the individual samples. Interpretation of these features is speculative in the absence of specific knowledge of the thermal and mechanical history and particulate state of each sample. A possible explanation for some of these bands is the appearance of forbidden modes of vibration of the corundum lattice due to mechanical damage of the crystallite surfaces as a result of particle-size reduction. Barker's work with single crystals[2] shows that forbidden modes could be made to appear in reflection by grinding with coarse diamond dust, and could be weakened or removed by optical polishing with fine diamond dust or etching with molten boric oxide and lead oxide. He reports three prominent forbidden modes and perhaps a dozen weaker ones, some of which are related to surface-damage distortions of the crystal lattice. These forbidden modes are manifested as additional, weaker minima in the reflection spectra. The superposition of several of these at different frequencies could account for some of the variable features of the powder spectra, and they could be introduced by the thermal and mechanical steps in the processing of the particles. One alumina powder might have been produced by pulverization from coarse grains, tending to enhance forbidden modes, while another alumina might have been produced by thermal decomposition of an alumina hydrate, where the high temperatures involved could have an opposite annealing effect[6].

CONCLUSION

An infrared spectrum characteristic of particulate alpha-aluminas has been established for qualitative analytical purposes. The intensity, more so than the shape, of this spectrum is influenced by particle size. This powder spectrum cannot be viewed as a simple absorption spectrum because scattering reflection losses at the crystallite surfaces dominate the overall form of the spectrum. Fundamental vibrational frequencies of the crystal do not of necessity coincide with the minima in its powder transmission spectrum, and therefore variations in this spectrum cannot be interpreted directly. These conclusions apply specifically to alpha-alumina and cannot necessarily be generalized to all other powdered inorganic materials. However, the optical effects discussed here in the case of alumina are believed to play a similar role, to a greater or lesser extent, in the spectra of other powdered materials.

ACKNOWLEDGEMENTS

Particle-size distributions were determined by M. N. Haller of the Electron Microscopy Laboratory of the Mellon Institute. X-ray diffraction studies of these samples were performed under the direction of W. L. Baun of the Analytical Branch of the Air Force Materials Laboratory.

REFERENCES

1. N. T. McDevitt and W. L. Baun, Contamination of KBr Pellets by Plastic Mixing Vials, *Appl. Spectr.* **14**, (5), 135 (1960).
2. A. S. Barker, Infrared Lattice Vibrations and Dielectric Dispersion in Corundum, *Phys. Rev.* **132**, (4), 1474 (1963).
3. H. J. Haefele, Das Infrarotspektrum des Rubins, *Z. Naturforsch.* **18a**, 331 (1963).
4. W. G. Spitzer and D. A. Kleinman, Infrared Lattice Bands of Quartz, *Phys. Rev.* **121**, (5), 1324 (1961).
5. S. Bhagavantum and T. Venkatarayudu, Raman Effect in Relation to Crystal Structure, *Proc. Indian Acad. Sci.* **9A**, 224 (1939).
6. H.C. Stumpf, A.S. Russell, J.W. Newsome, and C.M. Tucker, Thermal Transformations of Aluminas and Alumina Hydrates, *Ind. Eng. Chem.* **42** (7), 1398 (1950).

Analysis of Acrylonitrile–Butadiene–Styrene (ABS) Plastics by Infrared Spectroscopy

B. D. Gesner

Bell Telephone Laboratories, Inc.
Murray Hill, New Jersey

ABS plastics are heterogeneous blends of styrene-acrylonitrile copolymers and modified polybutadiene rubbers. The degree of toughness of these versatile thermoplastics is almost entirely a function of the type and character of the rubber. The complex structure of ABS resins and the many additives used to improve the finished plastic make it difficult to analyze these materials without first breaking them down into their component parts. Infrared spectroscopy, however, allows a rapid method for comparing commercial resins. In addition, infrared calibration curves are extremely useful in analyzing the composition of ABS plastics. Analyses based on this technique provide information on which predictions about the physical properties of the molded plastic can be made.

INTRODUCTION

ABS resins are blends of styrene–acrylonitrile copolymers and suitably-treated polybutadiene rubbers.[1] * They belong to the styrene family of plastics, with acrylonitrile added for chemical resistance and butadiene for toughness. The heterogeneous nature of the composite and the microstructure of the rubber

* The rubber spheres in the ABS polymers in this study have been chemically treated by bonding or grafting some styrene–acrylonitrile copolymer directly to their surface.

are very important in determining the degree of toughness and the overall performance of these materials. Therefore information on the character and the amount of the rubber and acrylonitrile in ABS resins should be a valuable aid for predicting physical properties.

Our interest in devising a spectroscopic method for analyzing ABS resins comes about because of possible relationships between composition and physical properties and because of the multitude of ABS resins in today's market. We feel that any indirect method for predicting physical properties will benefit those of us who, because of a lack of equipment or because of a backlog of materials, are limited in the number of resins we can measure directly. We are, however, aware that the heterogeneity and the gelled nature of the rubber phase make ABS resins difficult to analyze by spectroscopic methods.

Compositional analyses of ABS plastics are not new. By elution and precipitation techniques[1] ABS resins can be broken down into their free and grafted phases. Subsequent analysis gives reasonable values for total composition. However, this method is tedious, and the number of operations sufficient to raise experimental error in compositional values to as high as 1.5%. Moore and Frazer[12] have used infrared analysis semiquantitatively to direct their studies on the characterization of ABS-type graft copolymers. In their procedure they employed the 6.1 μ line for butadiene concentration. We have found that for experimental materials this approach gives good results; for commercial materials, however, this technique is hampered by additive interference at the 6.1 μ line. Haslam and Willis[13] suggest that optical density ratios from infrared lines at 4.4, 6.3, and 10.4 μ can be used in conjunction with ratios from known copolymers to determine ABS compositions. With this method we obtained reasonably good results when the ratios of monomer units in the model copolymers were close to those in the ABS. From the relationships[13]

$$k' = \frac{A_{4.4\mu}}{A_{6.3\mu}} \frac{\text{wt.\% styrene}}{\text{wt.\% acrylonitrite}}$$

$$k'' = \frac{A_{10.4}}{A_{6.3}} \frac{\text{wt.\% styrene}}{\text{wt.\% butadiene}}$$

we calculated $k' = 3.9$ for a 70:30 styrene–acrylonitrile copolymer and $k'' = 4.8$ for an 85:15 styrene–butadiene copolymer. With these values and from the fact that for ABS resins the sum of the weight per cents for acrylonitrile, butadiene, and styrene is 100% compositional values for ABS resins were obtained with standard deviations of 1.5%.

The application of infrared spectroscopy to the study of polybutadiene alone and in mixtures is extensive. Spectral-line assignments for a variety of polybutadienes have been made by, among others, Binder[2] and Golub and Shipman.[3] A number of studies dealing with the use of certain infrared absorption lines for quantitative determination of polybutadiene microstructure have been presented.[4-7] Absorptions of 10.4 and 11.0 μ are favored for *trans*-1,4 and vinyl units, respectively. For *cis* structure, lines at 13.8,[4] 14.7[5]. 13..5,[6] and the area between 12–15. 75 μ have been used for polybutadienes examined in carbon disulfide. In spite of the variations in technique from one method to another, the results on various polybutadienes are quite similar.

Unlike polybutadiene, ABS resins must be handled as molded or cast films in infrared studies because solutions of these materials are possible only by chemical destruction of the gelled rubber, while liquid dispersions are restricted to solvents having strong absorptions in the infrared regions of interest. Field *et al.*[8], using the 10.1 μ line for vinyl units and the 10.4 μ line for *cis* and *trans* units, found the amounts of 1,2 and 1,4 structure for cast polybutadiene films in good agreement with values obtained by solution techniques.[4-7] Two polymers somewhat similar to ABS were also quantitatively analyzed as films using infrared spectroscopy. Wexler[9] prepared an absorbance–ratio–composition equation for butadiene–styrene copolymers using infrared bands at 6.10 and 6.25 μ, while Sands and Turner[10] prepared nomographs from which bound acrylonitrile in solid butadiene–acrylonitrile copolymer laminates was determined by taking absorbance ratios of the lines at 4.4 and 3.4 μ.

It is obvious from the infrared spectrum of ABS graft copolymers (Fig. 1) that the methods for determining microstructure would be difficult to apply

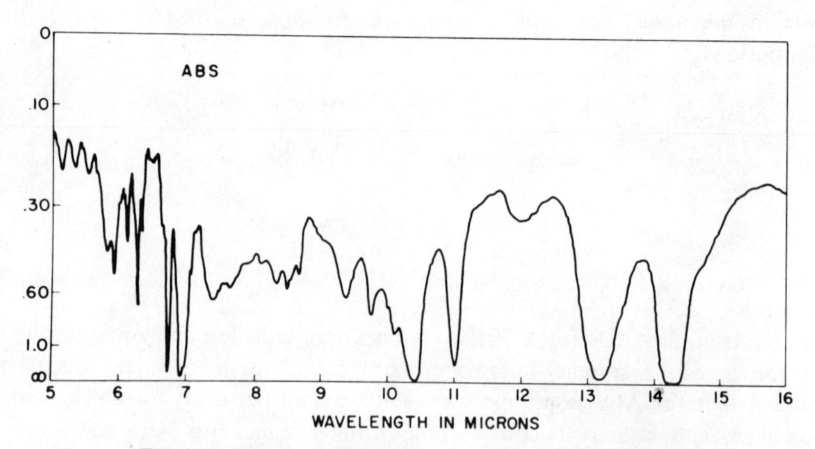

Fig. 1. Infrared spectrum of an experimental ABS resin.

in the case of grafted rubber. The intense lines arising because of the acrylonitrile and styrene components not only interfere with some, the *cis* lines at 13.5 and 14.7 μ, but also broaden other of the rubber characterizing frequencies, such as the *trans* 10.4 μ line. Variations of the absorbance ratio method of Field *et al.*,[8] however, might be applicable because the lines at 6.1, 10.1, and 10.4 μ, though broadened, do remain essentially unobstructed after grafting.

In this work we present the information we have obtained in examining ABS rubber microstructure on identifying source and on determining composition using variations of these spectroscopic methods.

EXPERIMENTAL

ABS plastic films for infrared analysis were compression-molded at 380°F and 5000 psi (ram pressure) in thicknesses that would allow the absorbance at 4.4 μ to remain within 0.2–0.8. Standard ABS resins were prepared by melt-mixing an ABS resin of known composition with varying proportions of a standard styrene–acrylonitrile copolymer.

Polybutadiene rubber was cast from carbon disulfide onto sodium chloride plates so that absorbance at 10.4 μ was less than 0.9. All polymers other than the base resins, which were supplied by various manufacturers and reported on previously,[1,11] were emulsion-polymerized with persulfate catalyst at from 50 to 60°C in a 1-gal stirred reactor. A Beckman IR8 spectrometer in the double-beam slow-scan mode was used to record the infrared spectra. The emulsion-polymerized ABS graft polymer analyzed by the phase-separation technique[1] contained 65 wt. % polybutadiene with M_v of 400,000 and 35 wt. % grafted styrene–acrylonitrile copolymer with M_v of 250,000. The system therefore contained approximately one graft per polybutadiene molecule.

RESULTS AND DISCUSSION

Microstructure

The absorbance ratios of the lines at 6.1, 10.1, and 10.4 μ (Table 1) for spectra of polybutadiene rubber before and after grafting are the same. We can interpret this result as meaning that the microstructure is unchanged by the grafting reaction. We feel that this is not surprising, since there is only about one graft per polybutadiene molecule. We would conclude that since so few olefin bonds are necessary here for grafting, infrared analysis cannot distinguish a change. Whether this solid-phase technique for examining micro-

TABLE I

	$A_{10.1}/A_{10.4}$	$A_{6.1}/A_{10.4}$
Rubber, untreated	0.09	0.09
Rubber, grafted	0.09	0.08

TABLE II

Olefin Optical Density Ratios for Some Commercial ABS Resins

ABS resin	Ratio $A_{6.1}/A_{10.4}$
I	0.40
II	0.00
III	0.25
IV	0.12

structure could provide information about *cis* structure in heavily-grafted and crosslinked systems remains to be seen.

Source

To identify the source of ABS the ratio of optical densities at 6.1 and 10.4 μ for six commercial ABS resins containing grafted rubber and having similar toughness properties were used, and were found to be sufficiently different to distinguish each plastic (Table II). We found, however, that plastics containing amide wax additives have increased absorbance at the 6.1 μ line and cannot be distinguished by this method.

We feel that even though this particular optical density ratio is comparison of *cis* and vinyl vs *trans* microstructure in the ABS resins, it serves no other purpose than to reduce the number of possible sources for ABS resins. We have successfully used comparisons of this type to identify plastic in defective parts from production lines incorporating resins from many different suppliers.

Composition

From the infrared spectra of standard ABS resins we have constructed calibration curves (Fig. 2) relating composition and corrected optical densities

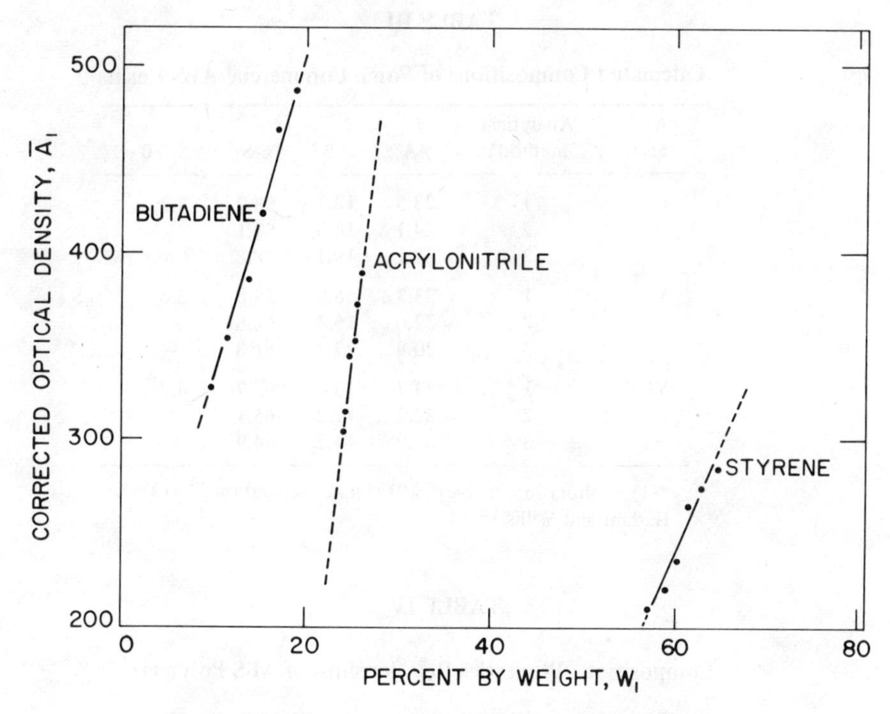

Fig. 2. Calibration curves for infrared analysis of ABS resins.

for the acrylonitrile line at 4.4 μ, the styrene line at 6.3 μ and the butadiene line at 10.4 μ. We define the corrected optical density $\bar{A}_1$ as the optical density for a particular line, A_1, divided by the sum of optical densities for all lines under consideration: $\bar{A}_1 = A_1 / \sum_{i=1}^{n} A_i$. The corrected optical density for the 4.4 μ line, $\bar{A}_{4.4}$, is therefore given by $\bar{A}_{4.4} = A_{4.4} / A_{4.4} + A_{6.3} + A_{10.4}$.

We have corrected the final values in Table III by assuming that the sum of the weight per cents of acrylonitrile, butadiene, and styrene is 100%. The corrected weight per cent $\bar{W}_1$ is then the weight per cent taken directly from Fig. 2, W_1, divided by the sum of the weight per cents: $\bar{W}_1 = W_1 / \sum_{i=1}^{n} W_i$. We have found standard deviations by this method to range from 0.3% for acrylonitrile to 0.9% for butadiene.

For comparison we have listed in Table III the values obtained by the calibration-curve method, the phase-separation method[11] and Haslam and Willis' method.[13] Values from these three methods agree within the range of their respective standard deviations.

TABLE III

Calculated Compositions of Some Commercial ABS Resins

ABS resin	Analytical method*	%A	%B	%S	%S/%B
I	1	23.5	18.3	58.2	3.2
	2	24.1	18.8	57.1	–
	3	22.1	19.1	58.8	–
V	1	23.3	16.9	59.8	3.6
	2	22.0	15.4	62.6	–
	3	20.8	17.4	61.8	–
VI	1	23.7	13.6	62.7	4.7
	2	22.7	12.2	65.1	–
	3	20.9	14.2	64.9	–

*(1) Calibration curves. (2) Phase separation.[11] (3) Haslam and Willis.[14]

TABLE IV

Composition–Properties Relationships of ABS Polymers[14]

Property	Behavior as S/B increases (A constant)
Impact strength	Decreases
Tensile strength	Increases
Flexural strength	Increases
Flexural modulus	Increases

TABLE V

Physical Properties of Some Commercial ABS Resins

Property	ABS resin I	ABS resin V	ABS resin VI
Impact strength (ft-lbs/in. notch, 73°F)	4.3	4.0	3.7
Tensile strength (psi)	5,700	5,600	5,300
Flexural strength (psi)	9,500	9,100	9,000
Flexural modulus	324,000	333,000	363,000

Now we can compare physical properties approximated from the infrared-determined composition of these materials (Table III) and the Frazer[14] physical-property–composition trends (Table IV) with measured physical properties (Table V). It appears that the three resins in Table III have the same acrylonitrile content. According to Frazer's[14] generalization (Table IV) the resins line up I > V > VI in impact strength and VI > V > I in tensile strength, flexural strength, and modulus. From the measured physical properties of these materials (Table V) we find that all resins are ordered properly in impact strength and flexural modulus. The reversal in tensile and flexural strength ranking is probably due to the greater copolymer molecular weights in resins I and V.[14] The effect of increasing molecular weight is much the same as that of increasing polarity. The net result is an increase in strength of the materials. In these cases the molecular-weight effect must be predominant.

We feel fortunate that in this series the physical properties are much in line with compositional correlations. However, as Frazer[14] points out, the profound effects that the structure of the substrate and the resinous phases have on physical properties should not be overlooked when making such correlations. Indeed, it would be far better to consider both structure and composition before attempting to predict physical properties. Correlations from infrared spectra can be helpful nonetheless.

ACKNOWLEDGEMENT

We thank L. T. Pappalardo for his efforts in the preparation and characterization of experimental polymers and D. J. Boyle for supplying physical property data on the commercial resins.

REFERENCES

1. B. D. Gesner, *Polymer Preprints,* 8, 1482 (1967).
2. J. L. Binder, *J. Poly. Sci.* **1**, 47 (1963); **3**, 1587 (1965).
3. M. A. Golub and J. T. Shipman, *Spectrochim. Acta* **16**, 1165 (1960); **20**, 701 (1964).
4. R. R. Hampton, *Anal. Chem.* **21**, 923 (1949).
5. R. S. Silas, J. Yates, and U. Thornton, *Anal. Chem.* **31**, 529 (1959).
6. J. L. Binder, *Anal. Chem.* **26**, 1877 (1954).
7. D. Morero, A. Santambrogio, L. Porri, and F. Ciampelli, *la Chemica e l'Industria* **41**, 758 (1959).
8. J. E. Field, D. E. Woodford, and S. D. Gehman, *J. Appl. Phys.* **17**, 386 (1946).
9. A. S. Wexler, *Anal. Chem.* **36**, 1829 (1964).
10. J. D. Sands and G. S. Turner, *Anal. Chem.* **24**, 791 (1952).
11. P. G. Kelleher, D. J. Boyle, and B. D. Gesner, *J. Appl. Poly. Sci.* **11**, 1731 (1967); B. D. Gesner, *J. Poly. Sci.* **3**, 3825 (1965).

12. L. D. Moore and W. J. Frazer, *Polymer Preprints* 8, 1486 (1967).
13. J. Haslam and H. A. Willis, *Identification and Analysis of Plastics*, D. Van Nostrand Company, Princeton, New Jersey (1965), p. 193.
14. W. J. Frazer, *Chem. Ind. (London)* 33, 1339 (1966).

Design and Application of a High Temperature Infrared Cell for the Study of Polymeric Materials*

W. R. Feairheller

Monsanto Research Corporation
Dayton, Ohio

and

W. J. Crawford

Wright-Patterson Air Force Base
Ohio

The recent interest in high-temperature polymeric materials indicates that it would be desirable to study these materials by infrared spectroscopy at temperatures well above the 300°C upper limit of most presently-available infrared cells. A new cell has been designed and constructed which has been found useful up to 600°C, covers the infrared region from 3000 to 250 cm^{-1}, and permits a choice of atmospheres in the cell of either air, inert gas such as nitrogen or argon, or vacuum. An additional feature is that the cell windows are easily removed for cleaning and polishing. The results obtained with this cell from room temperature to 600°C on high-temperature polymers such as poly-2,2'-(*m*-phenylene)-5,5'- bibenzimidazole as well as several other polymeric systems will be discussed. Recording the spectra while maintaining the desired temperature in the cell has a number of advantages over the method of external heating and cooling and then recording the spectrum.

*This work was supported in part by the U.S. Air Force under Contract No. AF33(615)-1565.

The use of infrared spectroscopy in the study of polymeric systems is well known and well established. However, with the exception of pyrolysis techniques, much of the infrared work has either been limited to temperatures below 300°C, or involved heating samples in an external oven, cooling them, and recording the spectra.

It was concluded that it would be advantageous to have an infrared cell capable of higher temperatures for use with thermally stable materials. With this type of cell, rates of decomposition and oxidation at elevated temperatures as well as phase changes could be studied.

To serve as guidelines in the construction of the cell, four general requirements were considered:

1. All materials used in the cell should be stable to 600°C.
2. The windows used in the cell must be transparent over a wide spectral range and easily removable for polishing.
3. The atmosphere around the sample should be variable (air, oxygen, inert gases, or vacuum).
4. The cell should have a temperature controller and measuring device.

With the above considerations in mind the following ten specific design requirements were listed:

1. Cesium iodide windows should be used, since they are transparent to infrared radiation from 4000 cm^{-1} to 180 cm^{-1}, are not affected by thermal shock, are easily repolished, and melt at 621°C.
2. The cell should be capable of temperatures up to 600°C.
3. The cell should be fitted with seals to allow use of various atmospheres or vacuum.
4. Heating areas should have low mass to minimize thermal lag.
5. Heat transfer to the spectrometer should be minimized to protect the optics of the instrument.
6. A thermocouple should be provided to control and read out the sample temperature.
7. Sample area should be readily accessible when the cell is disassembled.
8. Sample area should be suitable for both solids and liquids (viscous materials), which means the sample compartment must be able to accept two windows.
9. Any fittings subject to high-temperature and/or mechanical stress should be easily replaceable.
10. Provision should be made to trap off-gases formed during degradation for vapor-phase chromatography and mass-spectrometric analysis.

Based on these requirements the cell (Figs. 1 and 2) was designed and built. Cesium iodide windows were used throughout. The front window is 32 mm in diameter by 3–4 mm thick, and the other three windows are 25 mm in diameter by 3–5 mm thick. Although the openings for these windows may seem small, it was found that the finished cell assembled without windows passed 98% of the available radiation, so that with well-polished windows the cell would show little loss of energy.

The outside of the cell body is made of stainless steel sections, silver-soldered together, and lathe-turned to final shape. A groove is cut in the front piece to hold an O-ring, which then seals the cell when the two halves are bolted together. A hose connection is provided in the rear part of the outside cell body for attachment to a vacuum line. A plug in the front half (see Fig. 1), used for vacuum studies, is replaced by a hose connection for flushing the cell with various atmospheres.

Figure 2 shows the stainless steel standoff fittings that support the ceramic ring insulator and also act as connectors for the heater and as a

Fig. 1. External view of high-temperature infrared cell.

Fig. 2. Internal view of completely assembled infrared cell.

feedthrough for the thermocouple. The inner cell holder is made of nickel. Three concentric heat baffles made of molybdenum foil surround the inner cell area when the cell is assembled. Nickel was chosen for the inner cell holder, largely due to its thermal conductivity; stainless steel was chosen for the outer cell body to reduce the heat flow to the instrument.

The heater is a Nichrome wire contained in a boron nitride chamber that completely surrounds the inner cell unit. The inner cell unit is supported by the ceramic insulator made from Grade H fired lava. The heating wire is connected to the standoff fittings, which are insulated from the cell body by Teflon sleeves. Wire from a Variac connects to the posts on the outside of the cell. Visible at the top of the photograph is the thermocouple wire and feedthrough. A hole, also visible in the picture, is drilled about 1/8 in. into the salt plate; the tip of the thermocouple is placed in it. Spanner wrenches are used to remove and replace the window retainers. The outside windows are sealed with Teflon O-rings and have remained sealed under a vacuum of 20 μ at elevated temperatures.

The cell has met all of the original design requirements, and very little trouble has been experienced using it up to 600°C.

The cell can be used by applying voltage with a Variac and connecting potentiometer to the thermocouple to measure the temperature. However, this arrangement was considered undesirable due to temperature fluctuations.

Mr. Dail Hurley, under University of Cincinnati contract at Wright-Patterson Air Force Base, designed a very simple, but extremely accurate, temperature controller. Figure 3 shows the arrangement of the component parts of the controller.

Figure 4 is a schematic diagram of the controller attached to the cell. The controller regulates the voltage applied to the heat cell. This is accomplished by setting the potentiometer to the desired temperature. When voltage is applied to the heating element the cell heats to the specified temperature. At that temperature the thermocouple activates the galvanometer, and a photoswitch trips the circuit to the Variac.

To demonstrate the ease of observing phase changes in polymers, data were obtained for low-density and high-density polyethylene.

Figure 5 shows the data recorded using a Model 225 infrared spectrophotometer at the various temperatures. The two bands are the 731 cm^{-1} and

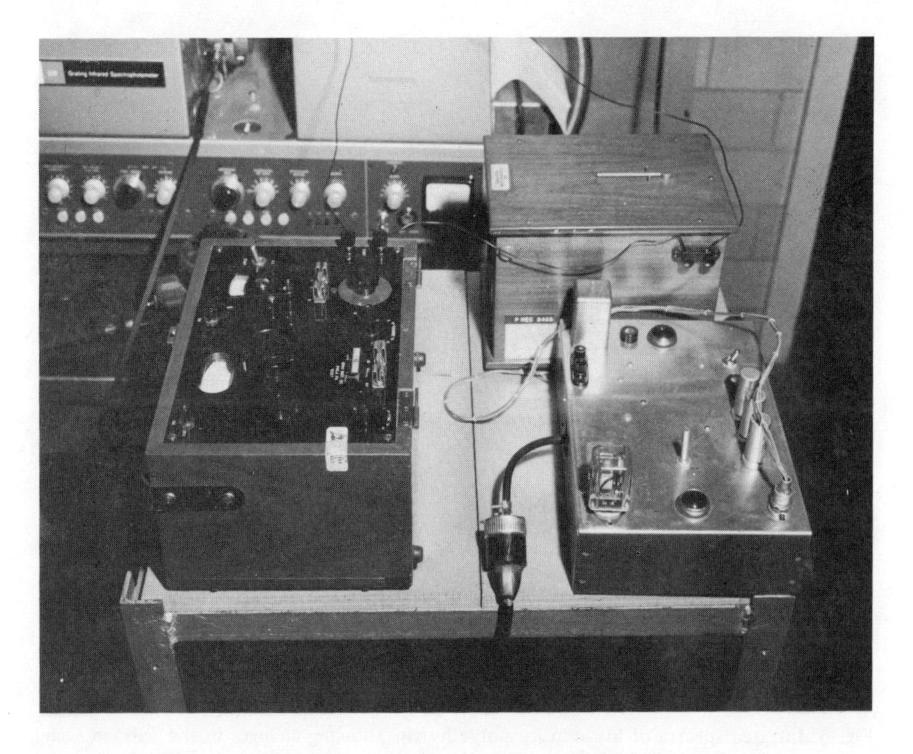

Fig. 3. Temperature controller for the high-temperature cell.

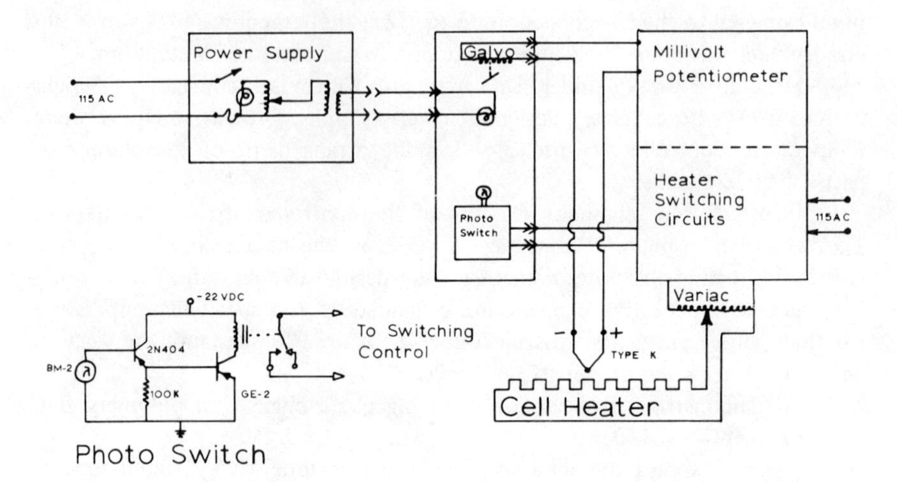

Fig. 4. Schematic of the temperature controller.

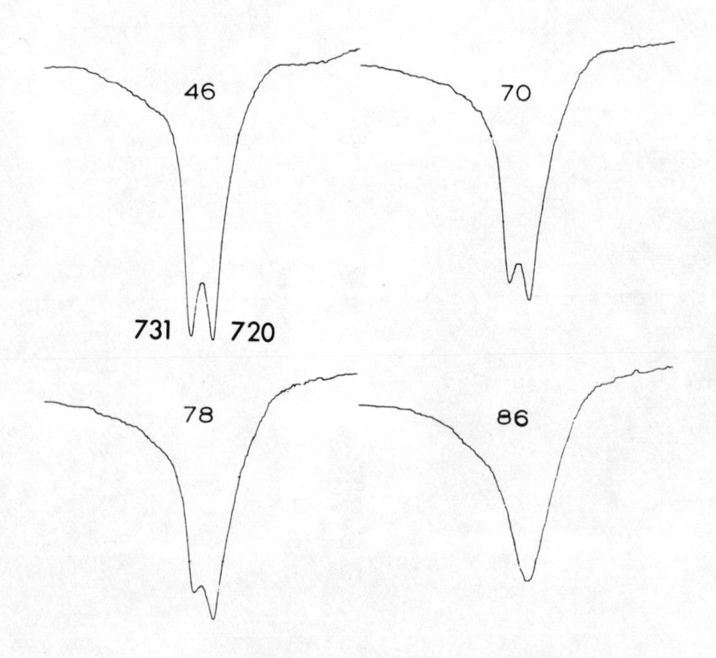

Fig. 5. Infrared spectra of low-density polyethylene showing changes in the 720 cm^{-1} and 731 cm^{-1} absorption bands with temperature.

720 cm^{-1} due to the $-CH_2-$ rocking mode of the polymer chain. Owing to the loss in crystallinity, the 731 cm^{-1} band decreases in intensity as the temperature increases. The spectrum was recorded by placing a film between cesium iodide plates and heating the cell at the specified temperatures.

Results of another series of tests on this same material in this region are shown in Figure 6. Due to the excellent control of the temperature controller

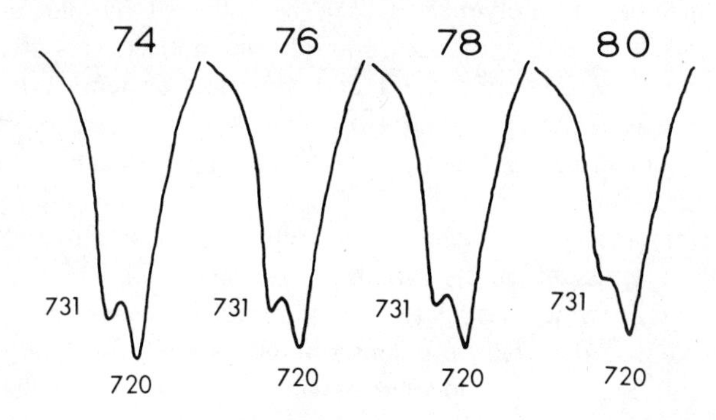

Fig. 6. Infrared spectra of the 720 cm^{-1} and 731 cm^{-1} bands of low-density polyethylene over the 74–80°C temperature range.

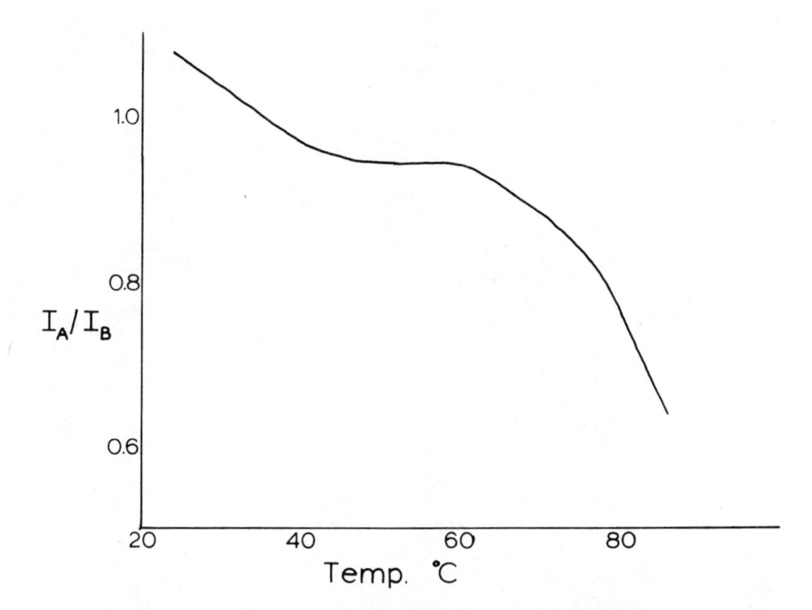

Fig. 7. Plot of the ratio of the intensity of the 731 cm^{-1} band to the intensity of the 720 cm^{-1} band against temperature for low-density polyethylene.

very small changes can be observed in the spectrum as the temperature is varied as little as 2°C. Thus it can be seen that the controller permits the temperature to be maintained within very close limits. This could become an important feature in the study of phase changes or other effects that would require very close temperature control.

Figures 7 and 8 demonstrate how easily the crystalline change in low-density and high-density polyethylene can be followed by plotting a curve. The graph represents a plot of the ratio of the intensities of the 731 cm^{-1} band to those of the 720 cm^{-1} band as a function of temperature. As the temperature increases, the curve starts to drop, showing a loss in crystallinity. This technique could be used effectively to investigate the phase changes in polymers.

One major problem that can be encountered in high-temperature cells is the effect of the emission of the sample on the spectrum. This is particularly true of situations in which very large temperature differences are required in the study of a material. This problem can be avoided by using a spectrophotometer with the chopper located between the source and the heat cell. However, if this type of spectrometer is unavailable, high-temperature effects

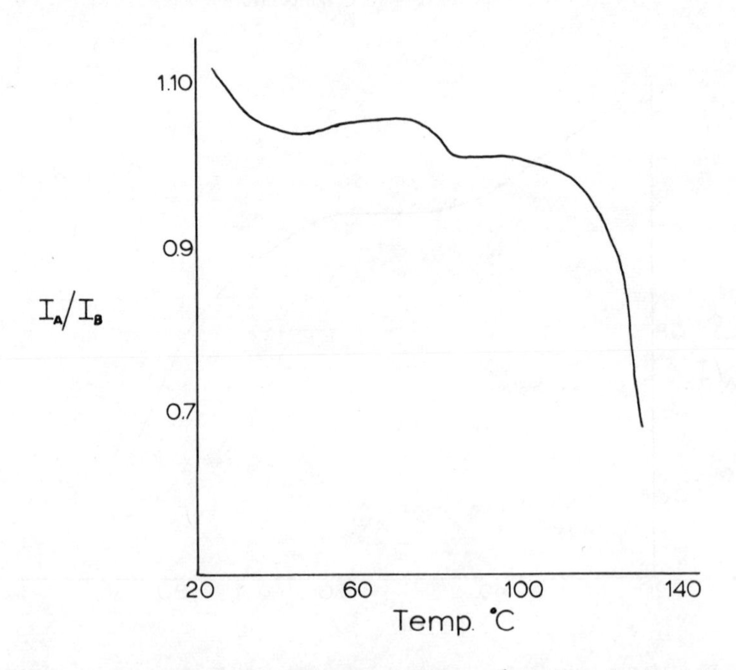

Fig. 8. Plot of the ratio of the intensity of the 731 cm^{-1} band to the intensity of the 720 cm^{-1} band against temperature for high-density polyethylene.

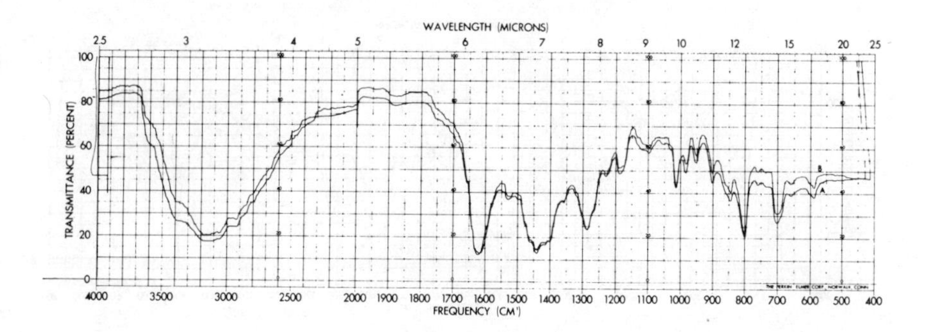

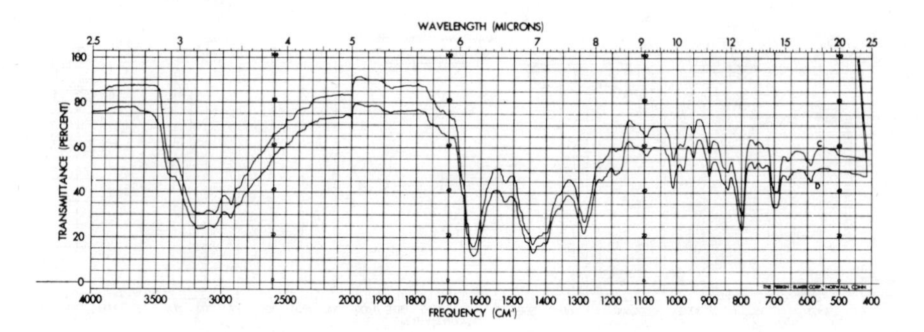

Fig. 9. Partial infrared spectra of poly-2,2′-(*m*-phenylene)-5,5′- bibenzimidazole heated in high-temperature cell. Nitrogen atmosphere. Film cast from dimethyl acetamide. (A) 25°C, (B) 53-64°C, (C) 111-124°C, (D) 171-197°C.

can still be minimized with a standard instrument. For example, a polymer can be evaluated at a temperature below the expected transition and above the transition. Any large differences observed in the spectra would be significant in terms of a phase or chemical change.

Effective use of the cell to study high-temperature polymeric systems is shown in the spectra of poly-2,2′-(*m*-phenylene)-5,5′-bibenzimidazole. The structure of this polymer is:

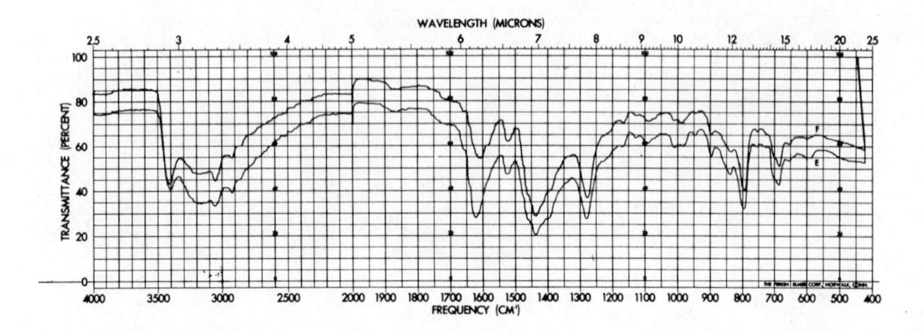

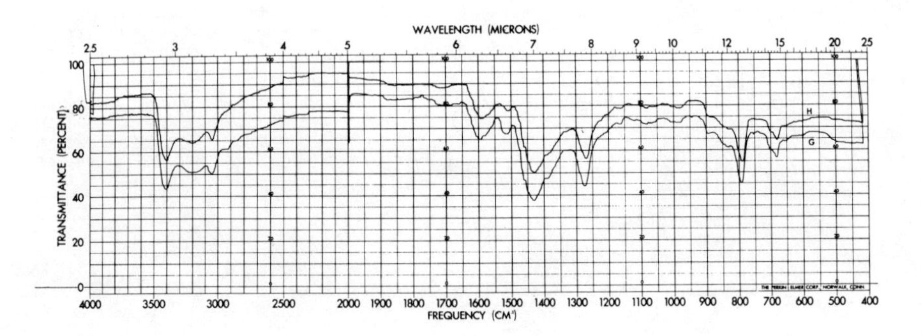

Fig. 10. Partial infrared spectra of poly-2,2′-(*m*-phenylene)-5,5' - bibenzimidazole heated in high-temperature cell. Nitrogen atmosphere. Film cast from dimethyl acetamide. (E) 268-294°C, (F) 356-373°C, (G) 457-460°C, (H) 533-540°C.

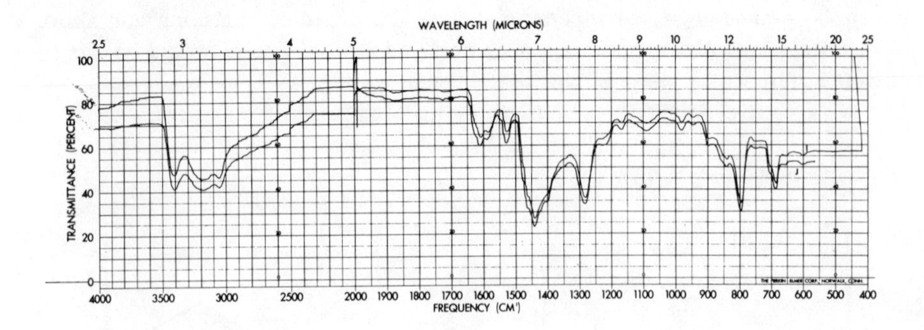

Fig. 11. Partial infrared spectra of poly-2,2′-(*m*-phenylene)-5,5′ - bibenzimidazole heated in high-temperature cell. Nitrogen atmosphere. Film cast from dimethyl acetamide. (I) 310-125°C, (J) 125-109°C.

Figures 9, 10, and 11 show the spectra in a nitrogen atmosphere of this material, obtained from a film cast from dimethyl acetamide, as it was heated from room temperature to 540°C and then as the material was cooled to room temperature.

Evidence of N,N-dimethyl acetamide is observed at 1630 cm^{-1} and 1010 cm^{-1} in the spectra until the temperature reaches 200°C. Above this temperature these bands are no longer visible. At about 260°C the absorption at 3400 cm^{-1} increases. Such behavior continues until a temperature of 526°C is reached. Upon cooling to room temperature from 526°C no further change is noted, showing that no reversible reaction occured as the polymer cooled. As a result, it seems that the 3400 cm^{-1} band may indicate initial stages of degradation in the polymeric unit.

The cell has more than met the design requirements and has performed very satisfactorily. The examples presented demonstrate the ease of using the heat cell and indicate that spectral research with polymers at high temperature is now feasible.

Spectroscopic Studies of Hydrogen Bonding

C. N. R. Rao and A. S. N. Murthy

Department of Chemistry
Indian Institute of Technology
Kanpur, India

A review is given of the literature on spectroscopic studies of the hydrogen bond. Discussed in turn are self-association due to hydrogen bonding, hydrogen bonding between donors and acceptors, intramolecular hydrogen bonding, and bonding in aqueous systems. For each category the results obtained by IR, NMR, and electronic spectroscopy are discussed.

INTRODUCTION

Since the publication of the excellent book on the hydrogen bond by Pimentel and McClellan[1] a large number of publications on spectroscopic studies of hydrogen bonding have appeared in the literature. The present authors prepared a report on hydrogen bonding covering the period 1959–1963 for private circulation.[2] We have presently reviewed some of the important results obtained from spectroscopic studies of the hydrogen bond in recent years. In presenting this brief review we have found it convenient to classify the material on the basis of the type of hydrogen bond. Thus, we have discussed self-association due to hydrogen bonding, hydrogen bonding between donors (electron donors, or bases) and acceptors (proton donors or acids), and intramolecular hydrogen bonding. In each of these sections we have discussed the results obtained by infrared, NMR, and electronic spectroscopy.

SPECTROSCOPIC METHODS EMPLOYED TO STUDY HYDROGEN BONDING

Infrared spectroscopy provides the most straightforward method for the detection of hydrogen bonds. The formation of hydrogen bonds (X–H. . .Y) gives rise to shifts of the stretching mode (ν_{XH}) and its harmonics to lower frequencies and causes increase in the intensity (and the half-bandwidth) of the fundamental, ν_{XH}. The frequency shift, $\Delta\nu_{XH}$, has been correlated with various chemical and physical properties; the important correlations are between $\Delta\nu_{XH}$ and the hydrogen bond distance ($R_{X...Y}$), the enthalpy or free energy of formation of hydrogen bonds (ΔH° or ΔF°), and half-bandwidth or band intensity. Many workers have found it convenient to employ $\Delta\nu_{X-H}$ as a measure of base strength. The bending mode of the X–H bond is also shifted on hydrogen bonding and the shifts can be fruitfully correlated with other properties. The formation of hydrogen bonds also result in new vibrational modes, the most important of them being the stretching (ν_{σ}) and bending (ν_{β}) modes of the X. . .Y bond. Torsional modes are also found due to the restricted rotation around the hydrogen bond. With the availability of commercial far-infrared spectrometers going down to 10 cm^{-1} there has been increased activity in the study of these new vibrational modes caused by hydrogen bonding. Raman spectroscopy also provides valuable information complementary to that obtained from infrared spectroscopy, but there have been fewer Raman studies of hydrogen bonding, probably because Raman spectrometers are not readily available. With the advent of laser Raman spectroscopy there will undoubtedly be greater interest in hydrogen-bond studies in the future. There is no doubt that of all the spectroscopic methods employed to study the hydrogen bond, the most effective and popular method is infrared spectroscopy. Infrared studies of hydrogen bonding have been briefly reviewed in some of the recent texts.[3,4]

Since 1958 nuclear magnetic resonance spectroscopy has been increasingly employed for the study of hydrogen bonding. NMR measurements provide information on (1) chemical shifts of the hydrogen-bonded proton, (2) hydrogen-bond formation equilibria and exchange times (3) changes in relaxation times, and (4) position of hydrogen atoms in hydrogen-bonded crystals. Of these only the first two aspects are of relevance to this review. Hydrogen-bonding studies employing high- resolution NMR spectroscopy have been discussed in the texts by Pople *et al.*[5] as well as Emsley *et al.*[6]; recently Lippert[7] and Laszlo[8] have reviewed the NMR studies of hydrogen bonding.

Hydrogen bonding affects the electronic spectrum of a molecule if the chromophoric portion of the molecule is perturbed by the hydrogen bond. Although historically electronic spectroscopy has been employed for the study of the hydrogen bond for a longer period than either infrared or NMR spectroscopy, it is only recently that ultraviolet and visible absorption spectra have been effectively used for quantitative studies. The main effects of hydrogen bonding on electronic transitions are the changes in frequency and intensity. Thus the $n \rightarrow \pi^*$ and $\pi \rightarrow \pi^*$ transitions can be distinguished in terms of solvent effects; the solvent shifts are more prominent when they form hydrogen bonds with the chromophoric group. Many solvents can break hydrogen bonds already present due to self-association or intramolecular hydrogen bonding and thus cause variations in the spectra. In addition to electronic absorption spectroscopy, there have been studies in the last few years on the emission spectra associated with hydrogen-bonded systems; such studies throw some light on the hydrogen bonding in the excited states of molecules. While electronic spectra could be diagnostic in the identification of hydrogen bonds, it often becomes difficult to obtain quantitative information on thermodynamic quantities. Many compounds show appreciable association only at high concentrations, under which conditions the band intensities in the electronic spectra would be very large. There are some other systems which give rise to monomeric species only at extremely low concentrations. Difficulties are also encountered in the study of donor–acceptor hydrogen bonding, since the acceptors (proton donors) are often highly associated in the concentration regions normally employed for study. Furthermore, the absorption bands most suitable for investigating hydrogen bonding could have very low intensity or show their characteristic absorption bands in the far-ultraviolet region, where the use of solvents is rather limited. The study of hydrogen bonding by electronic spectroscopy has been briefly reviewed in the texts by Rao[9] and Jaffe and Orchin.[10] Dearden[11] has reviewed the published work up to about 1961.

SELF-ASSOCIATION DUE TO HYDROGEN BONDING

Self-association phenomena in hydroxylic derivatives, amines, and other systems have been studied effectively by employing infrared spectroscopy. Since most of the hydroxylic or amino compounds of common interest are highly associated even at moderate concentrations, one generally has to examine very dilute solutions to observe the monomer–dimer equilibrium. A method based on infrared band intensities for evaluating equilibrium constants of dimerization was first described by Liddel and Becker.[12] Several other methods of evaluating equilibrium constants have also been reported. Thus Geiseler and Stockel[13] have developed a graphical method for the determination of equilibrium constants of complexes of various sizes. Step-wise self-association constants from infrared data have been calculated using a rigorous procedure by Vinogradov.[14]

Ever since Arnold and Packard[15] suggested that the chemical shift of the OH proton of ethanol was dependent on temperature and concentration a number of workers have investigated self-association due to hydrogen bonding by NMR spectroscopy. In the NMR spectra unique signals are not seen, due to the free and associated XH bonds, but the variation of the observed (average) chemical shift with concentration can be interpreted fairly satisfactorily to provide information on the hydrogen-bond equilibrium.

Concentration and temperature dependence of band position and intensity in the electronic absorption spectra have been made use of to study self-association of aromatic hydroxylic derivatives, amines, etc.

Alcohols, Phenols, and Related Compounds

The early infrared studies of Liddel and Becker[12] showed that in aliphatic alcohols the hydrogen-bond energies are in the order: methanol > ethanol > *t*-butanol. 2,2,2-Trifluoroethanol shows the free OH stretching band even in the pure liquid,[16] and the enthalpy of dimerization is considerably less (–3.8 kcal/mole) than in ethanol (–5 kcal/mole).[17] Similarly, 2,2,3,3,-tetrafluoropropanol is less associated than *n*-propanol.[17] The enthalpy values for the dimerization of aliphatic alcohols do not show any simple relation to pK_a values. The $\Delta\nu_{OH}$ values in fluoroalcohols are greater even though the ΔH° values are lower. Generally speaking, there appears to be no linear relationship between $\Delta\nu_{OH}$ and ΔH° in the self-association of aliphatic alcohols.[17]

Cook and Reece,[18] as well as Singh and Rao,[17] have examined the self-association of sterically-hindered hydroxylic compounds in carbon tetrachloride solution by infrared spectroscopy. The highly hindered 2,2,4,4-tetramethyl-3-*iso*- propyl-3-pentanol shows no evidence of self-association; only the free OH stretching absorption is found even in the pure liquid. The slightly less hindered 2,4-dimethyl- 3-ethyl-3-pentanol shows some dimerization.

Maguire and West,[19] as well as Singh and Rao,[17] have studied the dimerization of several phenols by infrared spectroscopy, and the equilibrium constant generally seems to decrease with the increase in the acidity of the phenol. The steric effects of various groups on the formation of hydrogen bonds in phenol have been examined[20]; alkyl groups with the exception of the *t*-butyl group seem to have little effect on the formation of hydrogen bond in *ortho*-substituted phenols,[20] as further corraborated by the recent thermodynamic data.[17] 2,6-Di-*t*-butylphenol seems to be monomeric even in pure liquid.

The recent infrared studies of Bellamy and Pace[21] seem to show that the dimers of alcohols and phenols have an open-chain rather than a cyclic structure. The hydrogen bonds in trimers and higher polymers are apparently stronger than those in dimers.

Cardinaud[22,23] has determined the ratios of the energies of hydrogen bonds to those of deuterium bonds in several alcohols. The deuterium compounds are generally more associated. Singh and Rao,[24] however, find lower K_2 and $-\Delta H^\circ$ for the dimerization of phenol-OD compared to phenol. Hadzi *et al.*[25] have characterized the strengths of hydrogen bonds in the self-association of alcohols and phenols on the basis of $\Delta\delta_{OD}/\Delta\nu_{OD}$.

Far-infrared spectra of phenol and *m*-cresol have shown strong bands at 177 and 147 cm^{-1} respectively, due to the ν_σ stretching mode of the hydrogen bond.[26] In *m*- and *p*-nitro phenols this band is found at 125 and 99 cm^{-1}, respectively.[27] Lake and Thompson[28] have examined a homologous series of alcohols in the 50–400-cm^{-1} region by an interferometric technique and have reported the low-frequency O–H. . .O bands as well as the torsional modes in the monomers. The far-infrared spectra of the hydrogen bonds in phenols have been examined in detail by Jakobsen and Brasch,[29] who find the O–H. . .O vibration frequency in the range 98–187 cm^{-1}. These workers have been able to use the Lippincott–Schroeder potential function to predict the hydrogen-bond stretching vibrations with fair success. The hydrogen-bond stretching frequencies in cresols and chlorophenols have been obtained by Hurley *et al.*[30]; they find no correlation between ν_σ and $\Delta\nu_{OH}$. Based on normal vibration calculations Ginn and Wood[31] have proposed that the band at 150 cm^{-1} in solutions of phenol in carbon tetrachloride (143 cm^{-1} in phenol- OD) is probably due to phenol cyclic trimers or polymers.

Davis *et al.*[32] studied the proton magnetic resonance spectra of solutions of methanol, ethanol, *iso*-propanol, and *t*-butanol in carbon tetrachloride and evaluated the enthalpies of dimerization. The trend in $-\Delta H^\circ$ values is methanol > ethanol >*i*-propanol >*t*-butanol. Saunders and Hyne[33], from their NMR studies of *t*-butanol, have concluded that the trimer is the preferred species, although small quantities of tetramer or higher polymers may be present. Becker[34] has shown that the data of Saunders and Hyne can be interpreted in terms of dimers and polymers. In order to determine the importance of the electrical effects of the substituents in deciding the hydrogen bonding in alcohols, Rao *et al.*[35] have compared the proton resonance spectra of ethanol and 2,2,2-trifluoroethanol. The δ_M as well as the K and $-\Delta H^\circ$ for dimerization are less in 2,2,2-trifluoroethanol than in ethanol.[36]

Connor and Reid[37] have examined the concentration dependence of δ_{OH} of a few hydroxylic compounds; a limited correlation between the association shifts and the integrated intensities of the OH stretching vibrations has been found. Martin[38] has determined the enthalpy of dimerization for 3-ethyl-3-pentanol to be –5.8 kcal/mole. The highly-branched chain alcohol 2,2,4,4-tetramethyl-3-*iso*-propyl-3-pentanol failed to show evidence for self-association and remains a monomer at all concentrations.[17,36] The self-association of di-*t*-butylcarbinol has been studied by infrared and NMR spectroscopy[39], and the results seem to show the formation of linear dimers ($-\Delta H^\circ = 4.2$ kcal/mole).

A strong monomer–dimer–polymer equilibrium is found in phenol by a study of the NMR spectra as a function of dilution.[40] The formation of *cis-trans* dimers of *ortho*-halophenols in dilute solution has been studied by Allan and Reeves.[41] The effect of steric hindrance by *ortho*-substituents on the self-association of phenols has been examined by several workers.[17,42,43] The association shifts in the hindered phenols have been correlated with the size of the substituents.

Forbes and Templeton[44] have obtained evidence for different types of hydrogen bonding from ultraviolet absorption spectra of aromatic derivatives. Hydrogen bonding in phenols and anilines have been studied in detail by ultraviolet spectroscopy.[45] In the case of phenol the concentration dependence of the $\pi \rightarrow \pi^*$ band (230μ) was not appreciable below a concentration of 0.01 mole/liter, indicating negligible association in this concentration region. Ito[46] has investigated the association of phenols in nonpolar solvents at different temperatures employing the 270 mμ band. The absorption bands of the associated molecules were found at shorter wavelengths than those of the free molecules. Rao and Murthy[47] have found that the λ_{max} values of both the 210 mμ and 270 mμ bands of phenol decrease with increase in concentration and attain limiting values at fairly high concentrations. Dearden[48] has derived expressions to calculate the extent of self-association in phenol and aniline from the ultraviolet absorption data; association constants have been calculated and an explanation of spectral effects due to self-association has been offered.

The far-ultraviolet absorption of hydrogen-bonded methanol (~195 mμ) has been studied by Kaye and Poulson.[49] The vacuum ultraviolet spectra of solid ammonia and water have been compared with the vapor spectra.[50] The absorption coefficients of H_2O, HDO, and D_2O have been studied at different temperatures and the results interpreted in terms of hydrogen bonding.[51]

Carboxylic Acids

Dunken and Fink[52] have reported that the infrared and Raman spectra of acetic acid show, besides the monomer band at 1758 cm^{-1} and the ring dimer band at 1668 cm^{-1}, two intermediate bands at 1713 and 1727 cm^{-1} attributable to an open dimer structure. Normal coordinate analysis of monomeric and dimeric formic and acetic acids has been carried out by Nakamoto and Kishita[53,54] employing the Urey–Bradley force field. Nakamoto and co-workers have also carried out the normal coordinate analysis of the acid maleate ion[55] and the acid carbonate ion.[56]

The association of benzoic acid has been examined[57] in the vapor phase and in dilute solutions of nonpolar solvents, and the $-\Delta H^\circ$ values vary in the order vapor > cyclohexane > carbon tetrachloride > benzene. Recently the dimerization equilibria of a few substituted benzoic acids have been studied by Hanrahan and Bruce.[58] The far-infrared spectra of gaseous formic and acetic acids have been reported in the 65–300 cm^{-1} region and bands due to intermolecular vibrations assigned.[59] Carlson *et al.*[60] have recorded the far-infrared spectra of dimers of formic acid, formic acid-*d*, acetic acid, and CD_3COOH in the gas phase in the region 33–400 cm^{-1} and assigned all the infrared active hydrogen-bond vibrations. In the solid state formic acid showed some structure in the 250 cm^{-1} region, while acetic acid had only a single intense band at 198 cm^{-1}. Jakobsen *et al.*[61] have reassigned the bands in the far-infrared spectra of formic acid on the basis of normal vibration analysis; they have also made assignments of frequencies of solid polymeric forms of formic and acetic acids. Liquid formic acid seems to consist of hydrogen-bonded polymers, while liquid acetic acid contains mainly dimers.

Frequencies and intensities of several carboxylic acids (deuterated and/or O^{18} enriched) in carbon tetrachloride in the 33–500 cm^{-1} region have been determined by Statz and Lippert.[62] The hydrogen-bond vibrations decrease with increasing mass of the substituent. Ginn and Wood[63] have shown that the doublet observed in the far-infrared spectrum of acetic acid vapor is unlikely to be due to proton tunneling; the doublet persists in the deuterated species as well. Low-frequency Raman spectra have been employed to study association in formic and acetic acids in aqueous as well as in hydrocarbon solutions.[64] Miyazawa and Pitzer[65] have studied the infrared spectra of four isotopic species of formic acid in vapor phase as well as in a solid nitrogen matrix in the 800–400 cm^{-1} region. *A* normal coordinate analysis of the *cis* and *trans* isomers of monomer, dimer, and polymer has been made.

Bellamy and Pace[66] have examined the infrared spectra of α and β forms of anhydrous oxalic acid and of the dihydrate. The order of the strength of the hydrogen bond formed by the carboxyl group is dihydrate > α > β. The assignments for the lower-frequency vibrations of these molecules have been discussed. The unusual breadth and shape of the ν_{OH} band of dimeric monocarboxylic acids may arise in part from the presence of an equilibrium mixture of open-chain and cyclic forms.[67]

Davis and Pitzer[68] have carried out a detailed investigation of the proton magnetic resonance spectra of formic, acetic, and benzoic acids in carbon tetrachloride solution. In benzoic acid it has not been possible to extrapolate the chemical shift–concentration curve to zero mole fraction; this indicates an appreciable amount of dimerization even in the very-low-concentration region. The monomer chemical shifts were calculated assuming the known equilibrium constants of dimerization. Muller and Hughes[69] have made a careful study of the association of benzoic acid in benzene and could not confirm the previously-reported high value of the monomer chemical shift (δ_M; δ_M does not vary much with temperature, while δ_D/δ_M has a positive temperature coefficient. From an examination of the literature reports it becomes apparent that the study of hydrogen bonding of carboxylic acids by NMR spectroscopy may often yield results which are difficult to interpret.

Forbes and Knight[70] have studied the monomer–dimer equilibria of benzoic acid in cyclohexane and cyclohexane–ether solutions. They found that the intensity of the 230 mμ band increases with the increase in the acid concentration in cyclohexane or in cyclohexane containing small amounts of ether. If the proportion of ether was large, the concentration dependence of the intensity diminished, indicating competitive intermolecular hydrogen bonding. The concentration limit above which λ_{max} becomes approximately constant in cyclohexane solvent seems to be related to the strength of the hydrogen bond. Ito[46] has investigated the effect of concentration on the 280 mμ band of benzoic acid and its derivatives. Absorption bands associated with the dimeric molecules were found to appear at longer wavelengths than those of the monomeric species. The observation was also confirmed by studying the temperature effect on the absorption spectra. Rao and Murthy[47] have found that the λ_{max} of both the 230 mμ and 280 mμ bands of benzoic acid increase with concentration and attain limiting values around the same concentration as that reported by Forbes and Knight.[70] The strength of the intermolecular (dimeric) hydrogen bond in substituted benzoic acids, as estimated from the observed concentration dependence of the ultraviolet absorption spectra, has been found to vary with the nature and position of the substituent.[71]

Lippert and Oechssler[72] have recorded the electronic absorption spectra of acetic acid and halogenated acetic acids to determine the characteristic absorption of dimers, and the results have been discussed theoretically. Dimerization in thiolbenzoic acid is found to be negligible on the basis of the solvent dependence of ultraviolet absorption spectra.[73]

Amines and Related Compounds

Wolff[74] has recorded the infrared spectra of methylamine in carbon tetrachloride in the temperature range –80°C to +20°C and has assigned the bands at 3290 and 3200 cm^{-1} to cyclic and linear structures, respectively. The NH stretching frequencies of aniline have been studied by Moritz[75] in carbon tetrachloride and chloroform. The presence of new bands on the low-frequency side in chloroform solution has been attributed to a specific interaction between chloroform and aniline. Whetsel,[76] from his studies on several amines of varying strengths, concluded that the low-frequency shoulders observed in the spectra of chloroform solutions are due to hydrogen-bonded species. An investigation of the infrared spectrum of aniline by Murthy[77] showed that the NH_2 asymmetrical and symmetrical stretching vibration frequencies decreased with increase in concentration, but there was no systematic variation of either the apparent molar extinction coefficient or the integrated band intensity of the bands with concentration.

Lady and Whetsel[78] have recently examined the self-association of aniline in cyclohexane by employing the first overtone of the symmetrical stretching band and have been able to interpret the data on the basis of monomer–dimer–tetramer equilibria. Singh and Rao[17] have examined the association of di-*n*-butylamine and have generalized that the N–H. . .N bond is appreciably weaker than the O–H. . .O bond.

Sandorfy and co-workers[79-81] have investigated hydrogen bonding in various amine hydrohalides by infrared spectroscopy. In the solid state the hydrogen bonds are of the type $N^+–H–X^-$. The broad bands around 3000 cm^{-1} in primary amine salts are shifted to lower frequencies in secondary and tertiary derivatives. The shift to higher frequencies is in the order HCl, HBr, HI. The absence of band shifting in aqueous solutions has been interpreted as due to the formation of an $N^+–H–O$ type of bond.

Hydrogen bonding in pyrazole derivatives and imidazole has been investigated extensively.[82,83] Hydrogen bonding in pyrrolidine has been examined by Linnell *et al.*[84]. Pyrazolones with an aromatic character have been found to show tautomeric equilibria between the hydroxy pyrazole and the zwitterion of the corresponding amide form, and the molecules are associated through strong intermolecular hydrogen bonds.[85] The Raman and infrared spectra of cyclic amines in the liquid state has shown the presence of N–H. . .N bonds.[86] The far-infrared spectra of imidazole and its deuterated derivatives have been studied[87]; the highest frequency (140 cm^{-1}) band has been assigned to the hydrogen-bond stretching vibration.

Dimerization of primary and secondary amides[88] has been investigated in the N–H stretching and deformation regions. Quantitative studies of the self-association of *N*-methylacetamide and *N*-phenylurethane, which contain *trans* and *cis* peptide bonds, respectively, have been reported recently.[89] The band around 100 cm^{-1} the far-infrared spectra of *N*-methylacetamide was assigned to the CO. . .HN stretching vibration.[90] The dimerization of caprolactam has been studied by Lord and Porro[91]; Kyogoku *et al.*[92] have examined the association of 1-cyclohexyl uracil and 9-ethyladenine.

Infrared spectra of ammonia and hydrazoic acid suspended in solid nitrogen have been examined by Pimentel and co-workers,[93,94] who have made assignments of various important frequencies of the hydrogen-bonded species. The hydrogen bonding of hydrazine has been studied recently by Durig *et al.*[95]. The association of thiocyanic acid and its deutero derivatives have been reported[96]; the association seems to be negligible up to $10^{-2}M$.

Feeney and Sutcliffe[97] proposed a monomer–tetramer equilibrium in mono-and di-ethylamines and *iso*-butylamine from a study of the NMR spectra. A weak monomer–dimer–polymer equilibrium is found to be present in aniline.[40] NMR spectroscopy has been employed to study the association of pyrazole.[98] A double resonance study of the self-association of pyrrole and its association to pyridine has been reported.[99] LaPlanche *et al.*[100] have studied the hydrogen bonding of *N*-monosubstituted amides in inert and bonding solvents. Hydrogen bonding in methyl-substituted hydrazines has been studied by Cook and Schug.[101]

Mercaptans and Related Compounds

Bulanin *et al.*[102] have found evidence for the association of ethyl and propyl mercaptans in carbon tetrachloride and have identified a band due to the associated species. David and Hallam[103] find that the association in thiophenols occurs at concentrations above 1.0 M. There appears to be definite experimental evidence that thiols form much weaker hydrogen bonds than the corresponding hydroxylic compounds. Thus the hydrogen-bond energy varies in the order S–H. . .S < N–H. . .N < O–H. . .O.[40,17] Studies of ultraviolet absorption spectra of thiophenol, aniline, and phenol in ether and cyclohexane have shown that the association is least in thiophenol and greatest in phenol.[77] Hydrogen bonding occurs in thiolacetic and thiolbenzoic acids to a smaller extent than the corresponding carboxylic acids;[73,104] the bonded C=O and S–H peaks have been identified.

Forsen[105] observed an almost linear dependence of the SH-proton chemical shift of ethyl mercaptan with concentration and interpreted this linearity as due to the presence of monomers and dimers in solution. Rao *et al.*[40] have observed this linearity in thiophenol and interpreted it as due to the presence of a weak monomer–dimer equilibrium over the 0–1 mole fraction range. The enthalpy of dimerization has been found to be ∿1.2 kcal/mole by Singh and Rao.[17] Linear δ_{SH}–concentration relations in mercaptans have also been reported by Colebrook and Tarbell.[106] Marcus and

Miller[107] have obtained the monomer–dimer equilibrium constants in a number of mercaptans. An NMR study has shown that thiolbenzoic acid undergoes negligible self-association.[73]

Self-Association Involving C–H bonds

Becker[108] studied the intensity of the C–H stretching band (3018 cm^{-1}) of chloroform as a function of the concentration in carbon tetrachloride. The results are consistent with the idea of weak hydrogen bonding between chloroform molecules.

Jumper *et al.*[109] have studied the NMR spectra of chloroform in carbon tetrachloride and cyclohexane solutions as a function of concentration and determined the self-association constant assuming that chloroform associates to form only dimers. The hydrogen bonding in chloroform and bromoform has been examined by Singh and Rao[17] employing infrared spectroscopy and the enthalpies of dimerization were found to be –3.2 and –1.0 kcal/mole, respectively.

Solvent effect studies on the proton resonance spectra of phenylacetylene have shown that the ethynyl proton is self-associated through the interaction of π electrons;[110] in ether or amine solvents new hydrogen bonds are formed at the expense of the C–H. . .π hydrogen bonds.

HYDROGEN BONDING BETWEEN DONORS AND ACCEPTORS

Infrared and NMR Studies

Of the variety of hydrogen-bonded systems examined by infrared spectroscopy, those systems where there is hydrogen bonding between a donor (proton acceptor) and an acceptor (proton donor) comprise by far the largest group. In the last few years quantitative thermodynamic and spectroscopic data have been obtained from infrared spectroscopy on most of the typical donor–acceptor pairs.

The NMR studies of the donor–acceptor systems provide data on the variation of the chemical shift of the XH proton with donor concentration. NMR association shifts have been employed as measures of the basicities of the donors or acidities of the acceptors. There have been some quantitative studies on the thermodynamics of donor–acceptor hydrogen-bonding equilibria in recent years.

Alcohols, Phenols, and Acids

Becker[111] studied the hydrogen bonding of alcohols with the several donors in carbon tetrachloride solution by employing infrared spectroscopy, and reported the equilibrium constants and enthalpies for the formation of 1:1 complexes. Data for fifteen O–H. . .O complexes lie fairly close to the Badger–Bauer line correlating ΔH° and $\Delta\nu$. The ΔH° values also correlate well with the integrated intensities, B°, of the bonded peaks. While $\Delta\nu$ and B° are temperature-dependent, $\Delta\nu_{1/2}$ is temperature-independent. Pineau and Josien[112] have determined the enthalpy of 1:1 complex formation between butanol and a series of ketones. The half-bandwidth of the associated band (as well as the K) is found to be a linear function of the relative frequency displacement $\Delta\nu_g/\nu_g$ (where ν_g is the gas-phase frequency of the OH stretching vibration of butanol).

Hydrogen bonding between methanol and several donors has been studied by means of the first overtone of the O–H stretching vibration,[113] and a definite relationship has been found between the hydrogen-bond strength and $\Delta\nu_{OH}$ as well as the basicity of the donors. Krueger and Mettee[114] have determined the equilibrium constants and $\Delta\nu_{OH}$ for the interaction of methanol with organic halides and aromatic hydrocarbons and question the usefulness of $\Delta\nu_{OH}$ as a measure of donor basicity. The interaction of pyridine with a number of aliphatic alcohols has been studied quantitatively and linear K–$\Delta\nu_{OH}$ and K–$\Delta\nu_{1/2}$ relations have been found.[115]

The dissociation energy of gaseous methanol–ethylether complex has been determined.[116] The interaction of ammonia and other amines with methanol in the vapor phase has been studied, and the structure of the OH band interpreted in terms of the sum and difference bands with the low-frequency hydrogen-bond stretching mode.[117] The far-infrared spectrum of gaseous methanol-triethylamine complex[118] shows a new band around 142 cm^{-1}. Complexes of phenol with amines studied by Wood and co-workers[119,120] have shown that the observed structure was not due to the sum and difference combinations with the hydrogen-bond stretching mode. Quantitative hydrogen-bonding data between a number of aliphatic alcohols and benzophenone (and acetone) have been given by Singh *et al.*[121] who find that K, ΔH°, and $\Delta\nu_{OH}$ increase with the acidities of alcohols.

Dunken and Fritzsche[122–124] have determined the thermodynamic constants of 1:1 complex formation for phenol with several donors. Aksnes[125] has determined the 1:1 equilibrium constants for phenol-triethyl phosphate and phenol-N-diethylacetamide systems. Studies with phenol and eighteen organophosphorus donors have shown that ΔF°, ΔH°, and ΔS° vary linearly with $\Delta\nu_{OH}$ in such a way that an increase in enthalpy is counteracted by a decrease in entropy.[126]

A linear relationship has been shown to exist between log K and the pK_a values of the donors for the systems pyridines–phenol.[127] This relationship is not found to be applicable to aliphatic amines. The influence of various substituents on *N,N*-disubstituted amides on their hydrogen-bond association with phenol has been discussed.[128] It has been found that $\Delta\nu_{OH}$ in various types of donors is proportional to the corresponding enthalpies of association;[129] $\Delta\nu_{OH}$ decreases in the order: organophosphorus compounds > amides > esters > ketones > aldehydes > ethers > pyridines and tertiary amines. The frequency shifts $\Delta\nu_{C=O}$ or $\Delta\nu_{P=O}$ also follow the same order.

A linear relationship between the $\Delta H°$ and the ionization potential of the donor has been found in a few phenol–base systems.[130] Dialkylsulfides and selenides are much weaker bases toward phenol than the corresponding ethers; ethers show a good $\Delta\nu_{OH}$-$\Delta H°$ linear relation.[131] Linear relationships between $\Delta\nu_{OH}$ and $\Delta\nu_{1/2}$ as well as other thermodynamic quantities have been discussed by Fritzsche.[132] The equilibrium constants and $\Delta\nu_{OH}$ for the interaction of organophosphorus compounds with phenol, ethanol, and α-naphthol are correlated with Taft σ^* constants of substituents.[133] Hydrogen-bonding studies of phenol with sulfuryl derivatives and sulfoxides show linear relationship between $\Delta\nu_{OH}$ and $\Delta H°$; $\Delta H°$ varies linearly with $\Delta F°$. In sulfuryl compounds the $\Delta\nu_{OH}$ correlated with the SO stretching vibrations.[134,135]

Singh *et al.*[121] have examined the interaction of several substituted phenols with benzophenone and triethylamine quantitatively and found that H°, $\Delta\nu_{OH}$ and K increase with the increase in acidity of the phenol. The hydrogen bonding of substituted benzophenones with phenol also showed substituent effects on K, $\Delta\nu_{OH}$ and $\Delta H°$, which could be correlated with Hammett σ constants.

Hydrogen bonding between methanol and phenol with a number of nitriles was first investigated by Mitra.[136] Phenol–nitrile systems have been investigated recently by White and Thompson[137], who have correlated the K values with $\Delta\nu_{OH}$. These workers find an increase in the C≡N stretching frequency on hydrogen bonding and have offered an explanation; Allerhand and Schleyer[138] have examined the ability of nitriles as well as isonitriles as proton acceptors. The spectral shifts due to hydrogen bonding between proton donors and nitriles with widely differing substituents correlated very well with Taft σ^* constants.

Schleyer and West[139] have studied the proton-accepting abilities of various alkyl halides from the $\Delta\nu_{OH}$ values of phenol and methanol. The proton accepting ability was found to be in the order I > Br > Cl > F. Derivatives of sulfur, phosphorus, and arsenic are also found to be good proton acceptors. The recent accumulation of reliable thermodynamic data on hydrogen bonds has made apparent the invalidity of Badger–Bauer rule when applied to more than a limited series of closely-related compounds. The thermodynamic data for the interaction of phenol with alkyl halides have been determined in order to test the validity of Badger–Bauer rule.[140] Both the free energy and enthalpy of interaction are found to be in the order F > Cl > Br > I, the reverse of the spectral shift order. An explanation for the absence of a correlation between ΔH° and $\Delta\nu$ has been offered. ΔH° measures the total energy of the interaction A–H. . .B–Y, partially compensated by the weakening of the A–H and B–Y bonds, while $\Delta\nu$ measures the weakening of the A–H bond only. Thus, if the proton donor A–H is kept the same, $\Delta\nu$ and ΔH° may correlate for minor structural changes in the proton acceptor.

West[141] obtained evidence for the formation of weak intermolecular hydrogen bonds between phenols and olefins. A new O–H band characteristic of the hydrogen-bonded complex was found to be 60–130 cm^{-1} lower than the absorption due to the unassociated phenolic OH group. The relative basicities of some olefins have been measured by determining the frequency shifts on hydrogen bonding. The interaction energies between various hydroxy derivatives and π electrons of cyclohexene and *iso*-propylbenzene have been determined by Wada[142], who found no relation between ΔH° and $\Delta\nu$.

Bellamy *et al.*[143] have investigated the association of substituted phenols with various types of ethers. Systematic variations in the numbers and sizes of alkyl substituents of the phenols and ethers have been made, and the results show that with minor exceptions (particularly 2,6-di-*t*-butyl phenol) changes in the sizes of substituents do not materially affect the strength of the hydrogen bond as measured by $\Delta\nu$, but do lead to quite large alterations in the equilibrium constants. Recently Singh and Rao[144] have quantitatively studied the interaction of sterically-hindered alcohols and phenols with various donors and found that the enthalpies of hindered OH groups are larger and comparable with those of simple alcohols and phenols. The low K values are due to entropy factors. $\Delta\nu_{OH}$ and ΔH° are not related in these sterically-hindered systems. Furthermore, there is no linear free-energy relationship in the hydrogen bonding of *p*-substituted 2,6-di-*t*-butyl phenols.

As indicated in various publications referred to earlier in this section, a number of workers have correlated the $\Delta\nu_{OH}$ with ΔH° or equilibrium constants. Joesten and Drago[145] proposed the following linear relationship between $\Delta\nu_{OH}$ and ΔH° for phenol with a variety of donors (totalling 15 in number): $-\Delta H^\circ$ (kcal/mole) = $0.016\ \Delta\nu_{OH} + 0.63$. Singh *et al.*[121] recently re-examined the $\Delta\nu_{OH}$–ΔH° relation for phenol-donor systems in view of the large amount of reliable data (111 donors) available from recent literature. These data when subjected to least-square treatment gave the relation $-\Delta H^\circ$ (kcal/mole) = $0.01\ \Delta\nu_{OH} + 1.75$, with a correlation coefficient r of 0.92 and a standard deviation s of 0.65. Although the correlation is considered satisfactory, the data with very low (< 3 kcal/mole) or high (> 6 kcal/mole) values of $-\Delta H^\circ$ do not fit the above relation well. When the data for the $-\Delta H^\circ$ range 3–10 kcal/mole were subjected to least-square treatment the following expression was obtained:$-\Delta H^\circ$ (kcal/mole) = 0.010 $\Delta\nu_{OH}$ + 2.37. The relation is nearly the same as that of Joesten and Drago.[145] This is understandable, since linearity between $\Delta\nu_{OH}$ and ΔH° should be strictly obeyed only for systems of comparable hydrogen-bond energies or hydrogen-bond lengths (O–H. . Y). An ideal linear relation would be expected only for systems with medium hydrogen-bond energies ($-\Delta H^\circ$ values between ∿3 and ∿6 kcal/mole). A least-square treatment of the data on 65 donors with $-\Delta H^\circ$ in this range gave the expression: $-\Delta H^\circ$ (kcal/mole) = 0.012 $\Delta\nu_{OH}$ + 1.87, with an excellent correlation coefficient of 0.995 as expected. The data with substituted phenols show deviations from the $\Delta\nu_{OH}$–ΔH° plot for phenol. Aliphatic alcohols show different relations between $\Delta\nu_{OH}$ and ΔH°, the deviations being greatest with strong donors such as amines. Recently Epley and Drago[146] have obtained the relationship between $\Delta\nu_{OH}$ and ΔH° by determining ΔH° values from calorimetric measurements. Their new relationship is given as $-\Delta H^\circ$ (kcal/mole) = 0.011 $\Delta\nu_{OH}$ + 2.79. It is interesting that the above relation is almost exactly the same as that proposed by Singh *et al.*[121] based on infrared spectroscopic data for the hydrogen-bond enthalpy range 3–10 kcal.

The use of $\Delta\nu_{XH}$ as measure of the acidities of acceptors or the basicities of donors could lead to erroneous conclusions, since there is likely to be no simple relationship between $\Delta\nu_{XH}$ and ΔH° in widely-differing systems. In a related series of donors, however, the $\Delta\nu_{XH}$–ΔH° relations may be useful to predict hydrogen-bond strengths or basicities. In a related series of hydrogen-bonded systems ΔF° values are generally proportional to ΔH° and therefore one may find linear $\Delta F^\circ - \Delta\nu_{XH}$ relations. For the interaction of a series of acceptors with a donor, however, it appears that $\Delta\nu_{XH}$ may not increase proportionally with ΔH°; instead, there may be an inverse relationship, as pointed out by Bernstein.[147]

The O–H stretching band of carboxylic acids is perturbed considerably by the presence of donors, but a quantitative study of the interaction is generally difficult due to extensive self-association in these acids.[121] The displacement of the C=O band of cyclohexanone has been measured in monochloro and trichloroacetic acids and trifluoroacetic acid.[148] The interaction between the carboxylic proton and a few π donors has been studied.[149] Infrared spectra of solid 1:1 pyridine–benzoic acid complexes have been investigated in detail and the spectroscopic properties have been correlated with ΔpK_a of the complexes.[150]

The infrared spectra of the adducts of hydrochloric, hydrobromic, and trichloroacetic acids with triphenyl phosphine oxide, dimethyl sulfoxide, diphenyl sulfoxide, pyridine 1-oxide and 2-picoline 1-oxide have been investigated.[151] The spectra in some cases have been studied down to 150 cm^{-1} in order to find the low-frequency vibrations of the hydrogen bond.

Hydrogen-bond energies have been determined for the complexes formed between HBr, HCl, and HI with acetone, dioxane, and ether.[152] The infrared spectra of HCL–ether,[153] HF–bases,[154] DCL–ethers,[155] and HNO_3–ether[156] systems have been examined. A band at 175 cm^{-1} in the HNO_3–dimethyl ether has been attributed to the hydrogen-bond stretching frequency.[157]

Drinkard and Kivelson[158] measured the proton resonance shifts of the OH proton of methanol and water as functions of concentration in acetone and dimethyl sulfoxide and found the bond formed with carbonyl oxygen to be stronger than that formed with sulfuryl oxygen. Hydrogen-bonding studies between alcohols (methanol and *t*-butanol) and several donors have indicated that in methanol–donor systems the association shift decreases in the order: tri-*n*-butylamine > acetone > acetonitrile > benzene.[121] With acetone as the donor the association shifts decreased in the order: trifluoroethanol > *t*-butanol > aniline > thiophenol > chloroform > bromoform > iodoform. Triphenylcarbinol shows a smaller association shift with acetone than triphenylsilanol, probably because the latter is a stronger acid.[121] The interaction of 2-propanol with *N*-methylacetamide[159], methanol with methyl cyanide, and methyl *iso*cyanide[160] have been reported. The lower K and $-\Delta H^\circ$ values for the association of ethanol in benzene compared to the values in CCl_4[32] give strong indication of OH. . .π bonding in benzene solvent.

Grancher[161] has correlated the chemical shifts of the hydrogen-bonded hydroxyl proton of phenol with the corresponding infrared frequency shifts and discussed the possibility of using the resulting linear correlation to derive hydrogen-bond energies. Eyman and Drago[162] have proposed a correlation between the phenol-OH chemical shift and the $-\Delta H^\circ$ of hydrogen bonding based on their data on 30 phenol–base systems; $\delta_{obs} = 0.748\Delta H^\circ - 4.68$. Somers and Gutowsky[43] have obtained thermodynamic data on the interaction of sterically-hindered phenols with dioxan, and Takahasi and Li[163] with *N*-methylacetamide and *N,N*-dimethylacetamide. NMR dilution shift studies of some aliphatic carboxylic acids in benzene and pyridine have indicated the existence of O–H. . .π type bonds between the carboxylic hydrogen and π-electron system.[164] Davis and Pitzer[68] have interpreted the temperature-dependent monomer chemical shifts of benzoic acid protons in benzene in terms of a possible interaction between the carboxylic proton and the π-electron system of benzene.

Amines, Amides, and Related Compounds

Bellamy *et al.*[165] have found that the frequency shifts of the NH stretching band of pyrrole in a variety of solvents are linear with the corresponding shifts of many X–H stretching frequencies in the same solvents and concluded that the major factor involved in all the cases is hydrogen bonding. Dunken and Fritzsche[166] have determined the thermodynamic constants of 1:1 complexes formed between indole and several donors. Hydrogen bonding of pyrrole with several nitriles has been studied quantitatively by Mitra.[136]

Whetsel and Lady[167] have reported quantitative hydrogen-bonding data for 1:1 complexes of various amines with chloroform. The hydrogen-bonded complexes of aniline and *N*-methyl aniline with a variety of ethers have also been investigated by these workers.[168] Hydrogen bonding of indole with a variety of donors has been reported.[163,169] Singh *et al.*[121] have studied the hydrogen bonding between *N*-methyl aniline and a few donors and concluded that the N–H. . .Y hydrogen bonds are invariably weaker than the corresponding O–H. . .Y hydrogen bonds.

Infrared spectra of formamide recorded in a number of solvents have shown evidence for solute–solvent hydrogen bonding.[170] The formation of a complex between *N*-methylacetamide and benzene has been demonstrated by Suzuki *et al.*[171]. Hydrogen-bonding equilibria of *N*-methylacetamide and *N*-phenylurethane with a number of bases have been studied quantitatively by Bhaskar and Rao.[89] Hydrogen bonding between pairs of derivatives of the base constituents of nucleic acids and their analogs has been studied in chloroform solution by Pitha *et al.*[172] Kyogoku *et al.*[92] have carefully studied the hydrogen bonding between adenine and uracil derivatives in order to explain the stability associated with hydrogen bonds present in DNA.

NMR spectra of a series of amines in carbon tetrachloride, benzene, pyrrole, 2,6-dimethylbenzene, and 1,3,5-trimethylbenzene have indicated that amines are influenced by the π orbitals of the solvents.[173] Yamaguchi[174] has studied the concentration dependence of the amine-proton chemical shifts of aniline, toluidines, and xylidines in acetone and found them to be appreciable, indicating hydrogen-bond formation between the solvent and the solute. An equilibrium constant has been reported for the association of *N*-methyl aniline with acetone in carbon tetrachloride.[175] Schneider and Reeves[176] have demonstrated the formation of 1:1 complexes between acetylacetone and dimethylamine. Solute–solvent hydrogen bonding between pyrrole and pyridine and between pyrrole and oxygenated bases has been investigated by Freymann and Freymann.[177] The hydrogen-bonding interaction between free nucleosides (complementary bases) in dimethyl sulfoxide and chloroform has been investigated.[178]

Mercaptans

Qualitative infrared studies on the interaction of thiophenol with a number of donors showed evidence for hydrogen bonding[121]; quantitative studies were not possible due to the absence of separate bands due to hydrogen-bonded species.

Mathur *et al.*[179] have reported the proton magnetic resonance studies of hydrogen bonding between the S–H proton of thiophenol and dimethylformamide, pyridine, tributylphosphate, and benzene in carbon tetrachloride solution; the equilibrium constants and enthalpies for 1:1 complex formation have been evaluated. The enthalpy of hydrogen-bond formation has also been determined. Hydrogen bonding between some thiols and *N*-methylacetamide has been reported.[180]

Donor–Acceptor Interaction Involving C–H bonds

The proton-donating ability of acetylene and its derivatives has been studied by a number of workers employing infrared spectroscopy. The shifts of the C–H stretching and deformation bands of acetylene in a variety of donors have clearly indicated solute–solvent hydrogen bonding.[181–183] West and Kraihanzel[184] have determined the relative acidities of terminal acetylenes by means of C–H band shifts in reference bases. Both oxygen- and nitrogen-containing compounds have been studied as donors[185], and the nitrogen-containing compounds are found to be better bases than the oxygen-containing compounds. Recently Creswell and Barrow[186] have studied the association of acetylene with acetone and have reported 1:1 and 2:1 formation constants.

The C–H stretching frequency of chloroform shows a low-frequency component in the presence of ammonia; Cannon[187] has shown that this is due to the interaction between the lone-pair electrons on the nitrogen atom and the CH bond of chloroform. Whetsel and Kagarise[188] have studied the C=O stretching bands of acetone and cyclohexanone in cyclohexane–chloroform mixtures. The results have shown that chloroform forms 1:1 and 1:2 complexes with ketones. Allerhand and Schleyer[189] have investigated the ability of various CH bonds to act as proton donors in hydrogen bonding; strong donors like pyridine-d_5 and dimethyl sulfoxide-d_6 have been used. Pentachlorocyclopropane, pentachloroethane, Br_2CHCN, bromoform, iodoform, and many other compounds gave larger C–H shifts than chloroform.

There have been a number of papers related to the study of the proton-donating ability of haloforms employing NMR spectroscopy. Kaiser[190] has analyzed the NMR line shift of chloroform in dioxan in terms of an equilibrium between free chloroform and 1:1 and 2:1 chloroform–dioxan complexes. McClellan *et al.*[191] have reported the NMR shifts for mixtures of chloroform with dimethylsulfoxide, ethylene, and propylene carbonates and determined the equilibrium constants for the 1:1 and 2:1 chloroform–dimethylsulfoxide complexes and enthalpy for the 1:1 complex. Howard *et al.*[192] have obtained the equilibrium constants for the 1:1 complexes of chloroform with various donors. Association shifts of chloroform, bromoform, and iodoform in various donor solvents show that the proton-donating ability varies in the order: chloroform > bromoform > iodoform.[193] A similar trend has been found with acetone as a donor.[121] The interaction shifts of chloroform, bromoform, and iodoform in benzene have been found to be considerable, and these shifts may be interpreted in terms of a weak π-donor association.[36]

Creswell and Allred[194] have obtained thermodynamic data for the association of tetrahydrofuran with fluoroform, chloroform, bromoform, and iodoform, and the enthalpies vary in the order $CHCl_3 > CHBr_3 = CHF_3 > CHI_3$. Calorimetric and quantitative NMR studies have been carried out on the chloroform– pyridine system.[195]

The low-field shift δ_{CH} of phenylacetylene in donor solvents has been ascribed to the interaction with the nitrogen or the oxygen atoms of the solvent molecules; the high-field shift in benzene is ascribed to the interaction with the π-electron system of benzene.[196]

Electronic Spectroscopic Studies

Becker[197] obtained evidence for hydrogen bonding between ethanol and benzophenone from ultraviolet absorption spectroscopy. An isosbestic point was noticed in the benzophenone–chloroform system, indicating the presence of a hydrogen-bond equilibrium. Ito *et al.*[198] measured the solvent blue-shifts in a series of solvents for a few carbonyl systems and found hydroxylic solvents to produce large blue-shifts. Complementary infrared studies indicated that the ground electronic state of the solute is stabilized by hydrogen bonding. Balasubramanian and Rao[199] have investigated solvent effects on the $n \to \pi^*$ transitions of a number of chromophores. Isobestic points were observed in the $n \to \pi^*$ bands of ethylenetrithiocarbonate in heptane–alcohol and heptane–chloroform solvent systems, indicating the existence of equilibria between the hydrogen-bonded and free species of the solute. The blue-shifts decreased in the order: water, 2,2,2-trifluoroethanol, methanol, ethanol, *i*-propanol, *t*-butanol. Approximate hydrogen-bond energies for several donor–acceptor systems have been estimated on the basis of blue-shifts. Chandra and Basu[200] have determined equilibrium constants of pyridazine, diethylnitrosamine, and benzophenone with various alcohols and found them to be in the order primary > secondary > tertiary. Although the equilibrium constants of Chandra and Basu were determined at concentrations of alcohols where they would exist appreciably as dimers and polymers, these constants as well as some others similarly obtained by Singh *et al.*[121] show good agreement with those from infrared spectroscopy.

The proton-donor properties of phenols, carboxylic acids, mercaptans, and amines have been studied by measuring the solvent blue-shifts of the $n \rightarrow \pi^*$ transitions of various chromophores in these solvents.[121] The solvent shifts are found to be mainly determined by solute–solvent hydrogen bonding. In any series of solvents the blue-shift increases with the acidity of the proton donor. The effect of proton-donating solvents such as chloroform, methanol, *t*-butanol, and water on the $n \rightarrow \sigma^*$ transitions of *i*-propyl iodide has been studied.[201] Hydrogen bonding between ethylamines and several prototropic solvents has been studied by means of the solvent blue-shifts of $n \rightarrow \sigma^*$ transitions.[202] Equilibrium constants for triethylamine with various alcohols have been determined. Hydrogen bonding has been found to be an important factor in interpreting effects of hydroxylic solvents on the electronic absorption spectra of iodide ion.[203]

Baba and Suzuki[204] have determined the equilibrium constants for 1:1 complex formation of phenol and naphthols with dioxane. The hydrogen-bond energies for the ground and excited states of the solute molecules have been determined. Chandra and Banerjee[205] have derived an expression for determining the equilibrium constant between a donor and acceptor where the donor, the acceptor, and the complex have comparable absorption in the spectral region of interest. The thermodynamic data for the association of a series of amides with several proton donors has been determined,[206] and there appears to be no correlation between equilibrium constants and Taft σ^* constants.

The $\pi \rightarrow \pi^*$ red-shifts have been used in determining 1:1 equilibrium constants of tetrahydrofuran, dioxan, and ether with different phenols.[207] The spectra of *p*-nitrophenol and *m*- and *p*-nitroaniline in ether–hexane have shown a red-shift with the increasing proportion of ether; the equilibrium constants in these systems have been determined.[208]

The changes in the spectra of naphthols in aromatic solvents are explained to be due to O–H. . .π hydrogen bonding;[209] the equilibrium constants for such systems are reported. The interaction of chlorophenols with dioxan and triethylamine has been investigated.[210] Quantitative studies on the hydrogen-bonded complexes of phenol with trimethylamine *N*-oxide[211] and a nitrile *N*-oxide[212] have been reported.

The hydrogen-bonding capacity of the proton N–H group in various amines has been investigated by utilizing the $n \rightarrow \pi^*$ blue-shift phenomenon.[121,213] The equilibrium constant for hydrogen-bond formation of amines with donors are in the order primary > secondary > tertiary. Linnell,[214] from his studies on the $n \rightarrow \pi^*$ band of pyridazine, suggested the possibility of one molecule of water forming two hydrogen bonds with pyridazine.

INTRAMOLECULAR HYDROGEN BONDING

Infrared Studies

Intramolecular or chelated hydrogen bonds are readily identified by infrared spectroscopy[3]; intramolecular hydrogen bonds can be distinguished from the intermolecular type because the latter variety is concentration-dependent, while the former is not. Intramolecular hydrogen bonding in hydroxy, carboxy, amino, and other derivatives have been investigated extensively and the frequency shift due to hydrogen bonding found to be a good measure of the strength of the hydrogen bond.[3] Intramolecular hydrogen bonding has been found to be very useful in the study of the conformation of organic molecules. Applications of intramolecular hydrogen bonding to stereochemistry have recently been reviewed by Tichy.[215]

The conformation of mono-, di-, and tri-haloethanols have been examined by Krueger and Mettee,[216] who have recognized intramolecular O–H. . .X (X = halogen) hydrogen bonds in these compounds. The intramolecular hydrogen bonds in cyanoalcohols have been studied.[217,218] The *gauche* conformation of 2-cyanoethanol is stabilized by the intramolecular hydrogen bond.

The infrared spectra of nitroalcohols have indicated that the structures of these compounds are best represented in terms of intramolecular hydrogen bonds between the OH and NO_2 groups.[219] Kuhn *et al.*[220] have found the bonded OH band of β-nitroethanol at 3608 cm^{-1} compared to the free OH band at 3623 cm^{-1}; a *gauche* form of the compound is apparently stabilized by a weak hydrogen bond. Even though there was some controversy about intramolecular hydrogen bonding in β-nitroalcohols for some time, the recent infrared studies of Baitinger *et al.*[221] have established the presence of such bonding beyond doubt. The intramolecular hydrogen bonds of some nitroalcohols have been reported along with their conformations by Krueger and Mettee.[218]

Intramolecular hydrogen bonding in 1,3-diols has been investigated by Julia *et al.*[222]. The frequency shift $\Delta\nu_{OH}$ was in the range 30–110 cm^{-1}. The influence of conformation of the neighboring groups and the ring have been discussed. Schleyer[223] has examined the Thorpe–Ingold hypothesis of valency deviation by determining the spectral shifts due to intramolecular hydrogen bonding in propane-1,3-diols. Hydrogen bonding in triols has been investigated and the intensity ratio (free OH)/(bonded OH) is found to be less than that for diols.[224] Apparently, triols exist in a conformation which allows two intramolecular hydrogen bonds. The $\Delta\nu_{OH}$ due to intramolecular hydrogen bonding in 1,2-diols decreases with increasing azimuthal angle between the OH groups, while the reverse is true of 1,4-diols.[225] Based on the infrared data on intramolecular hydrogen bonding in 2,5-di-*t*-butyl-1,4-cyclohexane diol, a non-chair conformation has been proposed for the molecule.[226]

The interaction between the OH groups and the π electrons in ω-hydroxy-1-alkenes has been determined by Oki and Iwamura.[227] A number of benzyl alcohol derivatives were studied to determine the electronic effect of the substituents on the interaction between the OH group and π electrons,[228] substituted benzylimines were also studied. The substituent effects have been investigated in a series of ω-aryl alkanols.[229] Intramolecular hydrogen bonding involving double bonds, triple bonds, and cyclopropane rings has been studied by Schleyer *et al.*[230]. The O–H. . .π intramolecular hydrogen bonding in 2-phenyl ethanol[231] and acetylene alcohols[232] has been examined.

Baker and Kaeding[233] showed that in *ortho*-halophenols the $\Delta\nu_{OH}$ varies in the order F < Cl < Br < I. However, in 2,6- dihalophenols the order is Cl > Br > F > I. This anomalous order has been explained as due to the varying sizes of the halogens and to an orbital–orbital repulsive interaction. Measurements of intensity ratio I_{trans}/I_{cis} in *ortho*-halophenols at various temperatures have permitted the evaluation of the enthalpy on the assumption that I_{trans}/I_{cis} is proportional to the equilibrium constant.[234–236] The enthalpy values and hence the hydrogen-bond strengths vary in the order I < Cl < Br. *O*-fluorophenol shows only one OH band and seems to be present in the planar form. The enthalpies of the intramolecular hydrogen bonds in *ortho*-halophenols and their deuterated derivatives have been determined in the vapor phase by Lin and Fishman.[237]

The OH stretching band in *ortho*-nitrophenol is found at very low frequencies ($\Delta\nu_{OH}$ = 400 cm^{-1}) due to the high strength associated with the resonance-stabilized hydrogen bond. The intramolecular hydrogen bonds of *ortho*- nitrophenol and its derivatives have been examined by Dabrowska and Urbanski.[238] In 2-nitrophenol and 2,6-dinitrophenol[27] the O...O stretching band is found at about 100 cm^{-1}. Based on measurements of the frequencies and intensities of the fundamental as well as the first and second overtones of O–H stretching vibration in *ortho*-nitrophenol it has been shown that the strong hydrogen bonding causes increased mechanical anharmonicity.[239]

The studies of Baker and Shulgin[240] on the intramolecular hydrogen bonding in *ortho*-hydroxy aromatic Schiff bases have permitted the direct physical determination of Hammett σ constants of *ortho*, as well as *meta* and *para*, substituents. Nyquist[241] has studied the O–H out-of-plane deformation vibration in the 300–360 cm^{-1} region in various *ortho*-substituted phenols and suggested that this bending mode is a good "group frequency" in view of its reasonable intensity and unique broadness; some correlations with certain chemical parameters have been found. Baker *et al.*[242] have shown how the temperature dependence of the OH-band intensity can be used to calculate the enthalpies of intramolecular hydrogen bonds.

Baker and Shulgin[243] find evidence for intramolecular hydrogen bonds between hydroxyl groups and π electrons in compounds of the type 2-allylphenol and compare the O–H. . .π hydrogen bonds with other intramolecular hydrogen bonds in *ortho*-substituted phenols. Based on $\Delta\nu_{OH}$ the hydrogen-bond strengths were suggested to vary in the order $F < O < Cl \approx \pi < Br < I < S <:O < N$. Oki and Iwamura[244] have studied the intramolecular hydrogen bonding between OH and π electrons in 2-(ω-alkenyl)-phenols and 2-(ω-phenylalkyl)-phenols. From the variation of intensities with temperature the energetics of interaction for 2-allylphenol, 2-benzylphenol, 2-*iso*-propenylphenol, 2-hydroxybiphenyl, and benzyldimethyl carbinol have been determined.[245,246]

Friedman[247] has surveyed a number of simple compounds containing similar intramolecular O–H. . .O and O–H. . .N hydrogen bonds and has shown that the latter bonds are stronger. A detailed study of internal five- and six-member rings involving bonding of phenolic hydroxyl to azomethine nitrogen has also supported the above conclusions. Correlation between the O–H stretching frequency and the O–H. . X distance in intramolecular hydrogen-bonded systems has been examined by Dearden[248] and Matsui *et al.*[249]. Ogoshi and Nakamoto[250] have carried out the normal coordinate analysis of the enol forms of acetylacetone and hexafluoro acetylacetone.

Oki and Hirota[251,252] have studied intramolecular hydrogen bonding in α-keto-, α-alkoxy-, and α-aryloxy-carboxylic acids by infrared spectroscopy. An attempt has been made to establish the correct assignments of the C=O and O–H bonds for the *cis* and *trans* structures of alkoxy acids.[253] Intramolecular hydrogen bonding in *ortho*-methoxybenzoic acid, *ortho*- aryloxybenzoic acids, methoxynaphthoic acids, and methyl hydroxynaphthoates have been studied by Oki and co-workers.[254–256] Intramolecular hydrogen bonds have been found to occur between the carboxyl OH group and the nitrogen in pyridine-2- and quinoline-2-carboxylic acids.[257] Oki and Hirota[258] have reviewed the literature on intramolecular hydrogen bonding of carboxylic acids.

Intramolecular hydrogen bonding in *ortho*-substituted anilines has been investigated by a number of workers by infrared spectroscopy, and evidence for appreciable intramolecular hydrogen bonding is found only when nitro groups are present in both the 2 and 6 positions.[259] Solvent-effect studies on the N–H stretching frequencies of a number of 2-nitroanilines have indicated that intramolecular hydrogen bonding is present in 2,6-dinitroanilines but not in other 2-nitroanilines.[260] Intramolecular hydrogen bonding in secondary aromatic amides with *ortho*-nitro substituents has been indicated.[261] Hambly and O'Grady[262] have made a detailed study of intramolecular hydrogen bonding in several amines and monodeuterated 2-bromo-, 2-iodo-, and 2-cyano-anilines and showed evidence for extremely weak hydrogen bonding in these derivatives. The spectra of monodeuterated 2-alkoxy-, 2-fluoro-, and 2-phenyl-anilines did not show any evidence for intramolecular hydrogen bonding. Krueger[263] has found that the symmetrical stretch and first overtone bands of some *ortho*-substituted anilines show splitting. The results have been interpreted in terms of the double minimum potential.[264]

Krueger and Smith[265] have reported on the intramolecular N–H. . .halogen hydrogen bond in 5- and 6-member chelate rings; the bond strength increases in the order F < Cl < Br < I. Intramolecular hydrogen bonds in ethanolamine and its *O*- and *N*-methyl derivatives,[266] ethyleneamines, and other diamines[267] and *ortho*-phenylenediamines[268] have also been studied by Krueger and co-workers.

NMR Studies

Since intramolecular hydrogen bonding is concentration-independent, the shape of the NMR chemical shift–concentration curve can be used in detecting the presence of intramolecular hydrogen bonds. Porte *et al.*[269] have measured the OH chemical shifts for phenol, α-naphthol, 9-phenanthrol, and for chelated *ortho*- substituted derivatives and have shown that the $\Delta\delta$ values provide a measure of the strengths of intramolecular hydrogen bonds. The $\Delta\delta$ values are proportional to $\Delta\nu_{C=O}$ values determined from infrared spectroscopy. Reeves *et al.*[270] determined the nuclear shielding parameters for 24 intramolecularly-hydrogen-bonded phenols and naphthols and found that the $\Delta\delta$ values were approximately proportional to the $\Delta\nu_{OH}$ values. Based on the above characteristics Allan and Reeves[271] have examined the dilution shifts for the OH proton in *ortho*-halophenols and determined the equilibrium constants and enthalpies for the *cis-trans* equilibria.

The chemical shift of the OH proton and the C=O stretching frequency have been correlated in a number of hydrogen-bonded chelates.[272] C^{13}-chemical shifts of intramolecularly hydrogen-bonded C=O group have been examined[273]; the C^{13} resonance shifts down field due to the formation of the hydrogen bond. Evidence for intramolecular hydrogen bonding in tropolone has been found from NMR spectra.[274]

Methyl and ethyl diacetoacetates undergo complete enolization and form strong hydrogen bonds.[275] 3-Cinnamoyl pentane-2,4-dione shows the presence of two different chelated enols. Chemical shifts of the OH groups in a number of intramolecularly-hydrogen-bonded alcohols and other hydroxy derivatives have been reported by Shapetko *et al.*[276] Intramolecular O–H. . .π hydrogen bonding in norbornaols has been studied.[277] Intramolecular hydrogen bonding in anions of carboxylic acids has been investigated.[278,279]

Electronic Spectroscopic Studies

The ultraviolet absorption spectra of nitroparaffin derivatives containing an OH or NH group have failed to show the $n \rightarrow \pi^*$ band (at 260–270 mμ) of the NO_2 group.[280] This has been attributed to the formation of a six-membered chelate ring by internal hydrogen bonds between the NO_2 and OH or NH_2 groups. In the ultraviolet absorption spectra of isomeric nitroanilines the *ortho*-isomer showed a widened and displaced (toward longer wavelengths) band as compared to the other isomers.[281] The effect of the intramolecular hydrogen bond (in going from vapor to hexane solution) in compounds of the type A–C_6H_4–D, where A is an electron acceptor and D an electron dönor, on the charge-transfer bands has been investigated in a large number of compounds.[282]

A variety of methods have been discussed by Dearden and Forbes[284] to obtain information concerning intramolecular hydrogen bonding in phenols and anilines from ultraviolet spectra. One method is to note the spectral changes observed between the *ortho*-substituted phenols (or anilines) and the corresponding *meta*isomers. A bathochromic wavelength displacement of 5–12.5 mμ of the intense primary band is found in the *ortho*derivatives compared to the corresponding *meta* isomer. Another method is to note the absence of an appreciable spectral change on altering the solvent conditions. The absence of an appreciable spectral change in cyclohexane and ether or ethanol solution also provides support for the presence of a strong intramolecular hydrogen bond. The intramolecular hydrogen bonding in *ortho*- nitrophenol has been discussed with special reference to the effects of steric interactions on the absorption bands.[284] The effect of alkyl substituents on the O–H stretching band parallels the effect on λ_{max} values in the ultraviolet spectra; the λ_{max} is progressively displaced to longer wavelengths with increasing size of the alkyl substituent.

A study of the ultraviolet and infrared spectra of a number of 2,4-dinitrodiphenylamines in different solvents have shown that these compounds are nonplanar in many solvents and has also indicated the presence of an intramolecular hydrogen bond.[285] The hydrogen bond is rendered strong in these compounds by the presence of the *para*-nitro substituent, which increases the positive charge on the amino group. Pinchas[286] and Forbes[287] have studied the infrared, ultraviolet, and NMR spectra of *ortho*-nitrobenzaldehyde and related compounds and have concluded that the aldehydic C–H bond forms an intramolecular hydrogen bond with the oxygen of the nitro group.

Ultraviolet and infrared spectral data have shown that *ortho*- methoxy benzoic acid forms a stronger intramolecular hydrogen bond than *ortho*-fluoro benzoic acid.[71] The complete absence of the monomer C=O band in the infrared spectra of a dilute solution of *ortho*-fluorobenzoic acid suggests that the monomer is destabilized by the *ortho*-fluoro substituent; the possibility of the formation of an intramolecularly-bonded structure has thus been ruled out. Oki and Iwamura[288] have examined steric effects on the O–H. . .π interaction in 2-hydroxybiphenyl.

AQUEOUS SYSTEMS

Recently several workers have examined the infrared spectra of water and aqueous solutions in some detail to investigate hydrogen bonding in these systems. Buijs and Choppin[289] investigated the spectrum of water in the 1.16–1.25 μ region and suggested that these bands were due to the presence of three types of water molecules, those having zero, one, and two hydrogen bonds, respectively, to neighboring molecules. Luck[290] has investigated several of the near-infrared bands of H_2O, but his calculations are not very extensive. Luck[291] suggests that there is a considerable fraction of hydrogen-bonded species in water up to the vicinity of the critical point, basing his conclusion on experiments with water under high pressure. Thomas *et al.*[292] have examined the spectra of liquid H_2O, liquid D_2O, H_2O ice, and D_2O ice in the near-infrared region at different temperatures and have calculated the concentrations of unbonded, singly-hydrogen-bonded, and doubly-hydrogen-bonded water molecules in liquid H_2O and liquid D_2O. The results on D_2O seem to be consistent with their earlier calculations. Luck[293] has recently studied the temperature dependence of four overtone bands of water and has interpreted the results in terms of a simple cluster model. Worley and Klotz[294] have studied the near-infrared spectra of H_2O–D_2O solutions at different temperatures and have estimated ΔH° values of –2.4 kcal/mol for the O–H. . .O bond. Lippincott *et al.*[295] have recorded infrared spectra of water and ice at different temperatures and pressures and shown that at high pressures the hydrogen bonds responsible for the open structure of ice collapse. In the close-packed structure of dense ice, hydrogen bonding produces a minor effect on the vibrational modes.

Recent studies[296] on dioxan–water and pyridine–water mixtures have shown that the high-frequency shifts for the OH stretching band in either solvent are due to a reduction of hydrogen bonding present in pure water.

The low-frequency shift of the C–O stretching mode of dioxan in aqueous mixtures indicates that oxygen is the hydrogen-bonding site. Raman studies on dimethyl sulfoxide–H_2O mixtures[297] have shown that the intensities of Raman bands decrease continuously with increasing water content of the mixture. The S=O stretching frequency of dimethylsulfoxide is shifted to lower frequencies with increasing water content. The association of water with various bases and anions has been investigated by Mohr *et al.*[298] The effect of dissolved KBr, KOH, and KCl on the Raman spectrum of water has been studied by Busing and Hornig,[299] and bands due to H_3O^+ and to free and bonded OH^- peaks have been identified. The effects of temperature and electrolytes on the Raman spectra of water, in particular on the low-frequency band around 170 cm^{-1}, has been studied by Walrafen.[300,301]

Holmes *et al.*[302] have investigated the chemical behavior of water protons in a series of organic solvents by NMR and infrared techniques. The chemical shifts of the water proton and the O–H stretching frequencies in various solvents have been correlated with each other and with hydrogen-bond energies. The dimerization constants of water in these solvents were determined and the theory of chemical shifts due to hydrogen bonding has been discussed. Dioxan–water and pyridine–water mixtures have shown a marked concentration dependence of the proton resonance shifts, indicating the presence of strong interaction, probably hydrogen bonding.[303] The proton resonance of water has been studied at various temperatures in the presence of electrolytes. The results, along with the infrared data, are interpreted in terms of hydrogen bonding in these systems.[304] The hydrogen-bond energy between acetaldehyde and water has been determined by employing NMR spectroscopy.[305] The thermodynamics of hydrogen bonding of water with various donors in cyclohexane or carbon tetrachloride has been studied quantitatively.[306,307]

The O^{17} resonance in aqueous solutions of electrolytes has been reported.[308] Muller[309] has considered various structural models of liquid water to predict the number of hydrogen bonds broken during the melting of ice and the variation of the fraction of unbroken bonds with temperature; the latter is important in determining the position of the proton signal in aqueous systems.

CONCLUDING REMARKS

An examination of the recent literature reveals that the thermodynamic and spectroscopic data on most of the systems of interest have been reported. Most of the available spectroscopic and thermodynamic data have been reviewed and tabulated by Murthy and Rao.[310] There is, however, considerable scope for studying hydrogen bonding in aqueous systems, molecules of biological interest, and molecules in vapor phase. Use of far-infrared and laser Raman spectroscopy will certainly be of great value. Hydrogen bonding in electronically-excited states of molecules needs to be investigated in greater detail.

ACKNOWLEDGMENT

C. N. R. Rao is thankful to the Department of Chemistry, Purdue University, Lafayette, Indiana for a Visiting Professorship for the year 1967–68.

REFERENCES

1. G. C. Pimentel and A. L. McClellan, *The Hydrogen Bond*, W. H. Freeman and Company, San Francisco (1960).
2. A. S. N. Murthy and C. N. R. Rao, Hydrogen Bonding (1958–1963) – A Review, Chem. Tech. Report 1, Indian Institute of Technology, Kanpur, India, 1964, as reported in *Spectrochim. Acta.* **21**, 215 (1965) and *Chem. Abstr.* **63**, 15576g (1965).
3. C. N. R. Rao, *Chemical Applications of Infrared Spectroscopy*, Academic Press, New York (1963).
4. H. E. Hallam, in: *Infrared Spectroscopy and Molecular Structure*, (M. Davies, ed.), Elsevier Publishing Company, Amsterdam (1963).
5. J. A. Pople, H. J. Bernstein, and W. G. Schneider, *High Resolution Nuclear Magnetic Resonance*, The McGraw-Hill Book Co., New York (1959).
6. J. W. Emsley, J. Feeney, and L. H. Sutcliffe, *High Resolution Nuclear Magnetic Resonance Spectroscopy*, Vols. I and II, Pergamon Press, Oxford (1966).
7. E. Lippert, *Ber. Bunsen. Ges. Physik. Chem.* **67**, 267 (1963).
8. P. Laszlo, *in: Progress in NMR Spectroscopy*, Vol. 3. (J. W. Emsley, J. Feeney, and L. H. Sutcliffe (eds.), Pergamon Press, Oxford (1967).
9. C. N. R. Rao, *Ultraviolet and Visible Spectroscopy – Chemical Applications*, 2nd ed., Butterworths, London, and Plenum Press, New York (1967).
10. H. H. Jaffe and M. Orchin, *Theory and Applications of Ultraviolet Spectroscopy*, John Wiley & Sons, New York (1962).
11. J. C. Dearden, *Rev. Univ. Ind. Santander*, **4**, 229 (1962).
12. U. Liddel and E. D. Becker, *Spectrochim. Acta* **10**, 70 (1957).
13. G. Geiseler and E. Stockel, *Spectrochim. Acta* **17**, 1185 (1961).
14. S. N. Vinogradov, *Can. J. Chem.* **41**, 2719 (1963).
15. J. T. Arnold and M. E. Packard, *J. Chem. Phys.* **19**, 1608 (1951).
16. C. G. Cannon and B. C. Stace, *Spectrochim. Acta* **13**, 253 (1958).
17. S. Singh and C. N. R. Rao, *J. Phys. Chem.* **71**, 1074 (1967).
18. J. S. Cook and I. H. Reece, *Australian J. Chem.* **14**, 211 (1961).
19. M. M. Maguire and R. West, *Spectrochim. Acta* **17**, 369 (1961).
20. N. A. Puttnam, *J. Chem. Soc.* **1960**, 486.

21. L. J. Bellamy and R. J. Pace, *Spectrochim. Acta* **22**, 525 (1966).
22. R. Cardinaud, *Compt. Rend.* **249**, 1641 (1959).
23. R. Cardinaud, *Bull. Soc. Chim. France* **1960**, 629.
24. S. Singh and C. N. R. Rao, *Can. J. Chem.* **44**, 2611 (1966).
25. D. Hadzi, I. Petrov, and M. Zitko, *Proc. Intern. Meeting Mol. Spectry 4*th Bologna, **1959**, Vol 2 (1962), p. 794.
26. J. H. S. Green, W. Kynaston, and R. A. Gebbie, *Spectrochim. Acta* **19**, 807 (1963).
27. A. E. Stanevich, Opt. *Spectr.* **16**, 425 (1964).
28. R. F. Lake and H. W. Thompson, *Proc. Roy. Soc. (London)* **A291**, 469 (1966).
29. R. J. Jakobsen and J. W. Brasch, *Spectrochim. Acta* **21**, 1753 (1965).
30. W. J. Hurley, I. D. Kuntz, Jr., and G. E. Leroi, *J. Am. Chem. Soc.* **88**, 3199 (1966).
31. S. G. W. Ginn and J. L. Wood, *Spectrochim. Acta* **23A**, 611 (1967).
32. J. C. Davis, Jr., K. S. Pitzer, and C. N. R. Rao, *J. Phys. Chem.* **64**, 1714(1960).
33. M. Saunders and J. B. Hyne, *J. Chem. Phys.* **29**, 253 (1958).
34. E. D. Becker, *J. Chem. Phys.* **31**, 269 (1959).
35. B. D. N. Rao, P. Venkateswarlu, A. S. N. Murthy, and C. N. R. Rao, *Can. J. Chem.* **40**, 387 (1962).
36. A. S. N. Murthy, *Ph.D. Thesis*, Indian Institute of Science, Bangalore, India (1964).
37. T. M. Connor and C. Reid, *J. Mol. Spectry.* **7**, 32 (1961).
38. M. Martin, *J. Chim. Phys.* **59**, 736 (1962).
39. L. K. Patterson and R. M. Hammaker, *Spectrochim. Acta* **23A**, 2333 (1967).
40. B. D. N. Rao, P. Venkateswarlu, A. S. N. Murthy, and C. N. R. Rao, *Can. J. Chem.* **40**, 963 (1962).
41. E. A. Allan and L. W. Reeves, *J. Phys. Chem.* **67**, 591 (1963).
42. I. Yamaguchi, *Bull. Chem. Soc. Japan* **34**, 451 (1961).
43. B. G. Somers and H. S. Gutowsky, *J. Am. Chem. Soc.* **85**, 3065 (1963).
44. W. F. Forbes and J. F. Templeton, *Can. J. Chem.* **36**, 180 (1958).
45. J. C. Dearden and W. F. Forbes, *Can. J. Chem.* **38**, 896 (1960).
46. M. Ito, *J. Mol. Spectry.* **4**, 125 (1960).
47. C. N. R. Rao and A. S. N. Murthy, *J. Sci. Ind. Res. (India)* **20B**, 290 (1961).
48. J. C. Dearden, *Can. J. Chem.* **41**, 2683 (1963).
49. W. I. Kaye and R. Poulson, *Nature.* **193**, 675 (1962).
50. K. Dressler and O. Schnepp, *J. Chem. Phys.* **33**, 270 (1960).
51. J. Barrett and A. L. Mansell, *Nature* **187**, 138 (1960).
52. H. Dunken and P. Fink, *Z. Chem.* **2**, 117 (1962).
53. K. Nakamoto and S. Kishida, *J. Chem. Phys.* **41**, 1554 (1964).
54. S. Kishida and K. Nakamoto, *J. Chem. Phys.* **41**, 1558 (1964).
55. K. Nakamoto, Y. A. Sarma, and G. T. Behnke, *J. Chem. Phys.* **42**, 1662 (1965).
56. K. Nakamoto, Y. A. Sarma, and H. Ogoshi, *J. Chem. Phys.* **42**, 1177 (1965).
57. G. Allen, J. G. Watkinson, and K. H. Webb, *Spectrochim. Acta* **22**, 807 (1966).
58. E. S. Hanrahan and B. D. Bruce, *Spectrochim. Acta* **23A**, 2497 (1967).
59. Y. Nakai and K. Hirota, *Nippon Kagaku Zasshi* **81**, 881 (1960).
60. G. L. Carlson, R. E. Witkowski, and W. G. Fateley, *Spectrochim. Acta* **22**, 1117 (1966).
61. R. J. Jakobsen, Y. Mikawa, and J. W. Brasch, *Spectrochim. Acta* **23A**, 2199 (1967).
62. G. Statz and E. Lippert, *Ber. Bunsen. Ges. Physik. Chem.* **71**, 673 (1967).
63. S. G. W. Ginn and J. L. Wood, *J. Chem. Phys.* **46**, 2735 (1967).
64. P. Waldstein and L. A. Blatz, *J. Phys. Chem.* **71**, 2271 (1967).
65. T. Miyazawa and K. S. Pitzer, *J. Chem. Phys.* **30**, 1076 (1959).

66. L. J. Bellamy and R. J. Pace, *Spectrochim. Acta* **19**, 435 (1963).
67. L. J. Bellamy, R. F. Lake, and R. J. Pace, *Spectrochim. Acta* **19**, 443 (1963).
68. J. C. Davis, Jr., and K. S. Pitzer, *J. Phys. Chem.* **64**, 886 (1960).
69. N. Muller and O. R. Hughes, *J. Phys. Chem.* **70**, 3975 (1966).
70. W. F. Forbes and A. R. Knight, *Can. J. Chem.* **37**, 334 (1959).
71. W. F. Forbes, A. R. Knight, and D. L. Coffen, *Can. J. Chem.* **38**, 728 (1960).
72. E. Lippert and D. Oechssler, *Z. Physik. Chem. (Frankfurt)* **29**, 403 (1961).
73. A. S. N. Murthy, C. N. R. Rao, B. D. N. Rao, and P. Venkateswarlu, *Trans. Faraday Soc.* **58**, 855 (1962).
74. H. Wolff, *Z. Elektrochem.* **66**, 529 (1962).
75. A. G. Moritz, *Spectrochim. Acta* **17**, 365 (1961).
76. K. B. Whetsel, *Spectrochim. Acta* **17**, 614 (1961).
77. A. S. N. Murthy, *Indian J. Chem.* **3**, 143 (1965).
78. J. H. Lady and K. B. Whetsel, *J. Phys. Chem.* **68**, 1001 (1964).
79. B. Chenon and C. Sandorfy, *Can. J. Chem.* **36**, 1181 (1958).
80. C. Brissette and C. Sandorfy, *Can. J. Chem.* **38**, 34 (1960).
81. P. Sauvageau and C. Sandorfy, *Can. J. Chem.* **38**, 1901 (1960).
82. C. M. W. Anderson, J. L. Duncan, and F. J. C. Rossotti, *J. Chem. Soc.* **140**, 2165, 4201 (1961).
83. H. Zimmermann, *Z. Elektrochem.* **65**, 821 (1961).
84. R. H. Linnell, M. Aldo, and F. H. Raab, *J. Chem. Phys.* **36**, 1401 (1962).
85. S. Refn, *Spectrochim. Acta* **17**, 40 (1961).
86. Yu. N. Sheinker and E. M. Peresleni, *Zh. Fiz. Khim.* 32, 2112 (1958).
87. C. Perchard, A. M. Bellocq, and A. Novak, *J. Chim. Phys.* **62**, 1344 (1965).
88. P. G. Puranik and K. Venkataramiah, *Proc. Indian Acad. Sci.* **54A**, 69 (1961); *J. Mol. Spectry.* **3**, 486 (1959).
89. K. R. Bhaskar and C. N. R. Rao, *Biochim. Biophys. Acta* **136**, 561 (1967).
90. K. Itoh and T. Shimanouchi, *Biopolymers* **5**, 921 (1967).
91. R. C. Lord and T. J. Porro, *Z. Elektrochem.* **64**, 672 (1960).
92. Y. Kyogoku, R. C. Lord, and A. Rich, *J. Am. Chem. Soc.* **89**, 496 (1967).
93. G. C. Pimentel, M. O. Bulanin, and M. VanThiel, *J. Chem. Phys.* **36**, 500 (1962).
94. G. C. Pimentel, S. W. Charles, and K. Rosengren, *J. Chem. Phys.* **44**, 3029 (1966).
95. J. R. Durig, S. F. Bush, and E. E. Mercer, *J. Chem. Phys.* 44, 4238 (1966).
96. T. M. Barakat, N. Leggee, and A. D. E. Pullin, *Trans. Faraday Soc.* **59**, 1764 (1963).
97. J. Feeney and L. H. Sutcliffe, *Proc. Chem. Soc.* 118 (1961); *J. Chem. Soc.* **1962**, 1123.
98. A. Perotti and M. Cola, *NMR Chem. Proc. Symp. Calgiari, Italy.* (1964), p. 249.
99. J. A. Happe, *J. Phys. Chem.* **65**, 72 (1961).
100. L. A. LaPlanche, H. B. Thompson, and M. T. Rogers, *J. Phys. Chem.* **69**, 1482 (1965).
101. J. R. Cook and K. Schug, *J. Am. Chem. Soc.* **86**, 4271 (1964).
102. M. O. Bulanin, G. S. Denisov, and R. A. Pushkina, *Opt. i Spektroskopiya* **6**, 754 (1959).
103. J. G. David and H. E. Hallam, *Spectrochim. Acta* **21**, 841 (1965).
104. I. M. Ginzburg and L. A. Loginova, *Opt. i Spektroskopiya* **20**, 241 (1966).
105. S. Forsen, *Acta. Chem. Scand.* **13**, 1472 (1959).
106. L. D. Colebrook and D. S. Tarbell, *Proc. Natl. Acad. Sci. U. S.* **47**, 993 (1961).
107. S. H. Marcus and S. I. Miller, *J. Am. Chem. Soc.* **88**, 3719 (1966).
108. E. D. Becker, *Spectrochim. Acta* **15**, 743 (1959).

109. C. F. Jumper, M. T. Emerson, and R. B. Howard, *J. Chem. Phys.* **35**, 1911 (1961).
110. N. Nakagawa and S. Fujiwara, *Bull. Chem. Soc. Japan* **33**, 1634 (1960).
111. E. D. Becker, *Spectrochim. Acta* **17**, 436 (1961).
112. P. Pineau and M. L. Josien, *Proc. Intern. Meeting. Mol. Spectry 4th Bologna.* **2**, 924 (1959). (Pub. 1962).
113. Y. Sato, *Nippon Kagaku Zasshi* **79**, 538 (1958).
114. P. J. Krueger and H. D. Mettee, *Can. J. Chem.* **42**, 288 (1964).
115. J. Brandmuller and K. Seevogel, *Spectrochim. Acta* **20**, 453 (1964).
116. R. G. Inskeep, F. E. Dickson, and J. M. Kelliha, *J. Mol. Spectry.* **4**, 477 (1960).
117. D. J. Millen and J. Zabicky, *J. Chem. Soc.* **1965**, 3080.
118. S. G. W. Ginn and J. L. Wood, *Nature* No. 200, 467 (1963).
119. A. Hall and J. L. Wood, *Spectrochim. Acta* **23A**, 1257, 2657 (1967).
120. S. G. W. Ginn and J. L. Wood, *Spectrochim. Acta* **23A**, 611 (1967).
121. S. Singh, A. S. N. Murthy, and C. N. R. Rao, *Trans. Faraday Soc.* **62**, 1056 (1966).
122. H. Dunken and H. Fritzsche, *Z. Chem.* **1**, 127 (1961).
123. H. Dunken and H. Fritzsche, *Z. Chem.* **1**, 249 (1961).
124. H. Dunken and H. Fritzsche, *Z. Chem.* **2**, 345 (1962).
125. G. Aksnes, *Acta. Chem. Scand.* **14**, 1475 (1960).
126. G. Aksnes and T. Gramstad, *Acta. Chem. Scand.* **14**, 1485 (1960).
127. T. Gramstad, *Acta. Chem. Scand.* **16**, 807 (1962).
128. T. Gramstad and W. J. Fuglevik, *Acta. Chem. Scand.* **16**, 1369 (1962).
129. T. Gramstad, *Spectrochim. Acta* **19**, 497 (1963).
130. D. L. Powell and R. West, *Spectrochim. Acta* **20**, 983 (1964).
131. R. West, D. L. Powell, M. K. T. Lee, and L. S. Whatley, *J. Am. Chem. Soc.* **86**, 3227 (1964).
132. H. Fritzsche, *Ber. Bunsen. Ges. Physik. Chem.* **68**, 459 (1964).
133. T. Gramstad, *Spectrochim. Acta* **20**, 729 (1964).
134. P. Biscarini, G. Galloni, and S. Ghersetti, *Spectrochim. Acta* **20**, 267 (1964).
135. S. Ghersetti and A. Lusa, *Spectrochim. Acta* **21**, 1067 (1965).
136. S. S. Mitra, *J. Chem. Phys.* **36**, 3286 (1962).
137. S. C. White and H. W. Thompson, *Proc. Roy. Soc. (London)* **A291**, 460 (1966).
138. A. Allerhand and P. V. R. Schleyer, *J. Am. Chem. Soc.* **84**, 1322 (1962); **85**, 866 (1963).
139. P. V. R. Schleyer and R. West, *J. Am. Chem. Soc.* **81**, 3164 (1959).
140. R. West, D. L. Powell, L. S. Whatley, M. K. T. Lee, and P. V. R. Schleyer, *J. Am. Chem. Soc.* **84**, 3221 (1962).
141. R. West, *J. Am. Chem. Soc.* **81**, 1614 (1959).
142. S. Wada, *Bull. Chem. Soc. Japan* **35**, 707 (1962).
143. L. J. Bellamy, G. Eglinton, and J. F. Morman, *J. Chem. Soc.* **1961**, 4762.
144. S. Singh and C. N. R. Rao, *J. Am. Chem. Soc.* **88**, 2142 (1966).
145. M. D. Joesten and R. S. Drago, *J. Am. Chem. Soc.* **84**, 3817 (1962).
146. T. D. Epley and R. S. Drago, *J. Am. Chem. Soc.* **89**, 5770 (1967).
147. H. J. Bernstein, *J. Am. Chem. Soc.* **85**, 484 (1963).
148. G. S. Denisov, *Dokl. Akad. Nauk SSSR* **134**, 1131 (1960).
149. S. Wada, *Bull. Chem. Soc. Japan* **35**, 710 (1962).
150. S. L. Johnson and K. A. Rumon, *J. Phys. Chem.* **69**, 74 (1965).
151. D. Hadzi, *J. Chem. Soc.* **1962**, 5128.
152. I. M. Arefev and V. I. Malyshev, *Opt. i Spektroskopiya* **13**, 206 (1962).
153. J. E. Bertie and D. J. Millen, *J. Chem. Soc.* **1965**, 497.
154. J. E. Bertie and D. J. Millen, *J. Chem. Soc.* **1965**, 510.

155. J. E. Bertie and D. J. Millen, *J. Chem. Soc.* **1965**, 514.
156. D. J. Millen and O. A. Samsonov, *J. Chem. Soc.* **1965**, 3085.
157. G. L. Carlson, R. E. Witkowski, and W. G. Fateley, *Nature* **211**, 1289 (1966).
158. W. C. Drinkard and D. Kivelson, *J. Phys. Chem.* **62**, 1494 (1958).
159. F. Takahasi and N. C. Li, *J. Phys. Chem.* **68**, 2136 (1964).
160. A. Loewenstein and Y. Margalit, *J. Phys. Chem.* **69**, 4152 (1965).
161. I. Grancher, *Helv. Phys. Acta* **34**, 272 (1961).
162. D. P. Eyman and R. S. Drago, *J. Am. Chem. Soc.* **88**, 1617 (1966).
163. F. Takahasi and N. C. Li, *J. Phys. Chem.* **69**, 1622 (1965).
164. H. Shimizu, *Nippon Kagaku Zasshi* **81**, 1025 (1960).
165. L. J. Bellamy, H. E. Hallam, and R. L. Williams, *Trans. Faraday Soc.* **54**, 1120 (1958).
166. H. Dunken and H. Fritzsche, *Z. Chem.* **2**, 379 (1962).
167. K. B. Whetsel and J. H. Lady, *J. Phys. Chem.* **68**, 1010 (1964).
168. J. H. Lady and K. B. Whetsel, *J. Phys. Chem.* **71**, 1421 (1967).
169. T. Gramstad and W. J. Fuglevik, *Spectrochim. Acta* **21**, 503 (1965).
170. P. G. Puranik and K. Venkataramiah, *J. Mol. Spectry.* **3**, 486 (1959).
171. I. Suzuki, M. Tsuboi, T. Shimanouchi, and S. Mizushima, *J. Chem. Phys.* **31**, 1437 (1957).
172. J. Pitha, R. N. Jones, and P. Pithova, *Can. J. Chem.* **44**, 1045 (1966).
173. C. Griessner Prettre, *Compt. Rend.* **252**, 3238 (1961).
174. I. Yamaguchi, *Bull. Chem. Soc. Japan* **34**, 1606 (1961).
175. M. T. Chenon and N. Lumbroso-Bader, *J. Chim. Phys.* **62**, 1208 (1965).
176. W. G. Schneider and L. W. Reeves, *Ann. N. Y. Acad. Sci.* **70**, 858 (1958).
177. M. Freymann and R. Freymann, *Compt. Rend.* **248**, 677 (1959).
178. L. Katz and S. Penman, *J. Mol. Biol.* **15**, 220 (1966).
179. R. Mathur, E. D. Becker, R. B. Bradley, and N. C. Li, *J. Phys. Chem.* **67**, 2190 (1963).
180. R. Mathur, S. M. Wang, and N. C. Li, *J. Phys. Chem.* **68**, 2140 (1964).
181. S. Murahashi, B. Ryutani, and K. Hatada, *Bull. Chem. Soc. Japan* **32**, 1001 (1959).
182. E. V. Shuvalova, *Opt. i Spektroskopiya* **6**, 696 (1959).
183. D. N. Shigorin, M. M. Shemyakin, M. N. Kosolov, and T. S. Ryabchikov, *Stroenie Veshchestva i Spektroskopiya, Akad. Nauk SSSR* **1960**, 36.
184. R. West and C. S. Kraihanzel, *J. Am. Chem. Soc.* **83**, 765 (1961).
185. E. A. Gastilovich, D. N. Shigorin, E. P. Gracheva, I. A. Chekula, and M. F. Shostakovskii, *Opt. Spectr.* **10**, 312 (1961).
186. C. J. Creswell and G. M. Barrow, *Spectrochim. Acta* **22**, 839 (1966).
187. C. G. Cannon, *Spectrochim. Acta* **10**, 429 (1958).
188. K. B. Whetsel and R. E. Kagarise, *Spectrochim. Acta* **18**, 329 (1962).
189. A. Allerhand and P. V. R. Schleyer, *J. Am. Chem. Soc.* **85**, 1715 (1963).
190. R. Kaiser, *Can. J. Chem.* **41**, 430 (1963).
191. A. L. McClellan, S. W. Nicksic, and J. C. Guffy, *J. Mol. Spectry.* **11**, 340 (1963).
192. B. B. Howard, C. F. Jumper, and M. T. Emerson, *J. Mol. Spectry.* **10**, 117 (1963).
193. W. G. Paterson and D. M. Cameron, *Can. J. Chem.* **41**, 198 (1963).
194. C. J. Creswell and A. L. Allred, *J. Am. Chem. Soc.* **85**, 1723 (1963).
195. T. J. V. Findlay, J. S. Keniry, A. D. Kidman, and V. A. Pickles, *Trans. Faraday Soc.* **63**, 846 (1967).
196. N. Nakagawa and S. Fujiwara, *Bull. Chem. Soc. Japan* **33**, 1634 (1960).
197. R. S. Becker, *J. Mol. Spectry.* **3**, 1 (1959).
198. M. Ito, K. Inuzawa, and S. Iminishi, *J. Am. Chem. Soc.* **82**, 1317 (1960).

199. A. Balasubramanian and C. N. R. Rao, *Spectrochim. Acta* **18**, 1337 (1962).
200. A. K. Chandra and S. Basu, *Trans. Faraday Soc.* **56**, 632 (1960).
201. A. Balasubramanian, *Indian J. Chem.* **1**, 329 (1963).
202. D. P. Stevenson, *J. Am. Chem. Soc.* **84**, 2849 (1962).
203. S. Singh and C. N. R. Rao, *Trans. Faraday Soc.* **62**, 3310 (1966).
204. H. Baba and S. Suzuki, *J. Chem. Phys.* **35**, 1118 (1961).
205. A. K. Chandra and S. Banerjee, *J. Phys. Chem.* **66**, 952 (1962).
206. M. D. Joesten and R. S. Drago, *J. Am. Chem. Soc.* **84**, 2696 (1962).
207. B. B. Bhowmik and S. Basu, *Trans. Faraday Soc.* **59**, 2696 (1963).
208. K. Semba, *Bull. Chem. Soc. Japan* **34**, 722 (1961).
209. I. B. Ghosh and S. Basu, *Trans. Faraday Soc.* **61**, 2097 (1965).
210. P. Chiorboli, B. Fortunato, and A. Rastelli, *Ric. Sci. Rend., Ser. A.* **8**, 985 (1965).
211. T. Kubota, *J. Am. Chem. Soc.* **88**, 211 (1966).
212. T. Kubota, M. Yamakawa, M. Takasuka, K. Iwatani, H. Akazawa, and I. Tanaka, *J. Phys. Chem.* **71**, 3597 (1967).
213. B. B. Bhowmik and S. Basu, *Trans. Faraday Soc.* **58**, 48 (1962).
214. R. H. Linnell, *J. Chem. Phys.* **34**, 698 (1961).
215. M. Tichy, *Advances in Organic Chemistry*, Vol. 5 (R. A. Raphael, E. C. Taylor, and H. Wynberg, eds.), Interscience, New York (1965), p. 115.
216. P. J. Krueger and H. D. Mettee, *Can. J. Chem.* **42**, 326, 340 (1964).
217. N. Mori, S. Omura, H. Yamakawa, and Y. Tsuzuki, *Bull. Chem. Soc. Japan* **38**, 1627 (1965).
218. P. J. Krueger and H. D. Mettee, *Can. J. Chem.* **43**, 2888 (1965).
219. T. Urbanskii, *Bull. Acad. Polon. Sci. Ser.* **4**, 87, 381 (1966).
220. M. Kuhn, W. Luttke, and R. Mecke, *Z. Anal. Chem.* **57**, 680 (1963).
221. W. F. Baitinger, P. V. R. Schleyer, T. S. S. R. Murty, and L. Robinson, *Tetrahedron* **20**, 1635 (1964).
222. S. Julia, D. Varech, Th. Burer, and Hs. H. Gunthard, *Helv. Chim. Acta* **43**, 1623 (1960).
223. P. V. R. Schleyer, *J. Am. Chem. Soc.* **83**, 1368 (1961).
224. L. P. Kuhn and R. E. Bowman, *Spectrochim. Acta* **17**, 650 (1961).
225. L. P. Kuhn, P. V. R. Schleyer, W. F. Baitinger, and L. Eberson, *J. Am. Chem. Soc.* **86**, 650 (1964).
226. R. D. Stolow, *J. Am. Chem. Soc.* **83**, 2592 (1961).
227. M. Oki and H. Iwamura, *Bull. Chem. Soc. Japan* **32**, 567 (1959).
228. M. Oki and H. Iwamura, *Bull. Chem. Soc. Japan* **32**, 955 (1959).
229. M. Oki and H. Iwamura, *Bull. Chem. Soc. Japan* **32**, 1135 (1959).
230. P. V. R. Schleyer, D. S. Trifan, and R. Bacskai, *J. Am. Chem. Soc.* **80**, 6691 (1958); *J. Am. Chem. Soc.* **90**, 327 (1968).
231. P. J. Krueger and H. D. Mettee, *Tetrahedron Letters* 1587 (1966).
232. F. Moll, *Arch. Pharm.* **299**, 429 (1966).
233. A. W. Baker and W. W. Kaeding, *J. Am. Chem. Soc.* **81**, 5904 (1959).
234. D. A. K. Jones and J. G. Watkinson, *Chem. Ind. (London)* **1960**, 661.
235. A. W. Baker and A. T. Shulgin, *Can. J. Chem.* **43**, 650 (1965).
236. D. A. K. Jones and J. G. Watkinson, *J. Chem. Soc.* **1964**, 2371.
237. T. S. Lin and E. Fishman, *Spectrochim. Acta* **23A**, 491 (1967).
238. U. Dabrowska and T. Urbanski, *Spectrochim. Acta* **21**, 1765 (1965).
239. G. Durocher and C. Sandorfy, *J. Mol. Spectry.* **15**, 22 (1965).
240. A. W. Baker and A. T. Shulgin, *J. Am. Chem. Soc.* **81**, 1523 (1959).
241. R. A. Nyquist, *Spectrochim. Acta* **19**, 1655 (1963).

242. A. W. Baker, H. O. Kerlinger, and A. T. Shulgin, *Spectrochim. Acta* **20**, 1467 (1964).
243. A. W. Baker and A. T. Shulgin, *J. Am. Chem. Soc.* **80**, 5358 (1958).
244. M. Oki and H. Iwamura, *Bull. Chem. Soc. Japan* **33**, 681 (1960).
245. M. Oki and H. Iwamura, *Bull. Chem. Soc. Japan* **33**, 717 (1960).
246. M. Oki and H. Iwamura, *Bull. Chem. Soc. Japan* **33**, 427 (1960).
247. H. H. Friedman, *J. Am. Chem. Soc.* **83**, 2900 (1961).
248. J. C. Dearden, *Nature* **206**, 1147 (1965).
249. Y. Matsui, M. Takasuka, and T. Kubota, *Shionogi Kenkyusho Nempo* **15**, 125 (1965); *Chem. Abstr.* **64**, 10589c (1966).
250. H. Ogoshi and K. Nakamoto, *J. Chem. Phys.* **45**, 3113 (1966).
251. M. Oki and M. Hirota, *Nippon Kaguka Zasshi* **81**, 855 (1960).
252. M. Oki and M. Hirota, *Spectrochim. Acta* **17**, 583 (1961).
253. M. Oki and M. Hirota, *Bull. Chem. Soc. Japan* **34**, 374 (1961).
254. M. Oki and M. Hirota, *Bull. Chem. Soc. Japan* **37**, 213 (1964).
255. M. Oki and M. Hirota, *Bill. Chem. Soc. Japan* **37**, 209 (1964).
256. M. Oki, M. Hirota, and S. Hirofuji, *Spectrochim. Acta* **22**, 1537 (1966).
257. M. Oki, M. Hirota, and Y. Morimoto, *Bull. Chem. Soc. Japan* **39**, 1620 (1966).
258. M. Oki and M. Hirota, *Nippon Kaguka Zasshi* **86**, 115 (1965).
259. L. K. Dyall and A. N. Hambly, *Chem. Ind. (London)*, 262 (1958).
260. L. K. Dyall, *Spectrochim. Acta* **17**, 291 (1961).
261. A. G. Moritz, *Spectrochim. Acta* **16**, 1176 (1960).
262. A. N. Hambly and B. V. O'Grady, *Australian J. Chem.* **15**, 626 (1962).
263. P. J. Krueger, *Can. J. Chem.* **41**, 363 (1963).
264. P. J. Krueger, *Can. J. Chem.* **42**, 201 (1964).
265. P. J. Krueger and D. W. Smith, *Developments in Applied Spectroscopy*, Vol 4, Plenum Press, New York (1965) p. 197.
266. P. J. Krueger and H. D. Mettee, *Can. J. Chem.* **43**, 2970 (1965).
267. P. J. Krueger, *Can. J. Chem.* **45**, 2143 (1967).
268. P. J. Krueger, *Can. J. Chem.* **45**, 2135 (1967).
269. A. L. Porte, H. S. Gutowsky, and I. M. Hunsberger, *J. Am. Chem. Soc.* **82**, 5057 (1960).
270. L. W. Reeves, E. A. Allan, and K. O. Stromme, *Can. J. Chem.* **38**, 1249 (1960).
271. E. A. Allan and L. W. Reeves, *J. Phys. Chem.* **66**, 613 (1962).
272. R. W. Hay and P. P. Williams, *J. Chem. Soc.* **1964**, 2270.
273. G. E. Maciel and G. B. Savitsky, *J. Phys. Chem.* **68**, 437 (1964).
274. Y. Ikegami, T. Ikenoue, and S. Seto, *Kogyo Kagaku Zasshi* **68**, 1415 (1965); *Chem. Abstr.* **64**, 190a (1966).
275. S. Forsen and M. Nilsson, *Acta. Chem. Scand.* **14**, 1333 (1960).
276. N. N. Shapetko, D. N. Shigorin, A. P. Skoldinov, T. S. Ryabchikova, and L. N. Reshetova, *Opt. Spektroskopiya* **17**, 459 (1964).
277. D. C. Kleinfelter, *J. Am. Chem. Soc.* **89**, 1734 (1967).
278. L. Eberson and S. Forsen, *J. Phys. Chem.* **64**, 767 (1960).
279. B. L. Silver, Z. Luz, S. Peller, and J. Reuben, *J. Phys. Chem.* **70**, 1434 (1966).
280. T. Urbanski, *Tetrahedron* **6**, 1 (1959).
281. A. E. Lutskii and V. T. Alekseeva, *Zh. Obsch. Khim.* **29**, 2992 (1959).
282. A. E. Lutskii, *Molekul. Spektroskopiya, Leningr. Gos. Univ. Sb. Statei* **1960**, 190.
283. J. C. Dearden and W. F. Forbes, *Can. J. Chem.* **38**, 1837 (1960).
284. J. C. Dearden and W. F. Forbes, *Can. J. Chem.* **38**, 1852 (1960).

285. A. Balasubramanian, J. B. Capindale, and W. F. Forbes, *Can. J. Chem.* **42**, 2674 (1964).
286. S. Pinchas, *J. Phys. Chem.* **67**, 1862 (1963).
287. W. F. Forbes, *Can. J. Chem.* **40**, 1891 (1962).
288. M. Oki and H. Iwamura, *J. Am. Chem. Soc.* **89**, 576 (1967).
289. K. Buijs and G. R. Choppin, *J. Chem. Phys.* **39**, 2035 (1963).
290. W. Luck, *Z. Elektrochem.* **67**, 186 (1963).
291. W. Luck, *Z. Elektrochem.* **68**, 895 (1964).
292. M. R. Thomas, H. A. Scheraga, and E. E. Schrier, *J. Phys. Chem.* **69**, 3722 (1965).
293. W. Luck, *Ber. Bunsen. Ges. Physik. Chem.* **69**, 627 (1965).
294. J. D. Worley and I. M. Klotz, *J. Chem. Phys.* **45**, 2868 (1966).
295. E. R. Lippincott, C. E. Weir, and A. V. Valkenburg, Jr., *J. Chem. Phys.* **32**, 612 (1960).
296. A. Fratiello and J. P. Luongo, *J. Am. Chem. Soc.* **85**, 3072 (1963).
297. J. J. Lindberg and G. Majani, *Acta Chem. Scand.* **17**, 1477 (1963).
298. S. C. Mohr, W. D. Wilk, and G. M. Barrow, *J. Am. Chem. Soc.* **87**, 3048 (1965).
299. W. R. Busing and D. F. Hornig, *J. Phys. Chem.* **65**, 284 (1961).
300. G. E. Walrafen, *J. Chem. Phys.* **40**, 3249 (1964).
301. G. E. Walrafen, *J. Chem. Phys.* **44**, 1546 (1966).
302. J. R. Holmes, D. Kivelson, and W. C. Drinkard, *J. Am. Chem. Soc.* **84**, 4677 (1962).
303. A. Fratiello and D. C. Douglass, *J. Mol. Spectry.* **11**, 465 (1963).
304. K. A. Hartmann, Jr., *J. Phys. Chem.* **70**, 270 (1966).
305. Y. Fujiwara and S. Fujiwara, *Bull. Chem. Soc. Japan* **36**, 574 (1963).
306. F. Takahasi and N. C. Li, *J. Am. Chem. Soc.* **88**, 1117 (1966).
307. N. Muller and P. Simon, *J. Phys. Chem.* **71**, 568 (1967).
308. Z. Luz and G. Yagil, *J. Phys. Chem.* **70**, 554 (1966).
309. N. Muller, *J. Chem. Phys.* **43**, 2555 (1965).
310. A. S. N. Murthy and C. N. R. Rao, *Appl. Spectr. Rev.* **2**, 69 (1968).

Vibrational and Electronic Raman Effect

J. A. Koningstein

E. W. R. Steacie Building
Department of Chemistry
Carleton University
Ottawa, Canada

The theory for overtones and combination bands in the vibrational Raman case has been formulated by introducing the concept of vibronic coupling. One of the main features is that the scattering tensor governing these types of processes is antisymmetrical. This is also the case for electronic Raman transitions of rare-earth ions, while it can also be shown that the intensities of electronic Raman transitions and those of overtones (vibrational Raman effect) are approximately equal in value.

Raman spectra of solids, liquids, and gases have served as important phenomena with respect to the assignment of the position of vibrational and rotational levels of molecules. The intensities of these Raman lines are related to a tensor. As there are three possible directions for polarization of the incident excitation radiation and again three possible directions of polarization for the scattered radiation, it follows that this tensor should contain nine elements. The elements a_{xx}, a_{yy}, and a_{zz} are on the diagonal of the tensor, while $a_{\rho\sigma}$ (where ρ and σ stand for x, y, and z) are all situated on off-diagonal positions.

The expression for an element of the tensor for the Raman effect was derived by Kramers and Heisenberg[1] before the advent of quantum mechanics. It was Dirac[2] who gave the full quantum-mechanical treatment, but in most cases the semiclassical expression derived by Placzek[3] is followed. All theo-

*Research supported by the National Research Council of Canada and the Defense Research Board.

rists, however, arrived at the same expression for $(a_{\rho\sigma})_{nm}$. This expression is written as

$$(a_{\rho\sigma})_{nm} = \frac{1}{h}\sum_r \frac{(M_\rho)_{nr}(M_\sigma)_{rm}}{\nu_{nr} - \nu_0} + \frac{[\rho \leftrightarrow \sigma]}{\nu_{rm} + \nu_0}$$

Here ν_0 is the frequency of the incident radiation and is far removed from that of an absorption band, n and m are the quantum numbers of the states which are involved in the transition, and $(M_\rho)_{nr}$ and $(M_\sigma)_{rm}$ are matrix elements of the electric dipole operators ρ and σ between the states n,r and r,m, respectively. The summation of r extends over all the excited states of the scattering molecule.

In Placzek's polarizability theory[3] the role of these virtual states is ignored. The polarizability of the molecule in the ground state (a_{nn}) is simply expanded to first order in the nuclear coordinate Q_a

$$a = a_0 + \left(\frac{\partial a}{\partial Q_a}\right)_0 Q_a + \cdots$$

In this paper I would like to make some general remarks with respect to the symmetry of the Raman tensor. It has been suggested that the scattering tensor for the vibrational Raman effect is a symmetrical tensor. Thus $a_{\rho\sigma} = a_{\sigma\rho}$. However, this condition is not easily recognized if one uses Placzek's theory. Also, from a purely theoretical point of view the omission of the intermediate states is not satisfactory.

The total wave function Ψ_{tot} of a molecule can, according to Born and Oppenheimer[4], be written as a product of the electronic, vibrational, and rotational wave functions. Here we shall restrict ourselves to the first two wave functions only. Is it possible to separate (completely) the electronic movements from the nuclear movements? This is not always so, and the vibronic coupling phenomenon plays an important role as the intensity-giving process for many a spectroscopic transition. However, the vibronic interactions can be considered as a rather weak effect, and in fact they can be considered as a perturbation of the Hamiltonian of the molecule. Thus

$$\mathscr{H} = \mathscr{H}^0 + \mathscr{H}' + \mathscr{H}''$$

where

$$\mathscr{H}' = \left(\frac{\partial \mathscr{H}^0}{\partial Q_a}\right)_0 Q_a \quad \text{and} \quad \mathscr{H}'' = \left(\frac{\partial^2 \mathscr{H}}{\partial Q_a\, \partial Q_b}\right)_0 Q_a Q_b$$

with Q_a and Q_b two normal coordinates of the molecule. H$'$ depends on one of them; in H$''$ the effects of two different types of nuclear motion of the molecule are realized.

The wave function of the molecule is written as: $\Psi_{tot} = \psi(r,Q)\,\varphi(Q)$, where $\psi(r,Q)$ is the electronic wave function and $\varphi(Q)$ is the vibrational one. One can now obtain $\psi(r,Q)$ by applying perturbation theory employing the Hamiltonian given by expression

$$\psi_g = \psi_g^0 + \sum_{l \neq g} h'_{lg}\psi_l^0 Q_a + \sum_{k,l \neq g} [h'_{kl}h'_{lg}Q_a^2 - h'_{gg}h'_{kg}Q_a^2 + h''_{kg}Q_aQ_b]\psi_k^0$$

$$h'_{pg} \int \psi_p^0(\mathbf{r}) \left| \left(\frac{\partial \mathscr{H}}{\partial Q}\right)_0 \right| \psi_g^0(\mathbf{r})\, d\tau \qquad h''_{pg} = \int \psi_p^0(\mathbf{r})\, |(\partial^2 \mathscr{H}/\partial Q_a\, \partial Q_b)|\, \psi_g^0(\mathbf{r})\, d\tau$$

The expression for the matrix element $(M_\rho)_{gi;er}$ now becomes

$$\begin{aligned}(M_\rho)_{gi;er} = 1 &+ \left[\sum_{g,e \neq l} \{h'_{le}M_\rho^0)_{gl} + h'_{lg}(M_\rho^0)_{le}\}\right] \langle \prod_{ab} gi|\, Q_a\, |er\rangle \\ &+ \left[\sum_{e,g \neq k} \{h''_{ke}(M_\rho^0)_{gk} + h''_{kg}(M_\rho^0)_{le}\}\right] \langle \prod_{ab} gi|\, Q_aQ_b\, |er\rangle \\ &+ \left[\sum_{e,g \neq l} \{h'_{lg}h'_{le}(M_\rho^0)_{ll}\}\right] \langle \prod_{ab} gi|\, Q_a^2\, |er\rangle\end{aligned}$$

It is seen that in the calculation of the scattering tensor $(a_{\rho\sigma})_{gi;gj}$ products occur of the type $(M_\rho)_{gi;er}(M_\sigma)_{er;gj}$ which contain products of the type $\langle \prod_{ab} gi|\, Q_a\, |er\rangle\langle er \mid gj\rangle$ This integral has to be divided by $\nu_{gi;er} - \nu_0$, where g and e are the quantum numbers of the electronic wave functions for ground and excited states, respectively, and (i,j) and r are quantum numbers of the vibrational levels associated with g and r, respectively. If the variation of $\nu_{ji;er} - \nu_0$ with the vibrational quantum numbers i and r is neglected, then a common denominator is found for the above matrix elements and closure can be invoked in the space of the vibrational functions so that

$$\langle \prod_{ab} gi|\, Q_a\, |er\rangle\langle er \mid gj\rangle = \langle gi|\, Q_a\, |gj\rangle$$

Because of the orthonormality of the vibrational wave functions, this matrix element is only different from zero if $j = i \pm 1$. This indicates that a Raman transition occurs between two vibrational levels of the normal coordinate Q_a separated by one vibrational quantum. The expression for $(a_{\rho\sigma})_{gi;gj}$ is now

$$\begin{aligned}(a_{\rho\sigma})_{gi;gi\pm 1} = \frac{1}{h} \sum_{r,l,g} &\{h'_{lg}[(M_\rho^0)_{ge}(M_\sigma^0)_{el} + (M_\rho^0)_{le}(M_\sigma^0)_{eg}] \\ &+ h'_{le}[(M_\rho^0)_{ge}(M_\sigma^0)_{lg} + (M_\sigma^0)_{eg}(M_\rho^0)_{gl}]\}\, \frac{\langle gi|\, Q_a\, |g,\, i+1\rangle}{\nu_{ge} - \nu_0} + \text{c.t.}\end{aligned}$$

Note that the replacement $\rho \leftrightarrow \sigma$ leaves (6) unaltered, and it is thus found that $a_{\rho\sigma} = a_{\sigma\rho}$ and the tensor for a first-order Raman effect both for the Stokes and anti-Stokes component is indeed symmetrical. This is also the case for "hot" bands as long as only one-quantum jump is involved. For a detailed analysis of the Raman selection rules and their relation to the constants h'_{pq}, etc., the reader is referred to the work by Albrecht[5], who was the first to employ the vibronic wave functions in the first-order Raman effects.

It is seen that in employing the full expansion (4) in the computation of a scattering tensor, matrix elements occur of the type $\langle gi| Q_a^2 |gj\rangle$ and $\langle gi| Q_a Q_b |gj\rangle$. The former matrix element unequal to zero if $j = i \pm 2$ and the latter if $j = i \pm 1$. Thus (1) a second-order Raman effect of the overtone type, and (2) a combination effect, are the processes with nonzero intensity. The expression for the scattering tensor for a combination type can be written as

$$(a_{\rho\sigma})_{g_i;g_i\pm 1} = \frac{1}{h} \sum_r \sum_{k,l \neq e,g} [h''_{kg}(M^0_\rho)_{ge}(M^0_\sigma)_{ek} + h''_{kg}(M^0_\rho)_{ke}(M^0_\sigma)_{eg} + h''_{ke}(M^0_\rho)_{ge}(M^0_\sigma)_{kg} + h''_{le}(M^0_\rho)_{gk}(M^0_\sigma)_{eg}]\langle \prod_{ab} gi| Q_a Q_b |gi \pm 1\rangle$$

It is seen that the replacement $\rho \leftrightarrow \sigma$ results in an identical expression, and thus $a_{\rho\sigma} = a_{\sigma\rho}$. A detailed calculation for the overtone Raman effect reveals that the scattering tensor is symmetric; however, if only the expansion of the electronic wave function to first order is taken in (4), then the tensor is asymmetric. For all other Raman processes which involve the excitation of one mode by more than two quanta or two modes by more than one quantum each, it is found that the scattering tensor is symmetric.

In the electronic Raman effect the transition originates in an electronic level (e.g., the ground state) and terminates in an electronic excited state. If we label these states with the quantum numbers n and m, then from (1) it can be concluded that the tensor is also antisymmetrical. For many molecules the first excited state is some ten thousands of cm^{-1} above the ground state, and thus the excitation of a Raman transition between these states shall be weak. A better case is that of the rare-earth ions in crystals. The electronic configurations of the lanthanides give rise to excited states which can be as low as a few hundred wave numbers above the ground state. If these ions are brought into a crystal then these states are split as a result of the crystal field which the ions experience, and electronic Raman effects may take place between the crystal-field components of the split ground manifolds of the lanthanides. The position of the excited states of the rare-earth ions are known, and an expression have been derived for the intensity of the absorption bands. It is thus possible to compute reasonably accurate values for the $(a_{\rho\sigma})_{nm}$ and $(a_{\rho\sigma})_{nn}$ (Rayleigh scattering), and we have recently been able to calculate absolute intensities and the degree of antisymmetry of the scattering tensor for electronic Raman transitions of the rare-earth ions[6]. It turns out that these transitions are weak, and by and large it can be said that second-order vibrational Raman transitions and electronic Raman effects of trivalent rare-earth ions are of equal intensity.

A Raman transition is allowed if in the product of the representations of initial and final states Γ_i and Γ_f, respectively, a species occurs to which a part of the scattering tensor also belongs. Antisymmetry of the tensor now suggests that $a_{\rho\sigma} - a_{\sigma\rho} \neq 0$ and thus we are now interested in the transformation properties of these combinations. In our theory on the electronic Raman effect it was pointed out that the antisymmetrical part transforms identically to the rotations R_z, R_y, and R_x. The species with these properties are tabulated, and it is found that in some cases $a_{\rho\sigma} - a_{\sigma\rho}$ transforms uniquely. If a Raman process is governed by only an antisymmetrical tensor, then the depolarization ratio for polarized light, ρ_s, for a system of randomly-oriented molecules or atoms is $\rho_s = \infty$. If both the symmetrical and antisymmetrical combinations belong to the same representation then $\frac{3}{4} < \rho_s < \infty$.

REFERENCES

1. H. A. Kramers and W. Heisenberg, *Z. Phys.31*, 681 (1925).
2. P. A. M. Dirac, *Proc. Roy. Soc. (London)* **114**, 710 (1927).
3. G. Placzek, *Handbuch der Radiologie*, Vol. VL, Part 2, (1934), p. 205.
4. M. Born and J. Oppenheimer, *Ann. Physik* **84**, 475 (1927).
5. A. C. Albrecht, *J. Chem. Phys.* **34**, 1476 (1961).
6. J. A. Koningstein and O. Sonnich Mortensen, *Phys. Rev.* **169**, 75 (1968).

Raman Spectroscopy of Polymeric Materials. Part I—Selected Commercial Polymers

Dorothy S. Cain[†] and Albert B. Harvey

Naval Research Laboratory
Washington, D.C.

The He–Ne laser has greatly reduced the problem of fluorescence which has in the past almost completely obscured conventionally excited (mercury arc) Raman spectra of all except the simplest and purest polymers. In many instances it is now possible to record Raman spectra of polymers without any prior purification whatever. Reported here are laser-excited Raman spectra of twelve selected commercial polymeric materials and their infrared counterparts. The complementary nature of Raman and infrared spectroscopy is apparent from vibrations such as C=C stretching (in polybutadiene) and S–S and C–S stretching (in polysulfide rubber) which give rise to strong Raman lines but produce weak infrared absorption bands. Conversely, the C=O group of polycarbonates and polyesters is a strong infrared absorber which produces a weak Raman line.

INTRODUCTION

The complementary techniques of infrared and Raman spectroscopy have not experienced equal application in the polymer field. The differences between the two techniques make it desirable to record both kinds of spectra, but difficulties encountered in sample handling and instrumentation have kept Raman spectroscopy from gaining the wide popularity enjoyed by infrared spectroscopy. Much of the trouble has been related to the Toronto mercury arc used as the exciting source, which, because of its low flux density and

† Present address, Chemistry Dept., University of South Carolina, Columbia, S.C. 29208.

proximity to the ultraviolet, demands relatively large samples, necessitates samples of high purity free from fluorescing materials, and poses problems of photodecomposition. Nielsen[1] reviewed the status of the Raman spectroscopy of polymers in 1963, and in his concluding remarks he mentioned the possibility of using lasers for Raman excitation. He said, "It is too early to make any predictions as to what role lasers may come to play in Raman spectroscopy; however, the role is likely to be important." The change to which he alluded is now taking place; several commercial instruments equipped with laser sources are available.

With laser excitation sample size is reduced because the illuminating flux is concentrated in a smaller volume. Wavelengths in the red and near-infrared range can be chosen, which reduces problems of photodecomposition and fluorescence in samples or small amounts of impurities therein. This innovation in Raman spectroscopy has opened the way to the acquisition of spectra from polymer samples previously considered unworkable. Schaufele[2] has demonstrated this capability for a few polymers, and the present work was undertaken to explore the possible applications of laser Raman spectroscopy to polymer chemistry. Compilations of infrared reference spectra of polymers have been published by this laboratory for some 200 commercial products.[3-5] Many of these samples have now been taken for study by the Raman effect.

I. EXPERIMENTAL TECHNIQUE

The samples examined were commercial plastics and gum elastomers for which no sample pretreatment or purification was attempted. Spectra were recorded on a Cary 81 Raman spectrophotometer equipped with a He–Ne laser source (6328 Å). Solid samples were simply placed in the beam for coaxial viewing as powders, chunks, or pellets. Infrared spectra obtained previously[4,5] are given for comparison. The wavenumber scales of the two spectra correspond up to 2000 cm^{-1}, at which point the infrared scale changes.

II. DISCUSSION

The major problem in obtaining Raman spectra of polymers has been and continues to be "fluorescence;" this is the name that has been applied to the phenomenon, but it is not fully understood. It is seen in spectra as a continuous background emission, which generally decreases in intensity toward higher Raman shifts. The purity of the sample seems to be the determining factor in the appearance of the fluorescence. Previously reported Raman spectra of polymers, obtained using conventional (mercury arc) exciting sources, have not been highly satisfactory, even when obtained from highly purified compounds. The types of polymers which have given the best spectra are those derived from easily purified starting materials. Polyethylene,[6-8] polypropylene,[6,9] and polystyrene[10] are examples of such materials, the spectra of which have been obtained by mercury-arc excitation.

Some workers have found that a purified sample which initially gives a good background can develop troublesome impurities on standing a few days.[1] With the laser the problem of background fluorescence is diminished from what it was with 4328 Å excitation, but it has not been eliminated. Some samples still have too much background emission to be studied, while others which appear to be poor when first inserted in the beam improve upon standing an hour or two. This "warming up" process is not peculiar to polymers, and it is reversible; i.e., a warmed-up sample which is removed from the beam reverts to giving poor background if it is reinserted. A better understanding of the causes of this phenomenon will be necessary to fully cope with it.

Figures 1–12 present spectra which have been selected from about 40 polymer samples examined. The top curve in each figure is the infrared

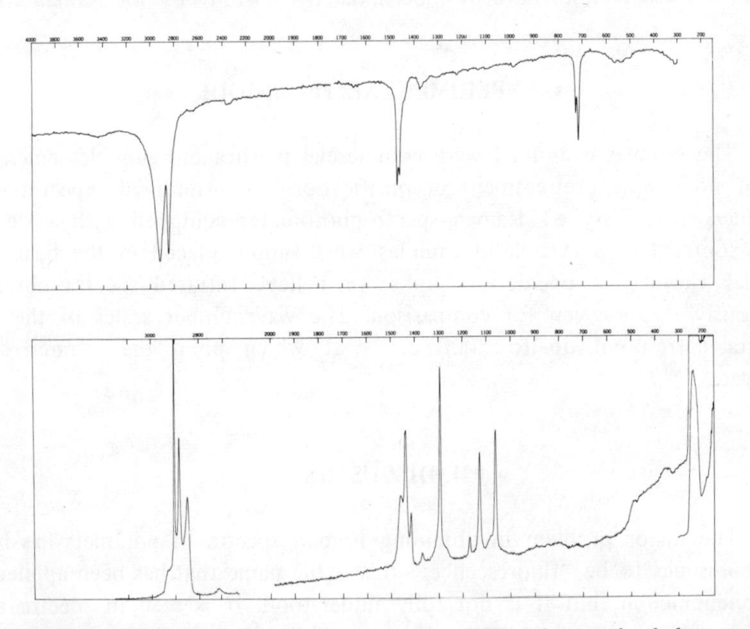

Fig. 1. Infrared (top) and Raman (bottom) spectra of Microthene polyethylene powder. Raman instrument settings: sensitivity, 100 × 2.5 (50 × 2.0 for inset from 150 to 300 cm^{-1}); period, 3.5 sec; slits, 10 cm × 6 cm^{-1}, single; scan speed, 1 cm^{-1}/sec.

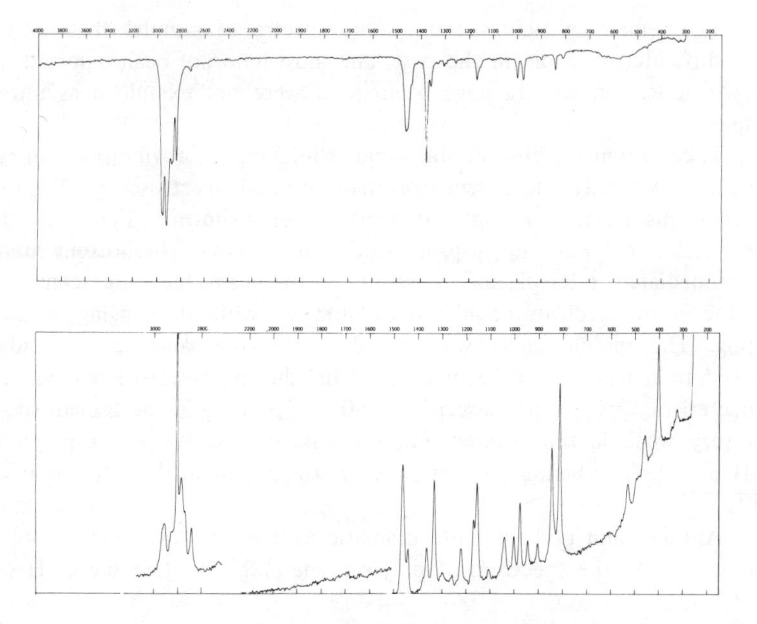

Fig. 2. Infrared (top) and Raman (bottom) spectra of Hercules polypropylene powder. Raman instrument settings: sensitivity, 100 × 3.0; period, 2 sec; slits, 10 cm × 6.5 cm^{-1}, single; scan speed, 2.5 cm^{-1}/sec.

spectrum, given for comparison. In all the Raman spectra bands at approximately 2865, 2905, and 2938 cm^{-1} are to be disregarded, as they are instrument ghosts.

Figures 1–3 illustrate the type of spectrum to be expected from laser-excited commercial samples of products which, using the mercury arc, gave only poor to fair spectra after tedious purification. It is interesting to note in these spectra some differences between Raman and infrared activity. Several fundamental vibrations appear in the Raman spectrum of polyethylene (Fig. 1) which are infrared inactive. It was not until the Raman spectrum was available that some combination bands seen in the infrared could be explained, since they are combinations of Raman and infrared-active vibrations.[1] The same type of situation exists for polypropylene (Fig. 2). One of the interesting features of the polystyrene Raman spectrum in Fig. 3 is the ring-breathing vibration seen near 1000 cm^{-1}, characteristic of monosubstituted aromatic rings. This vibration is weak in the infrared and can be useful for detecting phenyl groups from Raman spectra. The aromatic C–H stretch is also a good diagnostic line in Raman spectra, although this region is not always as helpful as in the infrared because the sensitivity of the detector diminishes toward longer wavelengths.

The remaining figures show some spectra of samples that have been more difficult to handle in the past, and most have not been reported. Those for which Raman spectra have been given were not examined as untreated samples.

The carbon–carbon double-bond stretching is a vibration for which Raman is generally more sensitive than infrared spectroscopy. Figures 4-6 illustrate this utility. A *cis*-polybutadiene rubber is shown in Fig. 4, which also demonstrates the warming-up process mentioned above. The bottom curve was run immediately after placing the sample in the beam; the fluorescence is high and the entire spectrum could not be obtained without changing the baseline setting. The middle curve was recorded after the same sample had been allowed to remain in the beam about 2 hr; the improvement is obvious. The characteristic C=C stretch is seen at 1660 cm^{-1}, strong in the Raman spectrum and very weak in the infrared. For this case of a stereospecific polymer the C–H out-of-plane bending vibration is very apparent in the infrared at $\sim$ 730 cm^{-1}.

Another and perhaps more dramatic example of this double-bond vibration is seen in the spectrum of polyisoprene (Fig. 5). Here the C–H out-of-

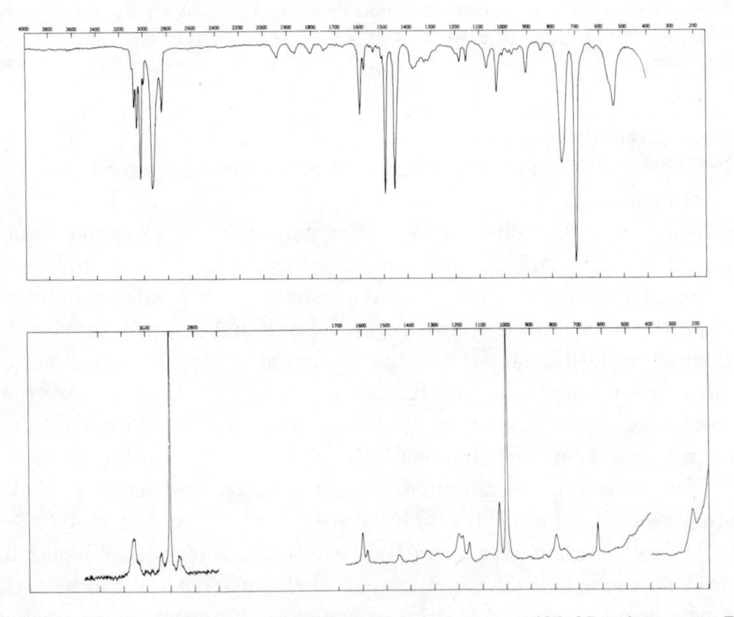

Fig. 3. Infrared (top) and Raman (bottom) spectra of Styron 690 27 polystyrene. Raman instrument settings: From 150 to 1700 cm^{-1} – sensitivity, 50 $\times$ 2.3; period, 2 sec; slits, 10 cm $\times$ 5 cm^{-1}, double (single to 400 cm^{-1}); scan speed, 1 cm^{-1}/sec. From 2700 to 3200 cm^{-1} – sensitivity, 100 $\times$ 3.0; slits, 10 cm $\times$ 5 cm^{-1}.

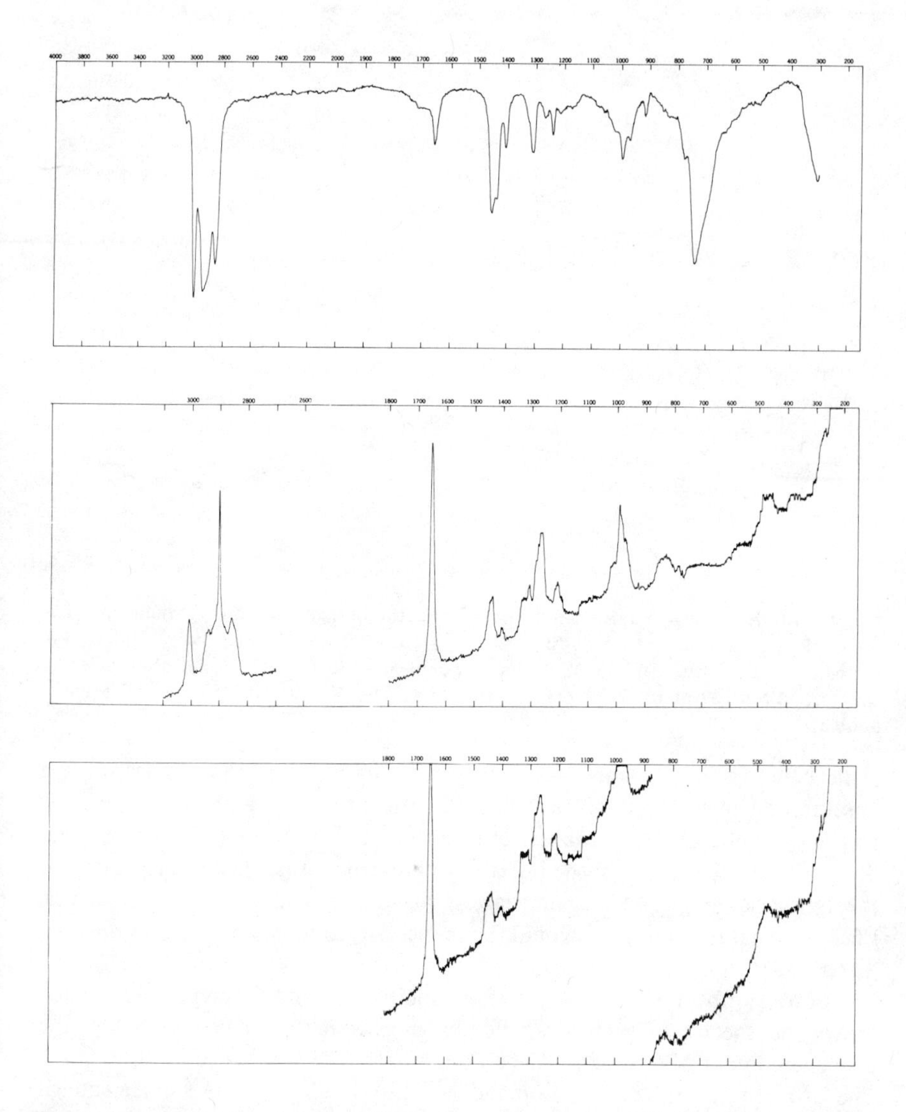

Fig. 4. Infrared (top) and Raman (bottom) spectra of Ameripol CB 220 *cis*-polybutadiene. The middle curve was obtained from a "warmed-up" sample, while the bottom curve was given by the same sample immediately after it was placed in the beam. Raman instrument settings: sensitivity, 100 × 3.0; period, 3.5 sec; slits, 10 cm × 5 cm^{-1}, double (10 cm × 4.5 cm^{-1} for middle curve); scan speed, 1 cm^{-1}/sec.

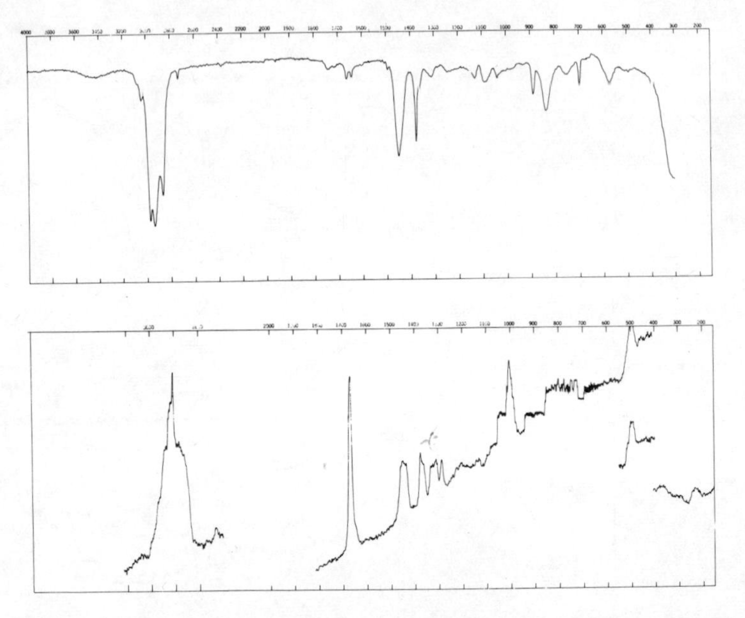

Fig. 5. Infrared (top) and Raman (bottom) spectra of Isoprene 350 synthetic polyisoprene. Raman instrument settings: From 150 to 1800 cm^{-1} – sensitivity, 100 X 3.0; period, 3.5 sec; slits, 10 cm X 6 cm^{-1}, double (single to 400 cm^{-1}); scan speed, 1 cm^{-1}/sec. From 2700 to 3100 cm^{-1} – sensitivity, 200 X 3.0; slits, 10 cm X 7.5 cm^{-1}, double.

plane bending does not give the help provided for the previous case, and the unsaturation might be doubtful from the infrared spectrum alone. This sample was a synthetic isoprene rubber, but natural rubbers were examined as well. Some Raman lines were seen, but the fluorescence was much worse than for the synthetic sample. This again suggests that the problem lies with impurities, since the natural product would be expected to contain more fluorescing materials.

Butyl rubber is a copolymer of isobutylene and isoprene, and Fig. 6 shows the spectra of such a sample. Judging from the Raman spectrum, the isoprene content is very small, since no C=C stretch is observed at 1660 cm^{-1}. The infrared spectra of the isoprene and Butyl rubbers were similar enough that the absence of a weak band near 1660 cm^{-1} was not very indicative, but the absence of what was a strong Raman line for the polyisoprene structure alone immediately points to the difference. This illustrates nicely the utility which Raman spectroscopy can have.

Another interesting group of polymers is the silicones. Work done on the structure of disoloxanes[11,12] has shown these molecules to be approximately linear, the Si–O–Si bond angle being about 155° as a result of back-bonding. The near-linearity creates an approximate center of inversion and leads to approximate mutual exclusion between Raman and infrared spectra. A similar situation evidently exists for the silicone polymers. Figure 7 presents spectra for polydimethylsiloxane, a silicone rubber. As in disiloxane, the strong Si–O–Si antisymmetrical stretching vibration seen in the infrared between 1000 and 1100 cm^{-1} is absent from the Raman spectrum, but the strong Raman line at about 490 cm^{-1} is probably the symmetrical Si–O–Si stretching vibration, which is weak in the infrared. The other prominent feature of the Raman spectrum is the C–Si stretch at $\sim$ 710 cm^{-1}, while the methyl groups are responsible for the other infrared bands.

The Raman spectrum of a mixed methyl phenyl vinyl silicone is seen in Fig. 8; this resembles that of the previous polymer, with the addition of Raman lines indicative of the phenyl substituents (ring-breathing vibration at 1000 cm^{-1}, ring mode at 1600 cm^{-1}, and C–H stretch at 3060 cm^{-1}). Unfortunately, the C=C stretch of the vinyl group did not appear in the Raman spectrum, although this might be expected to be weaker than other olefinic stretches.

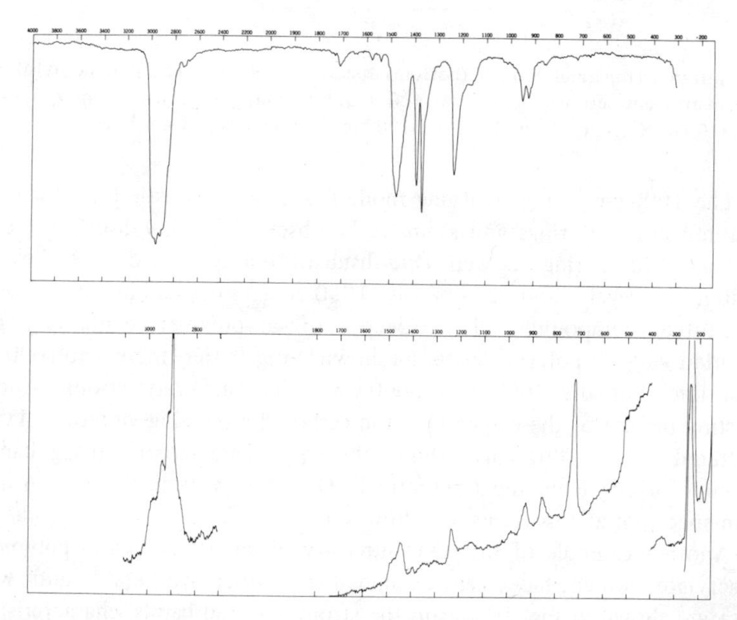

Fig. 6. Infrared (top) and Raman (bottom) spectra of Butyl HT 10-66 isobutylene–isoprene copolymer. Raman instrument settings: From 150 to 1700 cm^{-1} – sensitivity, 200 × 1.7; period, 3.5 sec; slits, 10 cm × 6 cm^{-1}, double (single to 400 cm^{-1}); scan speed, 1 cm^{-1}/sec. From 2700 to 3100 cm^{-1} – sensitivity, 200 × 3.0; slits, 10 cm × 8 cm^{-1}, double.

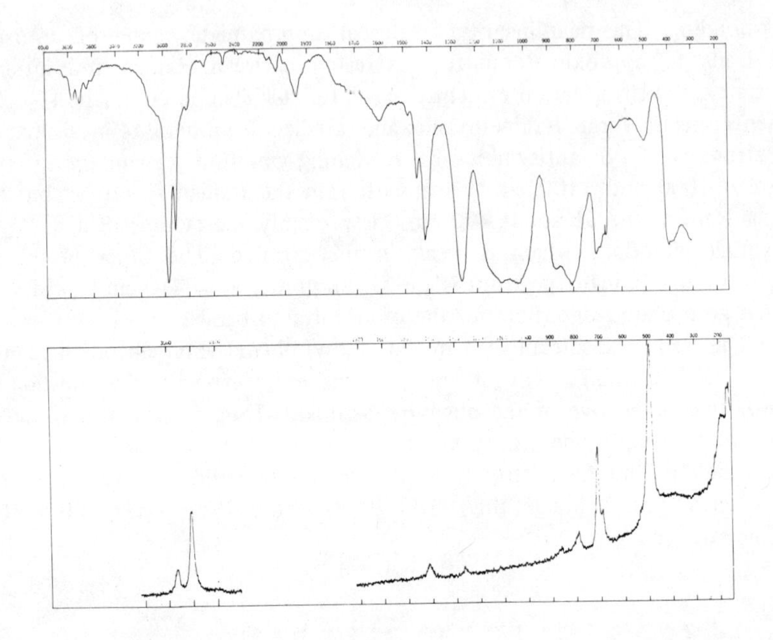

Fig. 7. Infrared (top) and Raman (bottom) spectra of Silastic 400 polydimethylsiloxane. Raman instrument settings: sensitivity, 100 X 3.0; period, 2 sec; slits, 10 cm X 5.5 cm^{-1}, double (10 cm X 7 cm^{-1} from 2700 to 3100 cm^{-1}); scan speed, 1 cm^{-1}/sec.

The 1000 cm^{-1} ring-breathing mode is a good diagnostic peak for monosubstituted aromatic rings and seems to be observed in *meta*-disubstituted and 1,3,5-trisubstituted rings as well. *Para*-disubstituted systems do not show this vibration, however, and a line at 1050 cm^{-1} occurs in many *ortho*-disubstituted compounds. The spectra of a polymer containing *para*-substituted rings, a polycarbonate, are shown in Fig. 9. Here there is no 1000 cm^{-1} Raman line, but the 1600 and 3060 cm^{-1} lines are very strong. Another interesting feature of these spectra is the carbonyl stretching vibration. Typical of infrared spectra, this curve shows the very characteristic strong band at 1780 cm^{-1} arising from the C=O stretch. Only a very weak line is seen in the Raman spectrum at this frequency, however.

Another example of this C=O intensity difference is seen in polymethyl methacrylate, which has been examined by other workers[13] and whose spectra are shown in Fig. 10. Again the strong infrared bands, characteristic of the C=O group and C–O stretch of the ester, have only weak Raman counterparts. This was a modified methyl methacrylate polymer, and the combination of Raman and infrared data suggests that the modification is the introduction of phenyl groups and unsaturation. The Raman spectrum,

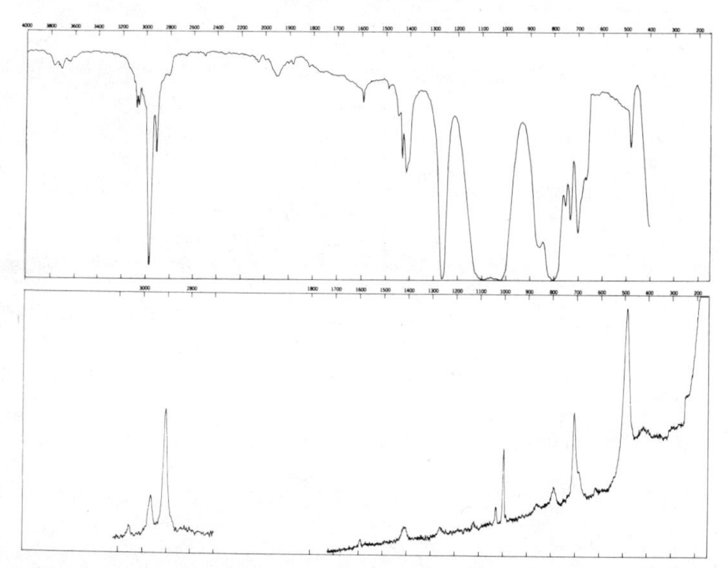

Fig. 8. Infrared (top) and Raman (bottom) spectra of Silastic 440, a mixed methyl phenyl vinyl silicone. Raman instrument settings: sensitivity, 100 $\times$ 3.0; period, 2 sec; slits, 10 cm $\times$ 7 cm^{-1}, double; scan speed, 1 cm^{-1}/sec.

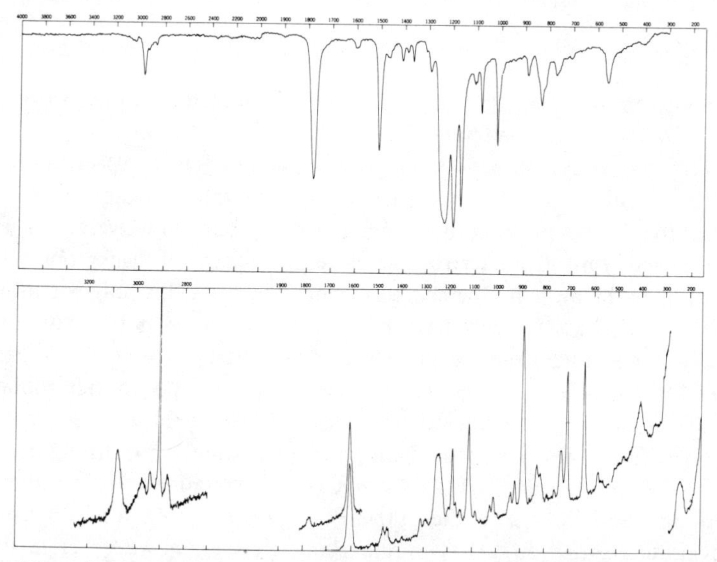

Fig. 9. Infrared (top) and Raman (bottom) spectra of Merlon M-90 polycarbonate. Raman instrument settings: From 150 to 1800 cm^{-1} – sensitivity, 100 $\times$ 2.4; period, 2 sec; slits, 10 cm $\times$ 4 cm^{-1}, double (single to 300 cm^{-1}); scan speed, 1 cm^{-1}/sec. From 2700 to 3200 cm^{-1} – sensitivity, 200 $\times$ 2.4; slits, 10 cm $\times$ 5 cm^{-1}, double.

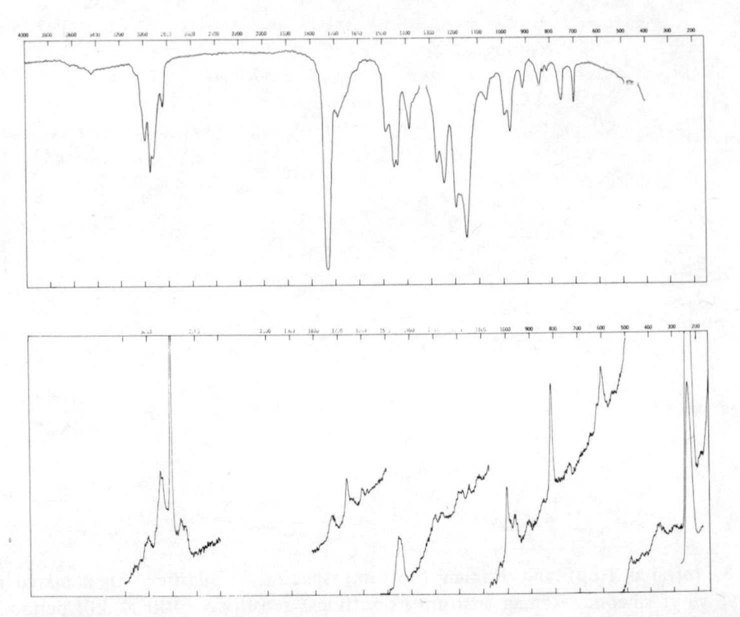

Fig. 10. Infrared (top) and Raman (bottom) spectra of Implex modified polymethyl methacrylate. Raman instrument settings: sensitivity, 100 × 3.0 (50 × 3.0 for inset from 150 to 300 cm^{-1}); period, 3.5 sec; slits, 10 cm × 7 cm^{-1}, double (single to 500 cm^{-1}); scan speed, 1 cm^{-1}/sec.

particularly, shows the 1000 and 1600 cm^{-1} lines of the monosubstituted ring and the 1670-cm^{-1} C=C stretch.

Polyoxymethylene is another polymer whose Raman spectrum has been studied. Matsui *et al.*,[14] using mercury-arc excitation, examined Carbowax samples that were purified by reprecipitation, but they were not able to obtain a spectrum from Delrin acetal resin because of background fluorescence. Figure 11 presents the spectrum obtained from this polymer using laser excitation. Band assignments have been made by Matsui and co-workers.[14,15]

An interesting example of the complementary nature of infrared and Raman spectra is seen in Fig. 12, the case of a liquid polysulfide rubber. The molecule contains two characteristic linkages, those of the ether groups and of the polysulfide units. The prominent infrared feature is the strong absorption in the 1000 cm^{-1} region arising from C–O–C vibrations; little indication is seen of the S–S linkage in the 500 cm^{-1} region expected for S–S stretching. Quite another situation exists in the Raman effect, however. Here the S–S stretch is observed at 510 cm^{-1} and the C–S stretch at 650 cm^{-1}, and these are the strongest lines. Conversely, little indication is seen of the ether groups so apparent in the infrared spectrum. This is a striking example of the necessity of having both spectra for the complete characterization of a molecule.

III. APPLICATIONS

The spectra presented illustrate the value of Raman spectroscopy to the polymer chemist. First, structural analysis is an area of interest. In addition to the information to be obtained merely from Raman activity, depolarization measurements on specific lines can give important symmetry and structural information. Recent work of Cornell and Koenig[16] illustrates the value of the laser in this application. Hendra and Willis[17] have used the polarized nature and directionality of the laser to investigate orientation effects of stretched polyethylene fibers. In addition, the far-infrared region is easily accessible in the Raman effect. Vibrations having frequencies in the range 50–300 cm^{-1} can be observed in the ordinary Raman spectrum, while special equipment is necessary to determine infrared absorption in this range. These vibrations may be of particular interest in polymer samples, since lattice modes of crystals, conformational transitions, and low-lying skeletal vibrations fall in this range.

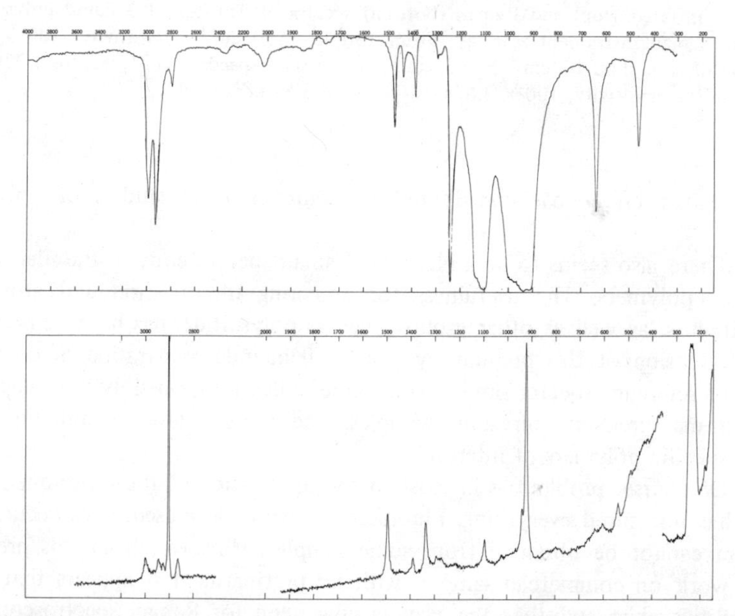

Fig. 11. Infrared (top) and Raman (bottom) spectra of Delrin polyoxymethylene. Raman instrument settings: sensitivity, 100 X 2.8; period, 2 sec; slits, 5 cm X 6 cm^{-1}, double (single to 350 cm^{-1}); scan speed, 1 cm^{-1}/sec.

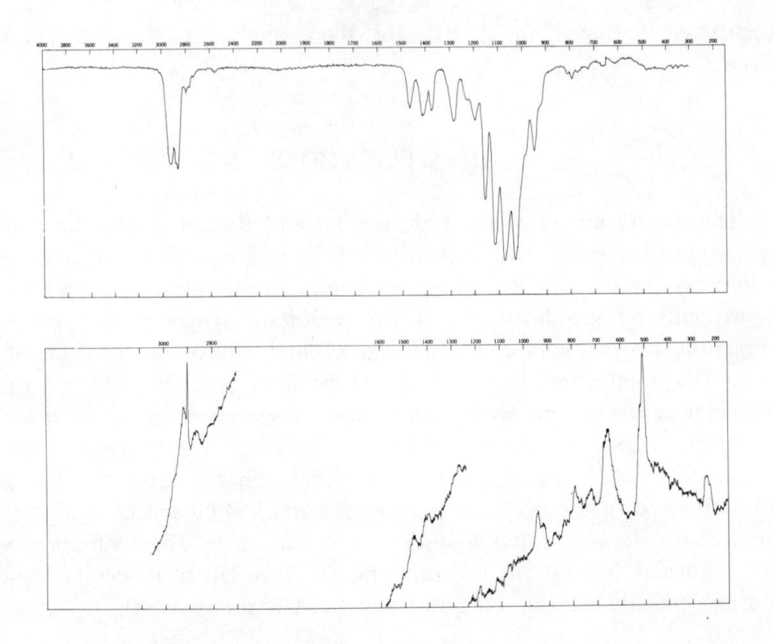

Fig. 12. Infrared (top) and Raman (bottom) spectra of Thiokol LP-3 liquid polysulfide rubber. Raman instrument settings: From 150 to 1600 cm^{-1} – sensitivity, 100×2.6; period, 3.5 sec; slits, 10 cm $\times$ 5.5 cm^{-1}, single; scan speed, 1 cm^{-1}/sec. from 2700 to 3100 cm^{-1} – sensitivity, 100×3.0; slits, 10 cm $\times$ 5.5 cm^{-1}, double.

The Raman effect has contributed to conformational studies of polypropylene.[18]

There also seems to be a place for Raman spectroscopy in the identification of polymers. The usefulness for detecting unsaturation and aromatic substituents, as well as other groups, such as polysulfide, has been demonstrated. Extension of this preliminary work will include preparation of compilations of reference spectra similar to infrared collections, possibly with depolarization measurements on some samples, and more detailed examination of some specific polymers of interest.

Of course, problems still exist in the application of the technique – the laser has not cured everything. Fluorescence, while decreased, does occur, and spectra cannot be obtained from some samples. However, from this preliminary work on commercial samples with no pretreatment it appears that new capabilities exist and that the way is now open for Raman spectroscopy to make valuable contributions in the field of polymer chemistry.

REFERENCES

1. J. R. Nielsen, *J. Polymer Sci., Part C,* **7**, 19 (1964).
2. R. F. Schaufele, *Trans. N. Y. Acad. Sci.* **30**, 69 (1967).
3. R. E. Kagarise and L. A. Weinberger, Infrared Spectra of Plastics and Resins, Naval Research Laboratory Report 4369 (May 26, 1954).
4. S. S. Stimler and R. E. Kagarise, Infrared Spectra of Plastics and Resins, Part 2 – Materials Developed Since 1954, Naval Research Laboratory Report 6392 (May 23, 1966).
5. D. S. Cain and S. S. Stimler, Infrared Spectra of Plastics and Resins, Part 3 – Related Polymeric Materials (Elastomers), Naval Research Laboratory Report 6503 (Feb. 28, 1967).
6. M. C. Tobin, *J. Opt. Soc. Am.* **49**, 850 (1959).
7. R. G. Brown, *J. Chem. Phys.* **38**, 221 (1963).
8. J. R. Nielsen and A. H. Woollett, *J. Chem. Phys.* **26**, 1391 (1957).
9. V. N. Nikitin and L. I. Maklakov, *Opt. Spectry. USSR* **17**, 242 (1964).
10. R. Signer and J. Weiler, *Helv. Chim. Acta* **15**, 649 (1932).
11. R. C. Lord, D. W. Robinson, and W. C. Schumb, *J. Am. Chem. Soc.* **78**, 1327 (1956).
12. R. F. Curl, Jr. and K. S. Pitzer, *J. Am. Chem. Soc.* **80**, 2371 (1958).
13. J. R. Ferraro, J. S. Ziomek, and G. Mack, *Spectrochim. Acta* 17, 802 (1961).
14. Y. Matsui, T. Kubota, H. Tadokoro, and T. Yoshihara, *J. Polymer Sci., Part A,* **3**, 2275 (1965).
15. H. Tadokoro, A. Kobayashi, Y. Kawaguchi, S. Sobajima, S. Murahashi, and Y. Matsui, *J. Chem. Phys.* **35**, 369 (1961).
16. S. W. Cornell and J. L. Koenig, Laser Excited Raman Scattering in Polystyrene, Case Western Reserve University., T. R. No. 84 (April 26, 1968).
17. P. J. Hendra and H. A. Willis, *Chem. Comm.* **1968**, 225.
18. H. Tadokoro, M. Kobayashi, M. Ukita, K. Yasufuku, S. Murahashi, and T. Torii, *J. Chem. Phys.* **42**, 1432 (1965).

Internal-Reflection Spectroscopy

Internal-Reflection Spectroscopy of Nonaqueous Solvent Systems: Halides in Liquid Sulfur Dioxide

D. F. Burow[†]

Department of Chemistry
Michigan State University
East Lansing, Michigan

Infrared spectra of several halide and pseudohalide salt solutions in liquid sulfur dioxide as well as that of the pure solvent were obtained using internal-reflection techniques. Perturbation of the solvent S–O symmetrical and asymmetrical stretch frequencies occurs in iodide, thiocyanate, and some bromide solutions. Solvation of anions in liquid sulfur dioxide is discussed in terms of a charge-transfer interaction.

INTRODUCTION

Examination of the chemical literature reveals that the majority of reactions studied occur in solution and that the course and rate of these reactions are very much solvent-dependent. An understanding of these phenomena can come only from a thorough knowledge of the structure of species in solution, the nature of their interaction with the solvent and with each other, and a quantitative evaluation of the parameters which influence these processes. Extensive studies of solvent effects have been carried out in water and a few other common solvents, but interpretation is often complicated by a number of factors such as hydrogen bonding and autoionization.

Liquid sulfur dioxide is an example of a dipolar, aprotic solvent with appreciable solvating properties, and is therefore a suitable medium in which

[†] Present address: Dept. of Chemistry, University of Toledo, Toledo, Ohio 43606.

to examine solvation processes in the absence of effects such as hydrogen bonding. This solvent system has been studied since the turn of the century[1-4], yet little is known concerning the nature of species in solution. Since the dielectric constant of liquid sulfur dioxide is low, its ionizing powers are expected to be quite limited, yet numerous "covalent" halides, which behave quite normally in other solvents, are electrolytes in liquid sulfur dioxide. This phenomenon has been attributed to an unusually strong solvation of halide ions in liquid SO_2.[3] Alkali, alkaline earth, and ammonium iodides and thiocyanates dissolve to produce intensely-colored solutions; solvates of the halides and psuedohalides can be isolated from solution. The molar ratio of SO_2 to salt in these solvates generally varies from one to four, but much higher ratios have been reported.[4] The infrared spectrum reveals a strong perturbation of the S–O vibrational modes in isolable iodide and thiocyanate solvates.[5] A better understanding of the solvation of simple salts in liquid sulfur dioxide should provide some insight into the origin of the electrolytic properties of "covalent" solutes, and yield a clearer concept of solvation processes in general. The infrared spectrum of liquid sulfur dioxide and that of several representative salt solutions have been examined to investigate the utility of such information in understanding the solution processes.

Application of conventional spectroscopic transmission methods to the study of low-temperature nonaqueous solvent systems presents several experimental difficulties. Cells of very short path length are a necessity. Some provision for cooling the solution or containing a liquid under pressure must be made. Solution manipulation must be carried out in rigorously anhydrous and oxygen-free conditions. The internal reflection method[6] is well recognized as a useful technique for the study of liquid systems and can be adapted to the peculiar requirements of this work. Radiation propagated in an optically-dense but transparent medium is totally reflected from the interface with a less-dense medium if the angle of incidence is greater than the critical angle. If the rarer medium absorbs at a particular frequency, total reflection is destroyed at that frequency. Measurement of the reflectivity of this interface as a function of frequency yields the internal-reflection spectrum. A detailed discussion of internal-reflection spectroscopy may be found in a monograph by Harrick[6] (also see the references cited therein).

EXPERIMENTAL

Infrared spectra were obtained using a 21-reflection, trapezoidal, internal-reflection element with a fixed 45° angle of incidence. The element was made of high-purity polycrystalline germanium and mounted in a Teflon cell (Fig. 1). The O-ring seals containing the sample permitted pressures

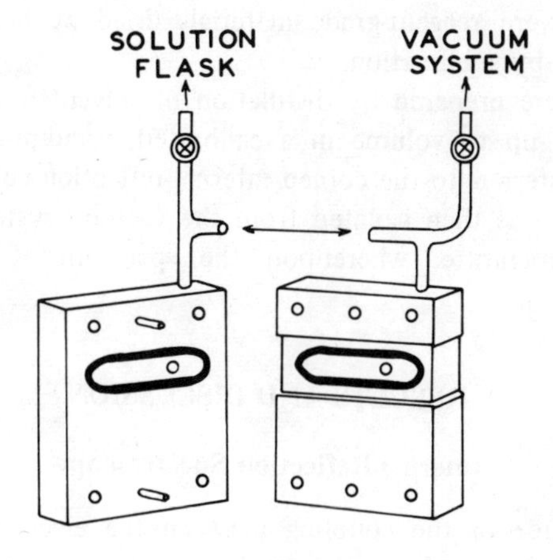

Fig. 1. Internal-reflection cell.

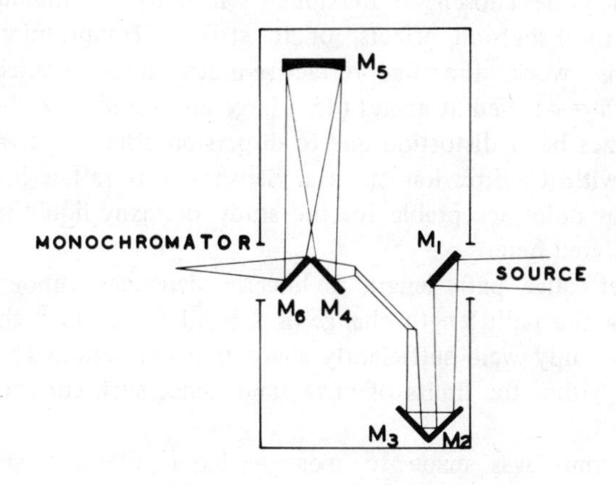

Fig. 2. Optical transfer unit.

between 3 atm and 10^{-4} torr to be maintained indefinitely within the closed cell. An optical transfer unit (Fig. 2) permitted use of the cell with a conventional spectrometer. Spectra were recorded using a Beckman IR-7 spectrometer in the single-beam mode.

All operations involving SO_2 purification, transfer, and solution make up were performed in a grease-free high-vacuum system fitted with Teflon stopcocks and connections. Sulfur dioxide (Matheson anhydrous, 99.98%) was refluxed over P_2O_5; degassed by multiple freeze, pump, melt cycles; passed through a mercury bubbler to remove SO_3; and finally passed through

P_2O_5. All salts were reagent-grade materials dried by heating *in vacuo* or, where convenient, by sublimation.

Solutions were prepared by distillation of solvent onto a known weight of salt and made up to volume in a calibrated, vacuum-jacketed flask. The solution was transferred to the cooled internal-reflection cell by its own vapor pressure. The cell was then isolated from the vacuum system and allowed to attain room temperature, whereupon the spectrum of the solution was recorded.

RESULTS AND DISCUSSION

Internal-Reflection Spectroscopy

The magnitude of the coupling between the electromagnetic field and the absorbing dipoles in the sample, and thus the intensity of the internal-reflection spectrum, is dependent upon a number of parameters.[6] These parameters may be chosen to maximize sensitivity, to minimize band distortion due to dispersion effects, or to strike a compromise between the two. In this work the high-refractive-index internal-reflection element (germanium, n = 4) used at angles (45°) large compared with the critical angle (19°) minimizes band distortion due to dispersion effects and facilitates direct comparison with transmission spectra. Sensitivity is rather low under these conditions but quite acceptable for the study of many liquid systems such as those encountered here.

Since effective path length is linearly dependent upon wavelength, a correction for the path length change as a band is traversed should be made. Bands in this study were sufficiently sharp that corrections for effective path lengths were within the limits of error, and hence such corrections have been neglected.

No attempt was made to measure band intensity, since a detailed knowledge of all the necessary parameters such as the interaction geometry and radiation polarization could not be readily obtained. Comparison of band shapes and peak positions among the various solutions is, nevertheless, quite meaningful, since all measurements were made under the same conditions. As a further check on the validity of these assumptions spectra of solutions of several alkali halide–SO_2 solvates in acetonitrile were obtained by both internal-reflection and transmission techniques. Peak positions in the two measurements differ by no more than 3 cm^{-1}. Band shapes as observed by the two methods, while not identical, possess the same features. Therefore it will be assumed that the internal-reflection spectra obtained in this study can be qualitatively compared with transmission spectra without the necessity for extensive data transformation.

Liquid Sulfur Dioxide

The infrared spectrum of liquid sulfur dioxide was examined from 800 to 4000 cm^{-1}. Two fundamental vibrational modes were observed at 1145 cm^{-1} (ν_1) and 1336 cm^{-1} (ν_3); the ν_2 region ($\sim$ 520 cm^{-1}) was not within the spectral range of the apparatus. No bands were observed at frequencies corresponding to overtone and combination modes; this is not unexpected, since they are very weak in the solid and in the Raman spectrum of the liquid.[7] Sketches of the bands along with spectral slit widths are shown in Fig. 3a. and 3b. Table 1 permits comparison of these observations with Raman[7,8] and infrared[9-11] data reported in the literature.

The symmetrical stretch (ν_1) was observed at 1145 cm^{-1} and is much

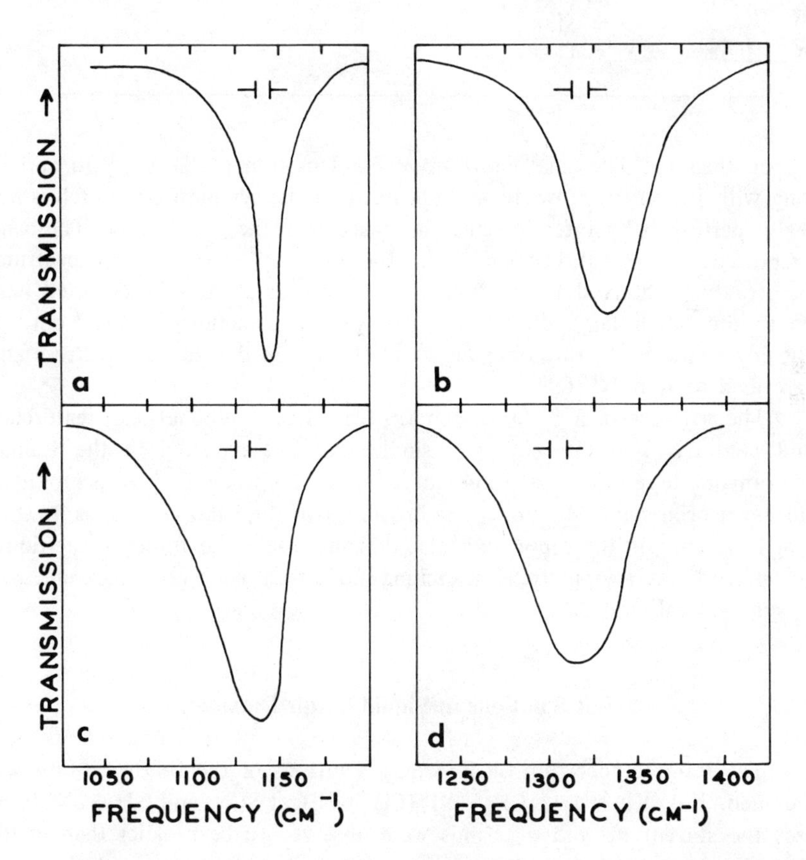

Fig. 3. Internal-reflection spectrum of liquid sulfur dioxide. (a) ν_1 of SO_2. (b) ν_3 of SO_2. (c) ν_1 of SO_2 containing 3 *M* KI. (d) ν_3 of SO_2 containing 3 *M* KI.

TABLE I

The Vibrational Spectrum of Sulfur Dioxide

	ν_1 (cm^{-1})	ν_2 (cm^{-1})	ν_3 (cm^{-1})	Ref.
Gas				
Infrared	1151	518	1362	11
Liquid				
Raman	1144	524	1336	8
	1145	524	1334	7
Infrared	1145	–	1336	Present work
Solid				
Raman	1144	523	1341	7
	1148	537		
Infrared	1144	521	1334	10
		535		

sharper than ν_3. The small shift upon condensation (1151 cm^{-1} in vapor), along with its very narrow shape, indicates that the symmetrical stretch is not greatly perturbed by intermolecular interactions in the liquid phase. The band is reported to occur at 1145 cm^{-1}[7] and 1144 cm^{-1}[8] in the Raman spectrum. The structure observed in the Raman spectrum[7] could not be detected here due to the rather large slit widths employed. The shoulder at 1127 cm^{-1} is probably the same feature observed at 1121 cm^{-1} in the Raman spectrum and is ascribed to ν_1 of $S^{16}O^{18}O$.

The asymmetrical S–O stretch (ν_3) band is a symmetrical, featureless band centered at 1336 cm^{-1}. A similar band is reported in the Raman spectrum of liquid SO_2 at 1334 cm^{-1}[7] and 1336 cm^{-1}[8]; in acetonitrile solution it occurs at 1340 cm^{-1}. The broadness of the band, as well as its shift from 1362 cm^{-1} in the vapor, indicates that intermolecular forces in the liquid state perturb the asymmetrical stretching mode to a much greater extent than the symmetrical mode.

Salt Solutions in Liquid Sulfur Dioxide

The infrared spectrum of 1 *M* SO_2 solutions of the following salts was examined: NaI, KI, NH_4I, $(CH_3)_4NI$, $(CH_3)_4NBr$, KSCN, and NH_4SCN. In all cases the solvent ν_1 and ν_3 bands were observed to be broader than in the pure solvent and markedly asymmetrical on the low-frequency side (Fig. 3c, d). No perturbations of solvent bands were observed in solutions of KBr and KCl, but since the solubility of these salts is markedly less than that of the salts in the previous set, no definite statement concerning the absence of a perturbation by this latter set is possible.

The concentration dependence of the infrared spectrum of solutions of KI (0.1 to 4.0 *M*) in liquid SO_2 (Fig. 4) reveals the probable existence of more than one SO_2 solvate band at the higher concentrations. This would indicate the presence of more than one solvated species in this concentration

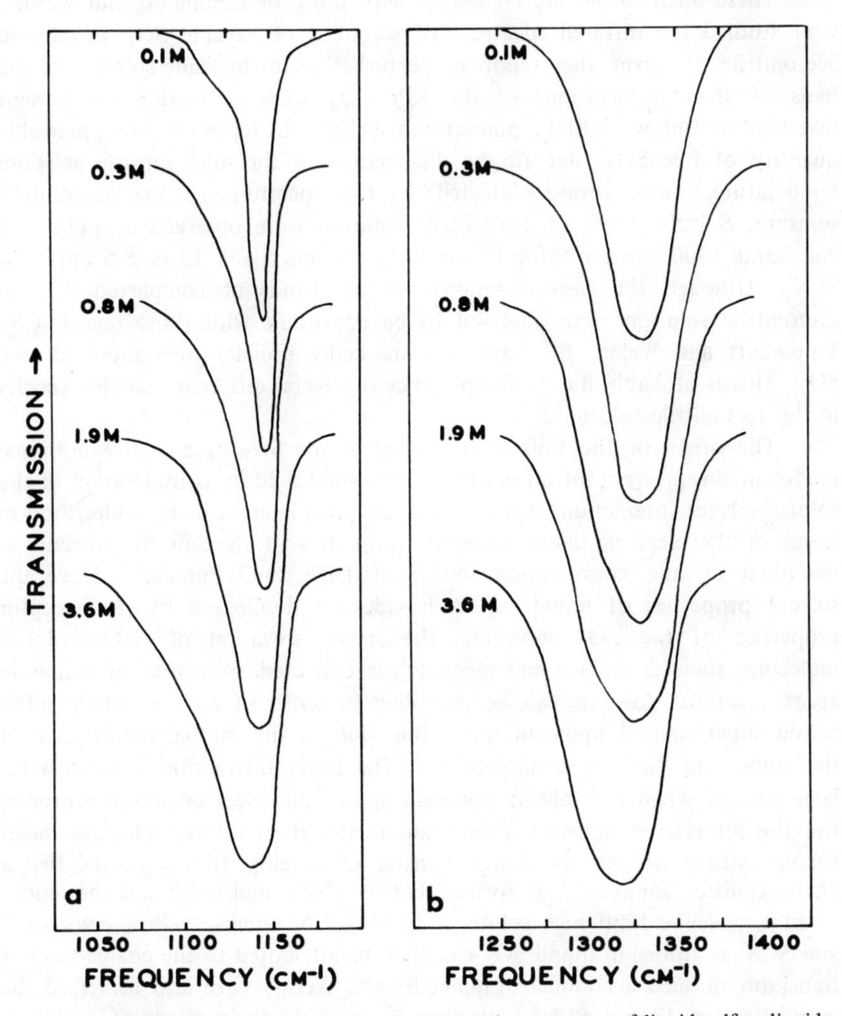

Fig. 4. The effects of dissolved potassium iodide on the spectrum of liquid sulfur dioxide. (a) ν_1 of SO_2 (b) ν_3 of SO_2.

range, but definite conclusions will require careful intensity measurements and band-shape analysis.

The magnitude of the perturbation as measured by the degree of asymmetry and broadening of the SO_2 bands is in the order $I^- \sim SCN^- > Br^-$. The cation has no measurable effect upon the perturbation; it is therefore concluded that a strong solvation of the anion occurs which results in a perturbation of the ν_1 and ν_3 vibrational modes of the solvent sphere molecules in liquid SO_2 solution.

These observations are consistent with those of Lippincott and Welsh[5], who studied the infrared spectrum of solutions of isolable SO_2 solvates in acetonitrile. However, they report no perturbation for bromide solvates on the basis of their observations of the KBr–SO_2 solvates. During the present investigation, the work of Lippincott and Welsh[5] was repeated. An appreciable quantity of free SO_2, due to the dissociation of the solid solvates at room temperature,[4] was always detected in the spectrum of the acetonitrile solution. Solvate bands in acetonitrile solution were observed at 1117 ± 5 cm^{-1} and 1300 ± 5 cm^{-1} for I^- and 1120 ± 5 cm^{-1} and 1305 ± 5 cm^{-1} for SCN^-. Although the peak positions for the bands of complexed SO_2 in acetonitrile solution were observed to be consistent with those reported by Lippincott and Welch, the bands are markedly broader than those of free SO_2. This is probably due to the presence of several different complex species in the acetonitrile solutions.

The origin of the shift in frequency of the SO_2 ν_1 and ν_3 vibrational modes in the presence of an anion can be established by consideration of the solute–solvent interaction. Aprotic solution phenomena can be understood in terms of the electron donor–acceptor properties of the solvent molecule as described by the coordination model of Drago and Purcell.[12] Since the solvent properties of liquid sulfur dioxide are dominated by the acceptor properties of the SO_2 molecule, the strong solvation of electron-donor molecules such as amines and alcohols[4] is expected. Solvation of anions in aprotic, dipolar solvents can be described in terms of an ion–dipole interaction superimposed upon an interaction due to the mutual polarization of the anion and the solvent molecule.[13] The latter interaction is greatest for large anions. When the solvent possesses appreciable electron-acceptor properties the interaction between a large anion and the solvent molecules might become strong enough for charge transfer to develop. It is suggested that a charge-transfer complex[14] is formed between SO_2 molecules and the anions of salts which exhibit high solubility in SO_2. The intense yellow color of I^- and SCN^- solutions in liquid SO_2 can then be attributed to the charge-transfer transition in such a complex. Lippincott and Welsh[5] have also described the interaction of I^- and SCN^- with SO_2 in the solvate in terms of a charge-transfer interaction.

Examination of the molecular-orbital description of SO_2 according to Walsh[15] indicates that the lowest-lying unoccupied molecular orbital is an S–O antibonding orbital ($b_1'' - \bar{\pi}_u$) constructed from the antisymmetrical overlap of the p atomic orbitals of O–S–O. This orbital is probably not pure $p\pi$ in character, but is made up of both $p\pi$ and $d\pi$ contributions by the sulfur.[16] In any case, an electron accepted by SO_2 in the charge-transfer process would occupy this orbital, thus reducing the S–O bond order. A reduction in the bond order would give rise to a concurrent decrease in the force constant of the S–O bond. This would be evident as a decrease in the frequency of both ν_1 and ν_3[17] upon formation of a charge-transfer complex with a large anion. Since Br^- is less polarizable than I^-, charge transfer would be less extensive, and the perturbation on the S–O stretching frequencies would be smaller, in agreement with observation. It must yet be determined whether the charge transfer occurs from the anion to a single solvent molecule or to a solvent cage as suggested for I^- in other solvents.[18,19]

The internal-reflection technique as a practical method of obtaining the infrared spectrum of low-temperature solvent systems has been established. The anions of salts which exhibit high solubility in liquid SO_2[1] markedly perturb the spectrum of the solvent. Since the liquid SO_2 solution spectra are qualitatively similar to those of the isolable solvate complexes, the solvate structure is probably comparable to solution species structure. Quantitative evaluation of the qualitative observations reported here is in progress.

REFERENCES

1. G. Jander, *Die Chemie in Wasserahnlichen Lösungsmitteln,* Springer-Verlag, Berlin (1949), p. 209.
2. L. F. Audrieth and J. Kleinberg, *Non-Aqueous Solvents*, John Wiley and Sons, New York, (1953), p. 210.
3. N. N. Lichtin, Ionization and Dissociation Equilibria in Solution in Liquid Sulfur Dioxide, in: *Progress in Physical Organic Chemistry*, Vol. 1, (S. G. Cohen, A. Streitwieser, and R. W. Taft, eds.), Interscience Publishers, New York (1963), p. 75.
4. T. C. Waddington, Liquid Sulfur Dioxide, in: *Non-Aqueous Solvent Systems* (T. C. Waddington, ed.), Academic Press, New York, (1965), p. 253.
5. E. R. Lippincott and F. E. Welsh, *Spectrochim. Acta* **17**, 123 (1961).
6. N. J. Harrick, *Internal Reflection Spectroscopy*, Interscience Publishers, New York (1967).
7. A. Anderson and R. Savoie, *Can. J. Chem.* **43**, 2271 (1965).
8. H. Gerding and W. J. Nijveld, *Nature* **137**, 1070 (1936).
9. R. N. Wiener and E. R. Nixon, *J. Chem. Phys.* **25**, 175 (1956).
10. P. A. Giguere and M. Faulk, *Can. J. Chem.* **34**, 1833 (1956).
11. R. D. Sheldon, A. H. Nielsen, and W. H. Fletcher, *J. Chem. Phys.* **21**, 2178 (1953).
12. R. S. Drago and K. F. Purcell, The Coordination Model for Nonaqueous Solvent Behavior, in: *Progress in Inorganic Chemistry*, Vol. 6, (F. A. Cotton, ed.), Interscience Publishers, New York (1964), p. 271.

13. A. J. Parker, *Quart. Rev.* **16**, 163 (1962).
14. R. S. Mulliken, *J. Am. Chem. Soc.* **74**, 811 (1952); *J. Phys. Chem.* **56**, 801 (1952).
15. A. D. Walsh, *J. Chem. Soc.* **1953**, 2266
16. L. Pauling, *J. Phys. Chem.* **56**, 361 (1952)
17. G. Herzberg, *Infrared and Raman Spectroscopy*, D. Van Nostrand Company, New York, (1945), p. 160.
18. M. Smith and M. C. R. Symons, *Discussions Faraday Soc.* **24**, 206 (1957).
19. G. Stein and A. Treinin, *Trans. Faraday Soc.* **55**, 1086 (1959); **55**, 1091 (1959); **56**, 1393 (1960).

Application of ATR to Low-Temperature Studies

John W. Cassels and Paul A. Wilks, Jr.

Wilks Scientific Corp.
South Norwalk, Connecticut

The recording of infrared spectra of many organic materials at reduced temperatures often results in more meaningful spectra due to the sharpening of the absorption bands. These effects can be observed both in transmission and ATR spectra. ATR has an advantage however, since spectra can be obtained on many samples opaque to transmission. An apparatus for obtaining ATR curves at temperatures approaching that of liquid nitrogen is described together with typical results.

The infrared examination of materials at temperatures as low as liquid nitrogen has been accomplished using direct-transmission techniques. Spectra thus obtained often show marked differences from spectra for the same materials recorded at room temperature. Absorption bands may be sharpened or even split into several bands by loss of rotational broadening, and in some cases new bands appear due to changes in the physical state of the sample.

The infrared cells used for low-temperature studies are generally of the demountable Dewar type with the actual cell inside an evacuable space fitted with windows to eliminate condensation on the cell.

As part of our applications program to extend the usefulness of multiple-internal-reflection techniques it was decided to examine the possibility of adapting existing equipment to accomplish low-temperature work. The Wilks Model 19 controlled-atmosphere internal-reflection attachment was modified to accept the standard multiple-internal-reflection sampling cells. The Model 19 (Fig. 1) is a single-beam attachment with an evacuable housing around it. The

Fig. 1. Model 19 controlled-atmosphere internal-reflection attachment.

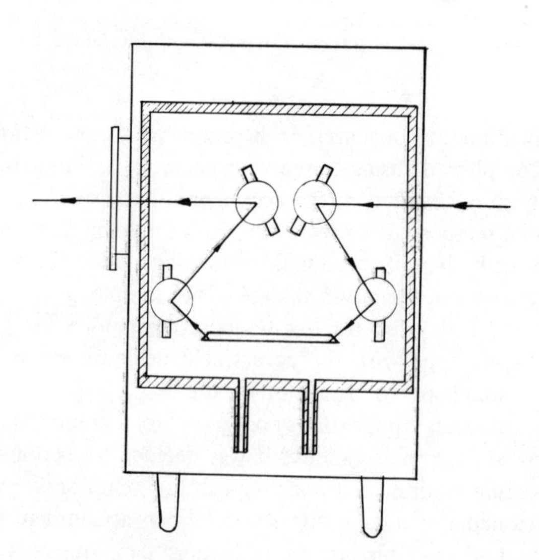

Fig. 2. Optical diagram of model 19.

housing is fitted with transmission windows. A combination of sampling cells was used to allow cooling of the internal-reflection crystal and to allow a solid sample to be held in contact with the crystal at the same time. The back of the standard solid holder was used to support the crystal, and the front portion of the liquid cell was used to clamp the crystal against the sample and also to present a cavity to one side of the crystal which could be filled with liquid nitrogen. Figure 2 is the optical diagram of the Model 19 with the crystal in place. Hose connectors were placed in the openings of the front plate of the liquid cell and Teflon tubing attached. A standard Dewar flask was used as a reservoir for liquid nitrogen, with one end of the Teflon tube placed down near the bottom of the flask. The tube attached to the second fitting of the liquid cell was connected to a vacuum pump so that the liquid nitrogen could be pulled across the internal-reflection crystal. An air bleed was placed in the vacuum line to control the amount and rate of cooling of the crystal. The top cover of the Model 19 was fitted with the tubing to facilitate flushing of the chamber. Figure 3 is the Model 19 in a Beckman IR-5A spectrophotometer with lines attached for cooling and flushing the equipment. A thermocouple was attached to the solid sample holder to monitor temperature. Figure 4 is a close up of the Model 19 in the instrument.

Passing liquid nitrogen across the crystal directly led to the problem of condensing water on the crystal even though the unit was flushed with dry carbon-dioxide-free helium. Apparently, some water came through with the nitrogen and condensed on the crystal, giving rise to absorption bands, The spectrum of the blank crystal at liquid-nitrogen temperature (Fig. 5) shows these bands to be shifted from the position normally found in liquid-nitrogen spectra. To overcome this water problem, a thin, anodized aluminum plate was placed

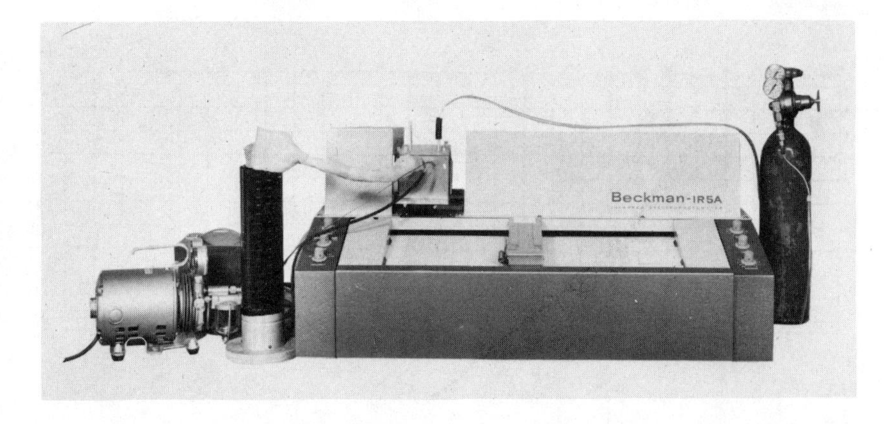

Fig. 3. Model 19 mounted in infrared spectrophotometer; set up for low-temperature work.

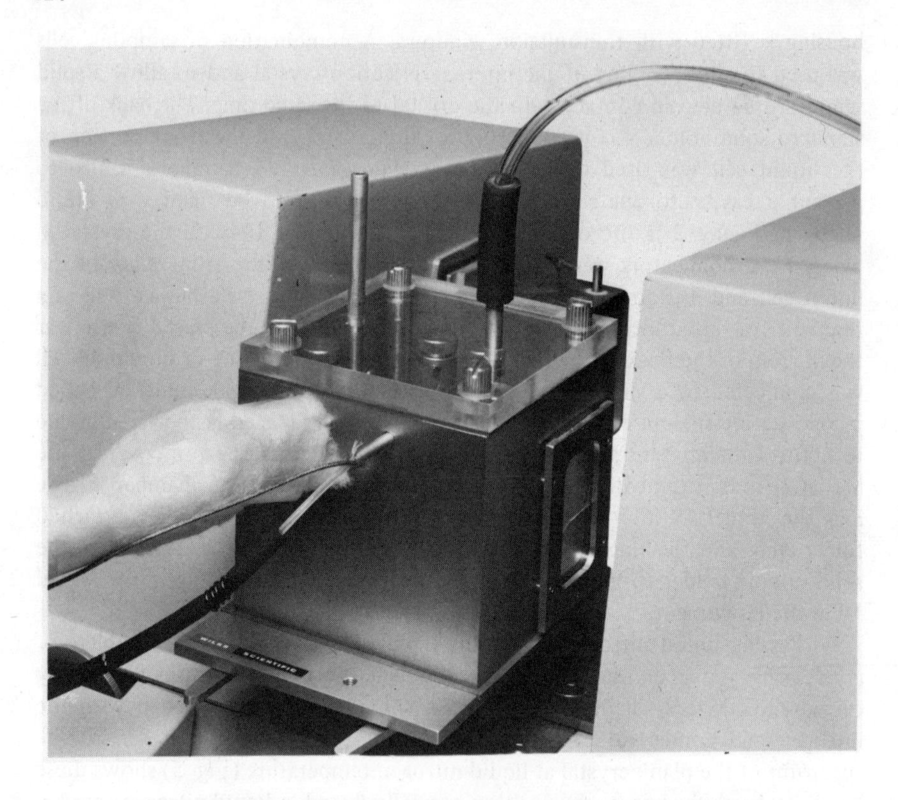

Fig. 4. Close-up view of Model 19 in the infrared spectrophotometer.

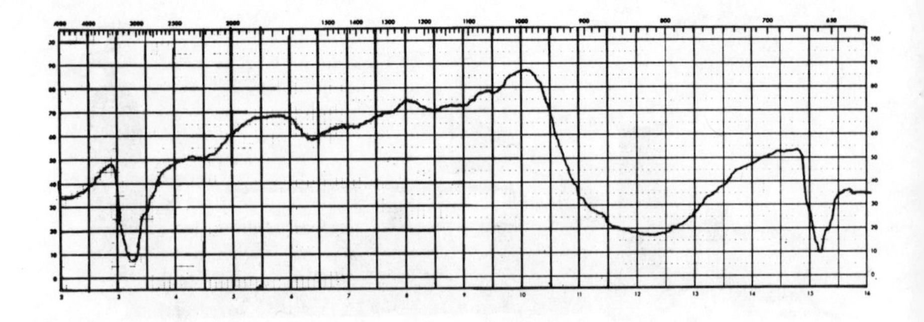

Fig. 5. Background scan taken without anodized aluminum plate in position.

between the crystal and the liquid-cell cavity. This kept water off the crystal and the anodized surface against the crystal; being nonconductive, it allowed energy to be transmitted through the crystal without energy loss.

Figure 6 is a comparison of the internal-reflection spectra of methyl oleate at room temperature and at liquid-nitrogen temperature. As can be seen, there are considerable spectral differences from liquid to solid phase. Glycerol monostearate (Fig. 7), which is a solid at room temperature, does not exhibit such a drastic spectral change, but there is some sharpening of some absorption bands. Polyethylene (Fig. 8) further illustrates this point. Note that the bands in the 13.7–14 μ area are changed considerably. With other solid materials pressed against the crystal occasionally it was noted that there was a general decrease in absorption-band intensities, most likely caused by a reduction in contact between sample and crystal due to shrinkage at low temperatures.

This technique appears to hold great promise in looking at samples at low temperatures directly and determining additional information about these samples which is not available from room-temperature examinations.

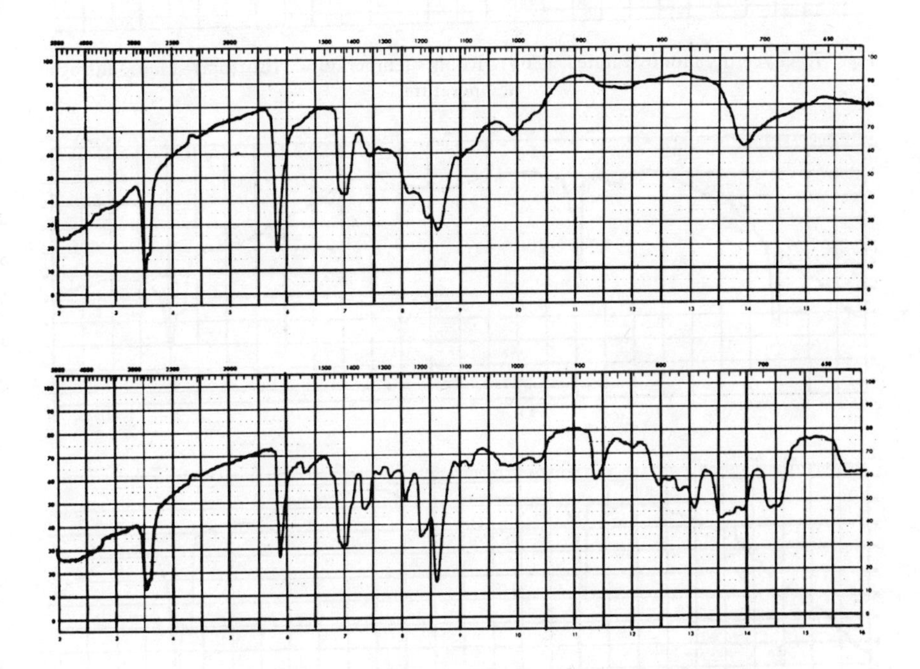

Fig. 6. Methyl oleate. (Top) Room temperature. (Bottom) Liquid-nitrogen temperature.

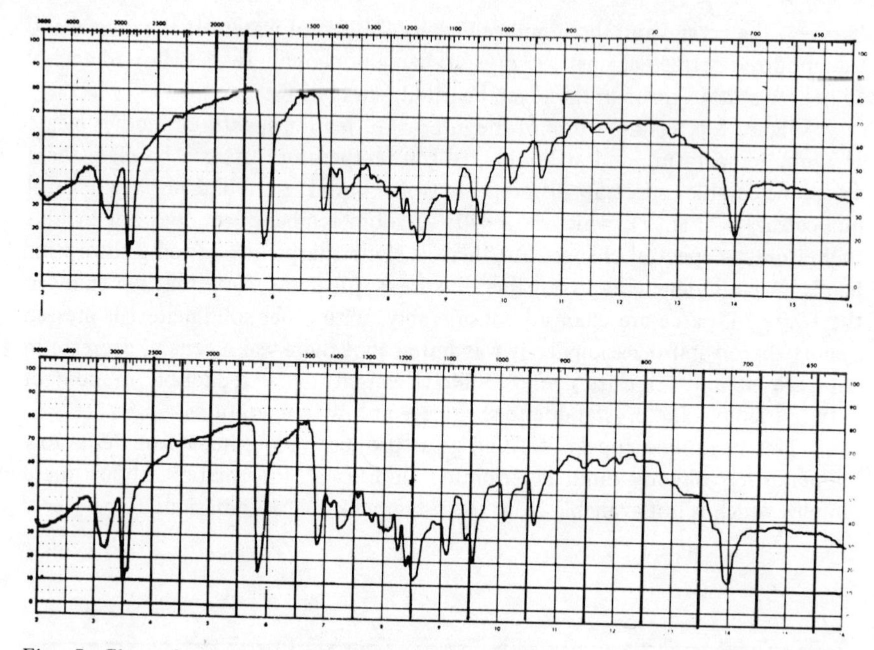

Fig. 7. Glycerol monostearate. (Top) Room temperature. (Bottom) Liquid-nitrogen temperature.

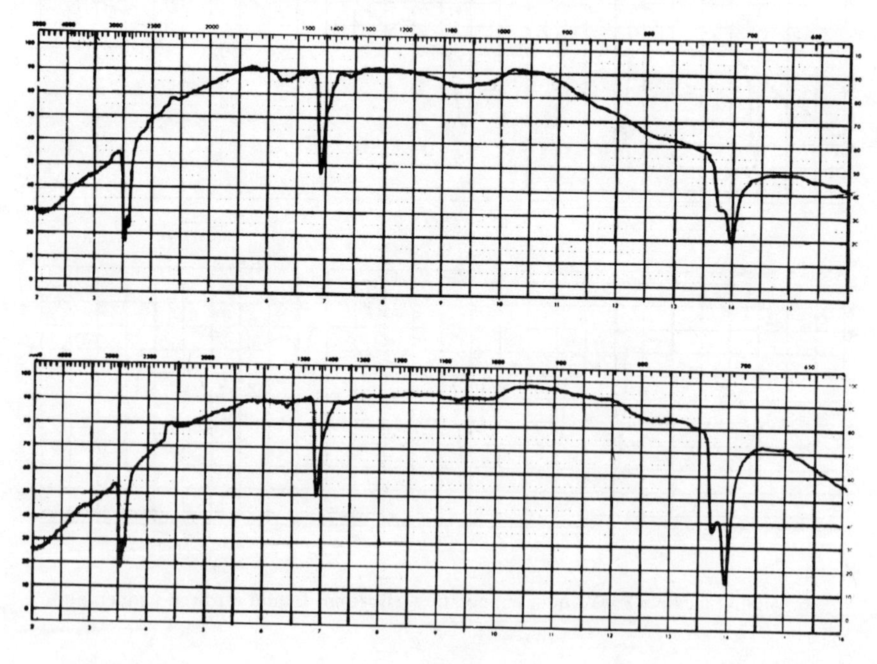

Fig. 8. Polyethylene. (Top) Room temperature. (Bottom) Liquid-nitrogen temperature.

REFERENCES

1. E. L. Wagner, and D. F. Horing, *J. Chem. Phys.* **18**, 298 (1950).
2. W. H. Duerig, and J. I. Modor, *Rev. Sci. Instr.* **23**, 421 (1952).
3. R. B. Holden, W. J. Taylor, and H. L. Johnson, *J. Opt. Soc. Am.* **40**, 757 (1950).
4. G. J. Janz and W. E. Fitzgerald, *Appl. Spectr.* **9**, 178 (1955).

Characterization of Commercial Formulations by Combining Several Chromatographic Techniques with Multiple-Internal-Reflection and Mass Spectrometry

T. S. Hermann, R. L. Levy, L. J. Leng, and A. A. Post

Midwest Research Institute
Kansas City, Missouri

Analytical chemists are frequently confronted with the problem of characterizing complex mixtures of compounds. The complex mixtures that are characterized and discussed in this paper are diverse commercial formulations. Although several approaches can be employed for the characterization of these formulations, this paper discusses one which we have found to be convenient and reliable. In this approach liquid–solid partition, thin–layer, and gas–liquid chromatography are employed to isolate and identify the components in a mixture, and multiple-internal-reflection and mass spectroscopy are employed to identify or confirm the identity of the components. This approach has been successfully employed to solve problems in our laboratories for several years. The advantages, disadvantages, and limitations of this approach are discussed in detail.

INTRODUCTION

Quite often a judiciously selected combination of techniques must be employed by the analytical chemist to characterize commercial formulations. Although many combinations of techniques are currently available, the combination which we have found to be most reliable for characterizing

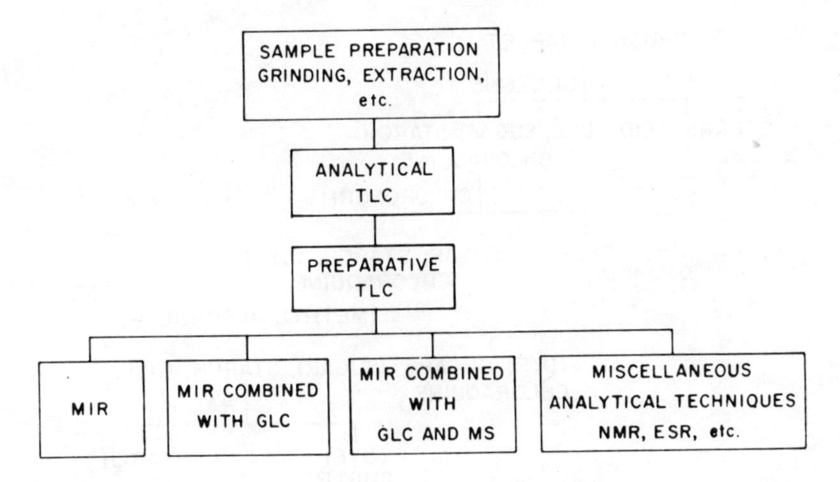

Fig. 1. General analytical philosophy.

unknown formulations is shown in Fig. 1. This approach is divided into seven parts: (1) sample preparation, (2) analytical and (3) preparative thin-layer chromatography, (4) infrared spectroscopy, (5) gas–liquid chromatography, (6) mass spectrometry, and (7) miscellaneous analytical techniques. The part labeled miscellaneous analytical techniques consists of elemental analysis, nuclear magnetic resonance, electron spin resonance, and ultraviolet–visible and emission spectroscopy. Although these techniques and several infrared transmission techniques are employed in nearly all characterizations, they will not be described in this paper. The object of this paper is to describe the manner in which multiple-internal-reflection spectroscopy can be employed in conjunction with chromatographic and mass-spectrometric techniques.

SAMPLE PREPARATION

The initial step in any analytical process is to make an overall study of the problem. Consequently, we must decide on the type of information that we wish to derive from the analysis, the amount of time and money to be expended, and other preliminary considerations. To demonstrate the steps that are usually taken after the preliminary considerations, we have selected a tablet that contains chloraquine as a model sample. One or more pills are ground in a mortar, and the resulting powder is separated into fractions with the extraction method shown in Fig. 2. In this extraction method the solvents increase in polarity in successive operations, thus extracting the relatively nonpolar compounds from the bulk sample. This order of using solvents is preferred

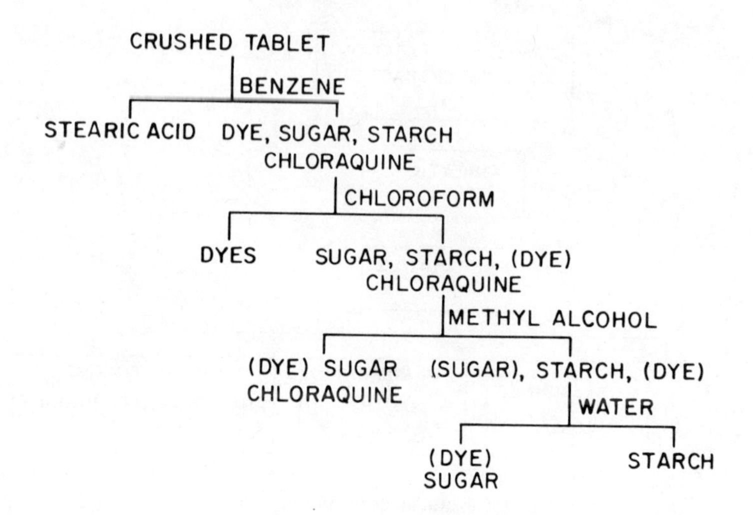

Fig. 2. Extraction method.

Fig. 3. Preparative thin-layer chromatoplates.

because the nonpolar components are usually present in relatively minute amounts, and it is conceivable that these compounds could be coextracted *in toto* with the larger amounts of polar compounds. Unfortunately, we have found that coextraction occurs to some extent regardless of the amount of care taken to avoid it, but to a lesser extent when the nonpolar solvents are used first. Consequently, another separation technique such as thin-layer chromatography (TLC) must be employed for further separation of the sample.

Analytical TLC is employed to further separate the components from the fractions, and it is also used to identify the functional groups that are present. By employing a great variety of chromogenic reagents to detect the compounds on developed chromatoplates the number of compounds and the nature of the functional groups on these compounds can be ascertained. It is obvious that a knowledge of the nature of the functional groups in these compounds is invaluable in the interpretation of spectra. In addition, this information is important in the choice of solvents for the preparative TLC technique. For preparative TLC relatively large amounts of the fractions are streaked onto a preparative chromatoplate and developed with the appropriate solvent system; the compounds are detected with a chromogenic reagent, the compounds in the TLC medium are removed from the chromatoplates, and then the isolated compounds are extracted from the TLC media into a volatile solvent such as methyl alcohol. A typical preparative TLC is shown in Fig. 3.

This combination of bulk-extraction and thin-layer chromatography has proven very reliable for the isolation of compounds from complex mixtures. Recovery of at least 90% of all components in amounts of 50 μg or more has been demonstrated on known formulations. The per cent recovery for components in amounts of less than 50 μg varies from 50 to 100%. These recoveries are satisfactory in characterizing unknown formulations. Without modification this method has been used by technicians to characterize several hundred diverse samples during the past few years. When the recoveries were less than desirable, modification of this method can easily be facilitated.

The major problem with this combination of techniques is the excessive amount of time that is expended to conduct the separations and functional group analysis. Time is especially a problem when poor resolution and tailing occurs for TLC, because at times a great many solvent systems have to be tested before an adequate one is found. Another serious problem that we have encountered is the interpretation of colors on TLC plates that the chromogenic reagents have not produced previously.

MULTIPLE-INTERNAL-REFLECTION SPECTROSCOPY

After each component of the formulation is isolated in methyl alcohol other techniques such as multiple-internal-reflection (MIR) spectroscopy can be

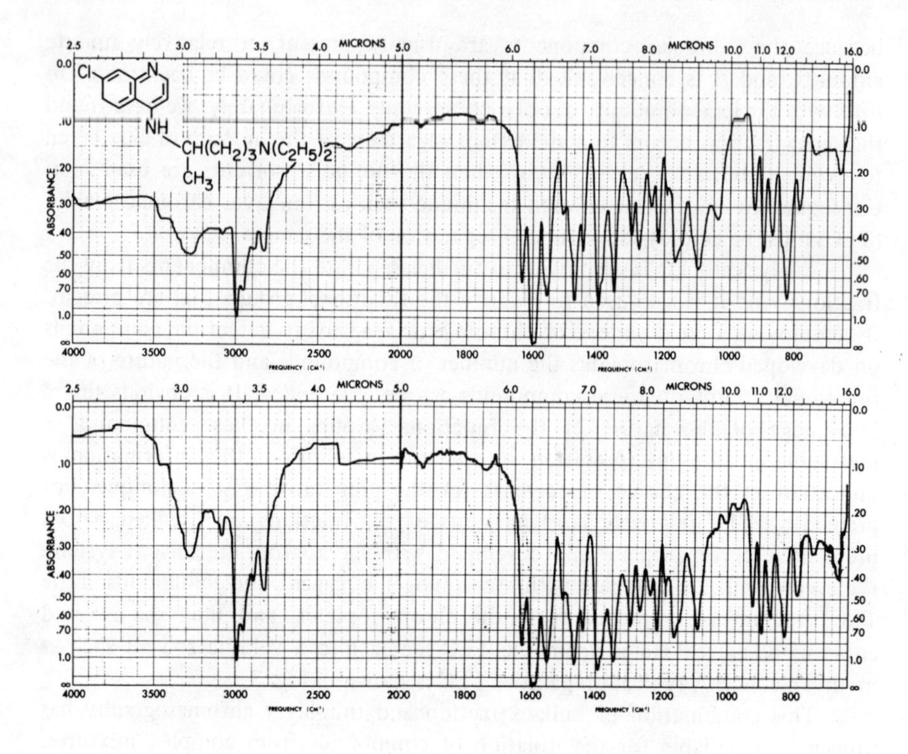

Fig. 4. Infrared spectra of chloroquine: (a) 30 μg of standard sample; (b) 30 μg of standard sample that had been passed through the separation method.

employed routinely for characterizations. Quite often the compound is simply evaporated onto an MIR plate and is scanned in the range of 4000–625 cm^{-1} on a Wilks Model 8B or a Beckman IR-12 spectrophotometer. This manner of obtaining infrared spectra is very convenient because it involves a minimum amount of sample handling, requires a relatively short time to perform, and solids and liquids can be prepared for scanning with the same technique. More than 10 μg of most materials will yield an interpretable spectrum without the need for scale expansion. To obtain an interpretable infrared spectrum for less than 10 μg of a material, scale expansion is required, and the samples must be spread on the center one-third of both analytical surfaces of a 50 x 20 x 0.75 mm MIR plate. The spectra of (a) about 30 μg of a standard sample and (b) about 30 μg of a standard sample that had been passed through the combined separation method are given in Fig. 4. These spectra show that most of the material is recovered unchanged and that little if any impurities have been introduced into the system.

When more than about 40 μg of material are present on an MIR plate made of KRS-5 a multiple-internal-reflection spectrum can be obtained with a Wilks Model 9 attachment on the Beckman IR-12 spectrophotometer. The plate can then be removed from its plate holder and inserted normal to the infrared beam to obtain its transmission spectrum. After several such experiments we have yet to observe an appreciable difference between MIR and transmission spectra except that MIR spectra are more intense.

MIR techniques possess the inherent advantages that the samples can be recovered for further analysis. This is not true, however, for all samples. We have found that organic compounds that possess nitrogen functional groups cannot be completely recovered from KRS-5 plates. In addition, these compounds etch the KRS-5 plates severely. Consequently, for these samples AgCl plates must be employed. Plates made of KRS-6, germanium, silicon, and other materials are also available for special applications.

GAS–LIQUID CHROMATOGRAPHY

At times preliminary examination of samples or information on their nature indicate that gas–liquid chromatography would be the most suitable method for isolating the compounds. A very simple device that has proven effective for collecting well-resolved peaks from a gas chromatograph is shown in Figure 5. This collector was fabricated from a piece of ordinary glass tubing and a hypodermic needle. The sample is condensed in the bulb, which is cooled in an acetone–dry ice bath. The sample can then be rinsed from the tube onto an MIR plate.

Other collectors, such as those shown in Fig. 6, have also been fabricated. For example, several 0.032 in. holes were drilled into MIR plates at various angles, and we attempted to collect chromatographic effluents within them.

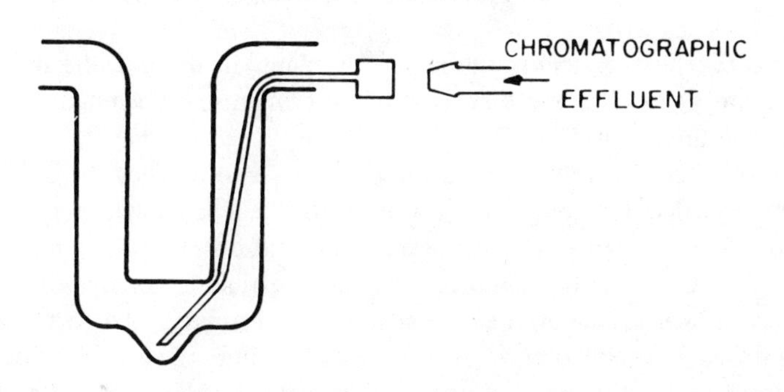

Fig. 5. Glass-tube fraction collector.

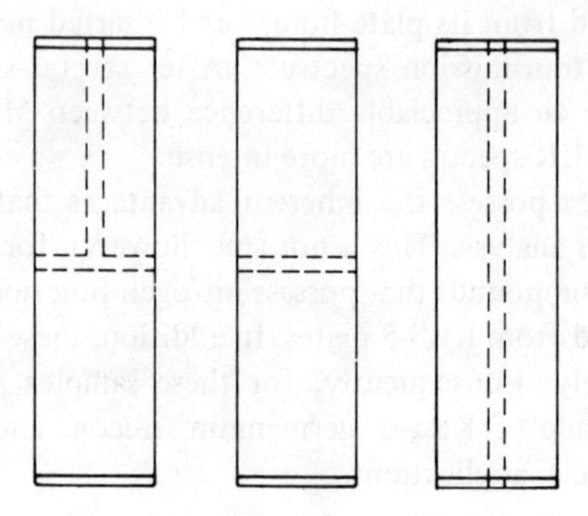

Fig. 6. Drilled multiple-internal-reflection plates.

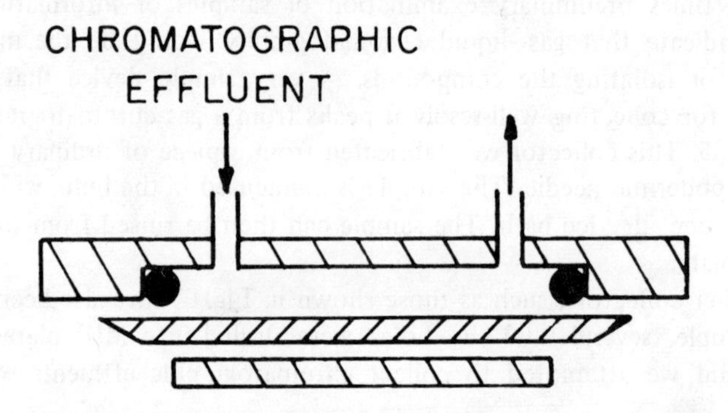

Fig. 7. Air-cooled fraction collector.

However, the heat gradient, the severe problems in drilling, the difficulty in cleaning the holes, and the high cost of experimenting precluded an extensive study of this approach.

Two other collectors are shown in Figs. 7 and 8. They are an air-cooled and a methyl-alcohol-cooled collector, respectively. These devices can be used to condense relatively large amounts of materials that possess boiling points greater than 150°C. Usually, the gas-chromatographic effluents condense in one spot, which is a severe deterrent to adequate sensitivity. To overcome this disadvantage, the collector shown in Figs. 9 and 10 was developed. The chromatographic effluents pass into a cell beneath the MIR plate. The top of the plate is cooled by passing cold methyl alcohol through the upper plate holder.

The bottom portion of the plate furthest from the entrance port is cooled in the same manner as the upper plate holder. The temperature of the bottom of the plate nearest the entrance port is controlled in the range –5°C to 50°C. The temperature of this portion of the plate will be optimized for each sample.

MASS SPECTROMETRY

At the present time we are using this cell to collect fractions from a Varian-Aerograph gas chromatograph that has been modified in our laboratories

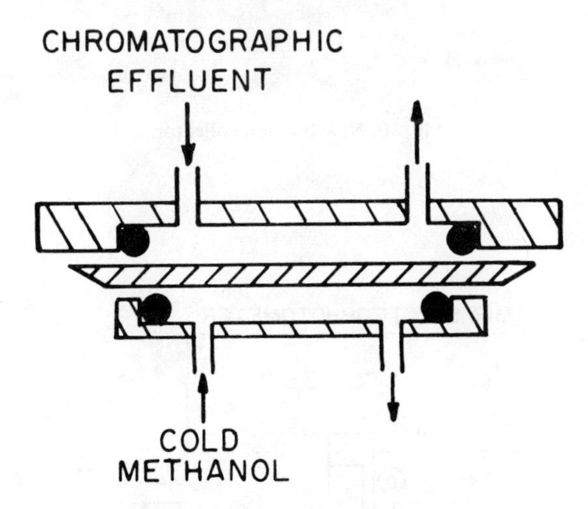

Fig. 8. Methyl-alcohol-cooled fraction collector.

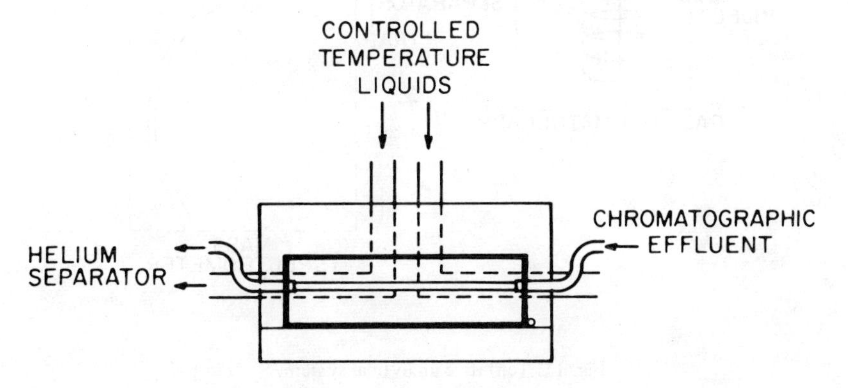

Fig. 9. Schematic of new fraction collector.

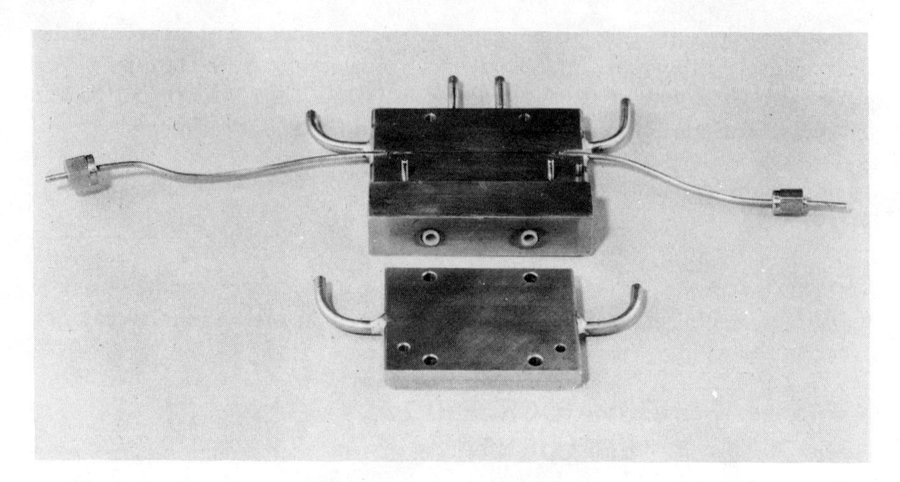

Fig. 10. New fraction collector.

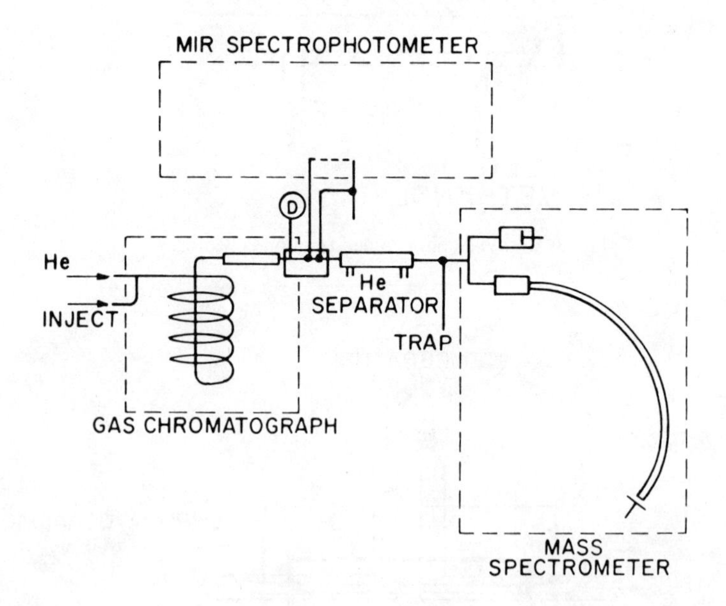

Fig. 11. Tentative analytical system.

for use with an Atlas-MAT CH-4B mass spectrometer. A schematic of the tentative system is shown in Fig. 11. In this approach the sample is separated in a gas chromatograph, and as the isolated compounds are eluted intermittently from the chromatograph each compound is successively condensed on an MIR plate and its spectrum obtained. Each compound is then vaporized and bled into the mass spectrometer. The MIR cell has several advantages over the conventional gas cells and microgas cells that are presently in use. Probably the most noteworthy advantage is sensitivity. Scott *et al*[1] have described a GC-MS-IR system that employed a microgas cell. The lower limit of detection was 25 μg for polar compounds and 100 μg for nonpolar compounds. The greatest disadvantages of using the MIR cell are condensation of materials between the chromatograph and the MIR cell, decomposition on the nonglass surfaces of the cell, and the ineffectiveness of the system for collecting compounds that boil below about 50°C. We are presently engaged in researching these problems and are quite confident of solving them.

REFERENCE

1. R.P.W. Scott, I.A. Fowlis, D. Welti, and T. Wilkins in *Gas Chromatography* 1966, A.B. Littlewood (ed.), Elsevier, New York (1967), pp. 318-333.

Nuclear Magnetic Resonance Spectroscopy

A Comparative NMR Study

P. D. Klimstra and R. H. Bible, Jr.

G. D. Searle & Co., Chicago, Illinois

The nuclear magnetic resonance positions of the C-18 and C-19 methyl group protons in a series of *A*- and *D*-ring substituted steroidal androstane derivatives were measured and found to be in excellent agreement with the values calculated using Zurcher's tables. In the *A*-ring hydroxylated derivatives the magnitude of the coupling between the proton on the carbon bearing the hydroxyl group and the protons on the adjacent carbon atoms was also found to be consistent with the proposed structures.

INTRODUCTION

In the course of synthesizing some steroidal androstane derivatives (Fig. 1 and Table II) for screening as potential pharmacological activity[1] the effect of *A*-ring and *D*-ring substituents on the nuclear magnetic resonance of the C-18 and C-19 methyl group protons was determined experimentally. This paper compares these observed values with those which were determined by the use of Zürcher's tables.[2] When the substituent in the *A*-ring was hydroxyl its spatial conformation was verified from the magnitude of the coupling constants.

EXPERIMENTAL

Samples

The steroidal derivatives used in this study were synthesized by chemical methods as described previously.[1]

TABLE I

Variant	A-60A	A-60
Filter bandwidth	2	2
RF field (mG)	0.04	0.16
Sweep time (sec)	250	250
Sweep width (Hz)	500	500
Spectrum amplitude	32	10

Instrumentation

The nuclear magnetic resonance spectra were recorded using a Varian A-60A and/or A-60 spectrometer. The settings shown in Table I were used for recording the spectra.

Sample Solutions

The solutions were prepared by dissolving 35–52 mg of sample in 0.4 ml of deuterated chloroform. The solutions were filtered through a cotton plug in the tip of a disposable pipet prior to determination of the spectra.

RESULTS AND DISCUSSION

The steroidal derivatives studied belong to the androstane series and possess the spatial conformation shown in Fig. 1. The four rings are fused in such a manner that a plane can be passed through all of the rings with the C-18 and C-19 angular methyl groups above and perpendicular to the plane.

The effects of the various substituents on the chemical shifts of the C-18 and C-19 protons are due primarily to the magnetic anisotrophy and the electric dipole moment of the substituents.[3] The magnitude of the effects vary both with the distances between the substituents and the angular methyl groups and with the relative orientations of the substituents and the angular methyl groups.[4]

Using model compounds, Zürcher[2] derived additive constants for the effects on the C-18 and C-19 methyl signals of various substituents at different locations on the steroid skeleton. These constants were used in this paper to predict the positions of the signals due to the angular methyl protons.

TABLE II

Calculated and Observed NMR Values of the C-18 and C-19 Methyl Group Protons

No.	*A*-ring*	*D*-ring (C-17)* R	R_1	Found (cps) C-18	C-19	Calculated (cps) C-18	C-19	Deviation (cps) C-18	C-19
I	"None"	"none"		41.5	47.5	41.5	47.5	–	–
II	1-"one"	"one"		52.5	71.0	51.5	70.5	–1.0	–0.5
III	2-"one"	"one"		52.0	47.0	52.0	47.5	0	+0.5
IV	3-"one"	"one"		54.0	63.0	53.0	62.5	–1.0	–0.5
V	4-"one"	"one"		52.5	46.5	52.5	46.5	0	0
VI	1-"one"	OH	H	44.5	70.0	44.5	70.0	0	0
VII	2-"one"	OH	H	44.0	46.0	44.5	46.0	0	+0.5
VIII	3-"one"	OH	H	46.0	62.0	46.0	62.8	0	+0.8
IX	4-"one"	OH	H	44.5	45.5	44.5	46.0	0	+0.5
X	1-"one"	OAc	H	47.5	70.0	47.5	69.5	0	–0.5
XI	2-"one"	OAc	H	47.0	45.0	48.0	46.0	+1.0	+1.0
XII	3-"one"	OAc	H	49.0	62.0	48.5	61.5	–0.5	–0.5
XIII	4-"one"	OAc	H	47.5	45.0	47.5	45.0	0	0
XIV	1-"one"	OH	CH_3	51.5	70.5	52.0	72.0	+0.5	+1.5
XV	2-"one"	OH	CH_3	51.0	45.5	50.5	46.5	–0.5	+1.0
XVI	3-"one"	OH	CH_3	53.0	62.5	53.0	62.5	0	0
XVII	4-"one"	OH	CH_3	51.5	46.0	52.0	46.5	+0.5	+0.5
XVIII	1α-OH	"one"		52.5	49.0	52.0	49.0	–0.5	0
XIX	2β-OH	"one"		52.0	63.5	52.0	63.5	0	0
XX	3β-OH	"one"		52.0	50.5	52.0	50.0	0	0
XXI	4β-OH	"one"		52.0	64.5	52.5	64.5	+0.5	0
XXII	1α-OH	OH	H	44.5	48.5	44.0	48.0	–0.5	–0.5
XXIII	2β-OH	OH	H	44.0	62.5	44.5	63.5	+0.5	+1.0
XXIV	3β-OH	OH	H	44.0	49.0	44.0	49.0	0	0
XXV	4β-OH	OH	H	44.0	63.5	44.0	63.5	0	0
XXVI	1α-OH	OAc	H	47.5	48.5	47.5	47.5	0	–1.0
XXVII	2β-OH	OAc	H	47.0	62.5	47.5	62.5	+0.5	0
XXVIII	3β-OH	OAc	H	47.0	49.5	47.0	49.0	0	+0.5
XXIX	1α-OH	OH	CH_3	51.5	49.0	51.0	49.0	–0.5	0
XXX	2β-OH	OH	CH_3	51.0	63.0	51.0	63.0	0	0
XXXI	3β-OH	OH	CH_3	51.0	48.0	51.0	48.5	0	+0.5
XXXII	4β-OH	OH	CH_3	51.0	64.0	51.0	64.0	0	0

*The "one" refers to a ketone functional group.

Table II summarizes the observed and calculated positions of the angular-methyl-group proton signals in the present study. It will be noted that the largest observed difference between the calculated and found values was ± 1.5 cps.

Zürcher has observed that many of the positions on the steroid nucleus are symmetrically disposed with respect to the protons on the methyl groups.[3] For example, the similarities of the relationships of the C-2 and C-4 substituents and the C-19 methyl can be seen from Fig. 1. Illustrations of these relationships can be seen by comparing the C-19 methyl proton signals in the 2- and 4-oxoderivatives (III and V; VII and IX; XI and XIII; XV and XVII), and the 2 β- and 4 β-axial hydroxyl derivatives (XIX and XXI; XXIII and XXV; XXX and XXXII).

In the hydroxyl derivatives the width of the signal due to the proton attached to the carbon atom which bears the hydroxyl group can also be used to define the spatial orientation of the hydroxyl function.[5] In the case of the axial hydroxyl compounds, such as the 2 β and 4 β derivatives, the proton attached to the carbon bearing the hydroxyl group is coupled only moderately (dihedral angles of approximately 60°) with the protons on the adjacent carbon atoms. On the other hand, in those analogs in which the hydroxyl group is equatorial, such as the 3β derivatives, the proton attached to the carbon atom bearing the hydroxyl group must be axial, and hence is coupled strongly with the axially-disposed protons on the adjacent carbon atoms (dihedral angle of approximately 180°) and coupled moderately with the neighboring equatorially-disposed protons. The net result is that the signal for the proton attached to the carbon atom bearing the hydroxyl group is very broad for the equatorial hydroxy analogs and reasonably narrow for the axial hydroxy derivatives. Table III summarizes the observed positions and widths of the signals due to the protons attached to the carbons bearing the hydroxyl groups in the various derivatives.

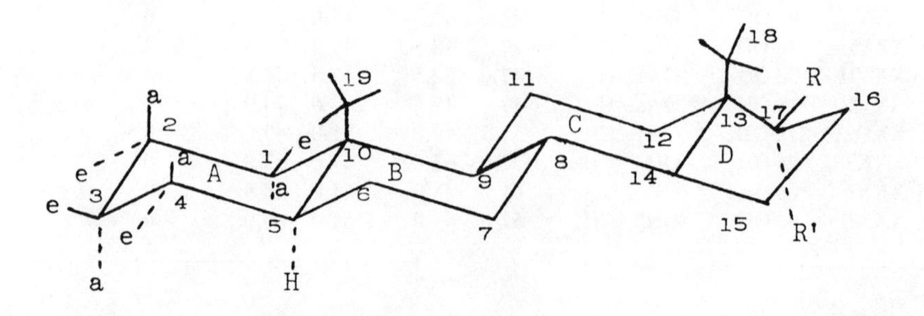

Fig. 1. Androstane molecule. Spatial conformation.

TABLE III
Hydroxyl Group Conformation Data*

A-ring	D-ring (C-17)			
(Hydroxyl position)†	*R*	*R′*	X[3] (cps)	Y[4] (cps)
C-1 (*a*)	"one"		220	5.5
C-2 (*a*)	"one"		248	7.0
C-3 (*e*)	"one"		213	16.0
C-4 (*a*)	"one"		231	6.0
C-1 (*a*)	OH	H	220	6.5
C-2 (*a*)	OH	H	249	7.5
C-3 (*e*)	OH	H	214	20.0
C-4 (*a*)	OH	H	229	7.0
C-1 (*a*)	OAc	H	221	6.0
C-2 (*a*)	OAc	H	247.5	8.0
C-3 (*e*)	OAc	H	213	17.0
C-1 (*a*)	OH	CH_3	222	7.0
C-2 (*a*)	OH	CH_3	248	8.0
C-3 (*e*)	OH	CH_3	215	17.5
C-4 (*a*)	OH	CH_3	228	7.0

*Refer to Fig. 1 for the basic skeletal structure of these substances.
†The (*a*) refers to axial conformation of the hydroxyl; the (*e*) refers to equatorial conformation of the hydroxyl.
‡Position of the proton on the carbon bearing the hydroxyl group in the *A*-ring.
§Results obtained after deuterium exchange to avoid broadening of the peak due to increased coupling with the hydroxyl proton.

CONCLUSION

The agreement between the calculated and found positions of the angular-methyl-proton signals in this series of steroids was excellent. The magnitude of the coupling between the proton bearing a hydroxyl group and the vicinal protons was also very useful in the assignment of configuration of the secondary alcohols.

ACKNOWLEDGMENT

The authors gratefully acknowledge the assistance of Mr. A. J. Damascus and Mrs. C. Dorn of the Analytical Division of G. D. Searle & Co.

REFERENCES

1. P. D. Klimstra, A. R. Zigman, R. E. Counsell, *J. Med. Chem.* **9**, 924 (1966).
2. R. F. Zürcher, *Helv. Chim. Acta* **46**, 2054 (1963).
3. R. F. Zürcher in *NMR in Chemistry*, B. Pesce (ed.), Academic Press, New York (1965) p. 45.
4. N. S. Bhacca and D. H. Williams, *Applications of NMR Spectroscopy in Organic Chemistry*, Holden-Day, San Francisco (1964), Chapter 2.
5. K. Tori and E. Kondo, *Tetrahedron Letters*, **1963** (10), 645.

An NMR Study of Acrylic Polymers

Daniel A. Netzel

Desoto, Inc.
DesPlaines, Illinois

Solvent chemical-shift and temperature effects on acrylic-copolymers were studied. The solvents used in this investigation were *o*-dichlorobenzene, bromoform, 1,1,2,2-tetrachloroethylene, and 1,1,2,2-tetrabromoethane. Bromoform was found to give the best resolved spectra at a temperature of 102°C. Analytical data obtained by integration of the NMR resonance bands for the arcylic copolymers were within a few percent of the initial monomer content.

INTRODUCTION

It is now widely recognized that nuclear magnetic resonance (NMR) spectrometry can be successfully applied to the characterization of polymeric systems.

Numerous NMR studies have been performed on acrylic polymers.[1-14] However, most of the studies involved characterizing the homopolymers of the acrylates. What has been reported are the effects of various polymerization conditions, such as catalyst, solvent, and temperature, on the resulting stereoregular polymer. Copolymers of the acrylates have been studied, but to a lesser degree.[15-18] Nothing has been reported on the effects of temperature and solvents on the NMR spectra of acrylic copolymers.

To obtain a reasonably good NMR spectra with high signal-to-noise ratio, one needs a fairly high concentration of a polymer in a suitable solvent. At room temperature high concentrations are limited by the high viscosity of the resulting polymer solution. High-viscosity solutions give NMR spectra in which the resonance lines are broadened. These broad lines are due to dipolar magnetic coupling between neighboring protons of adjacent molecules. There is now evidence that restricted segmental motion within the polymer may also cause an increase in resonance linewidth.[19] Segmental motion within a polymer molecule is dependent on the solvent.

Thus if a suitable solvent is chosen, i.e., one which is capable of breaking any intermolecular or intramolecular bonds which might restrict segmental motion, spectra may be obtained having narrow lines when the polymer solution is heated to a certain critical temperature. Above this temperature no apparent improvement of the spectrum will be obtained. This critical temperature will be different for different solvents for a given solute.

To determine the critical temperature and the best solvent for a given polymer, a trial-and-error process must be employed.

For qualitative identification of acrylic copolymers chemical-shift differences will be observed when comparing spectra of a given polymer in different solvents. These shifts can be confusing and can lead to erroneous conclusions regarding sample composition. However, solvent chemical-shift effects can be used advantageously in quantitative analysis of copolymers by separating overlapping resonance bands. The extent of the solvent shift is not necessarily the same for different protons within the same molecule for a given solvent.

The purpose of this paper is to provide qualitative information regarding the effects of solvent and temperature on acrylic copolymers and to utilize these effects to obtain quantitative data on the copolymer systems studied.

EXPERIMENTAL

Copolymer Preparation

The copolymers (70% methyl methacrylate and 30% ethyl methacrylate; and 70% methyl methacrylate and 30% ethyl acrylate) were prepared from their respective monomers on a weight basis using 2% benzoyl peroxide and 0.5% *t*-butyl perbenzoate as catalysts. Twenty per cent of the monomer charge containing the catalysts was added to 100% toluene and heated until reflux. The system was purged with nitrogen gas. The remaining 80% monomer charge was added over 1½ to 2 hr. The total system was refluxed for another 2 hr after final addition.

Solvents and Sample Preparation

The solvents chosen for this study were *o*-dichlorobenzene, 1,1,2,2-tetrachloroethylene, bromoform, and 1,1,2,2-tetrabromoethane. These solvents were chosen on the basis of their high boiling points, their ability to solubilize the copolymer, and the chemical shift of the solvent's hydrogen atoms. Hexamethyltrisiloxane (HMTS) was used as the reference material and set equal to 0 ppm. Chlorotriphenylmethane was used as a lock signal for 1,1,2,2-tetrachloroethylene. The samples were prepared by adding 1 ml of solvent to 300 mg of the copolymer. The resulting solution was filtered through a millipore 5 μ Teflon filter directly into a 5 mm thin-walled tube and sealed with a plastic pressure cap.

Instrument

The NMR spectra contained in this paper were recorded at 60 MHz with a JEOLCO C-60H NMR spectrometer. A variable-temperature probe was used

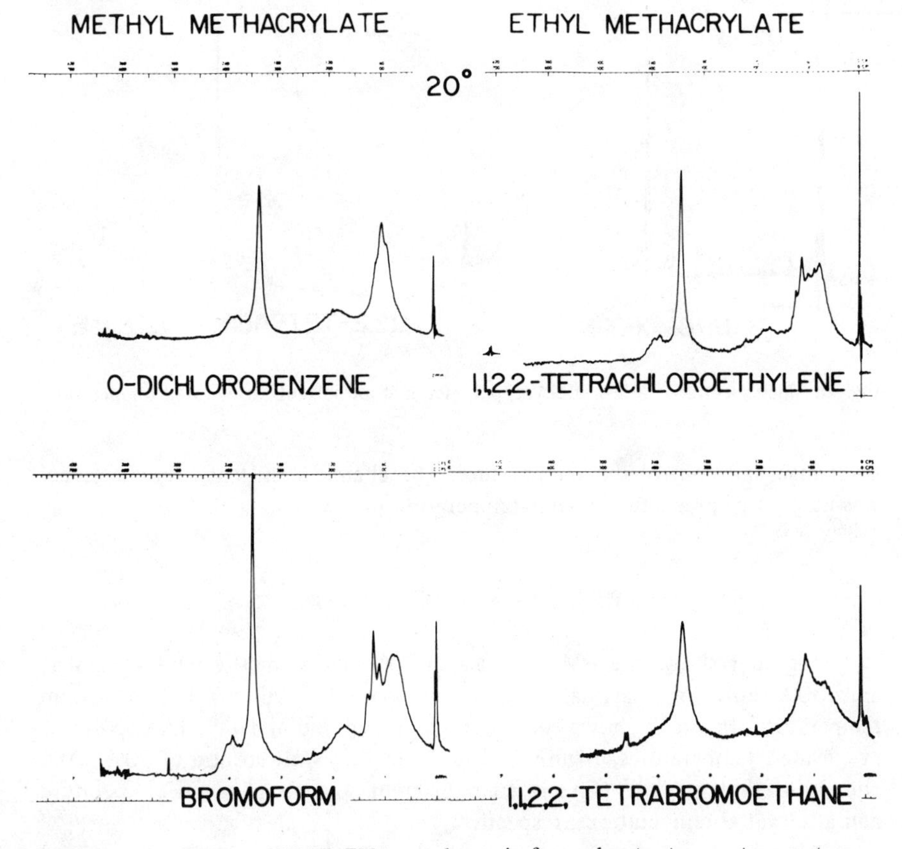

Fig. 1. NMR spectra of MMA–EMA copolymer in four solvents at room temperature.

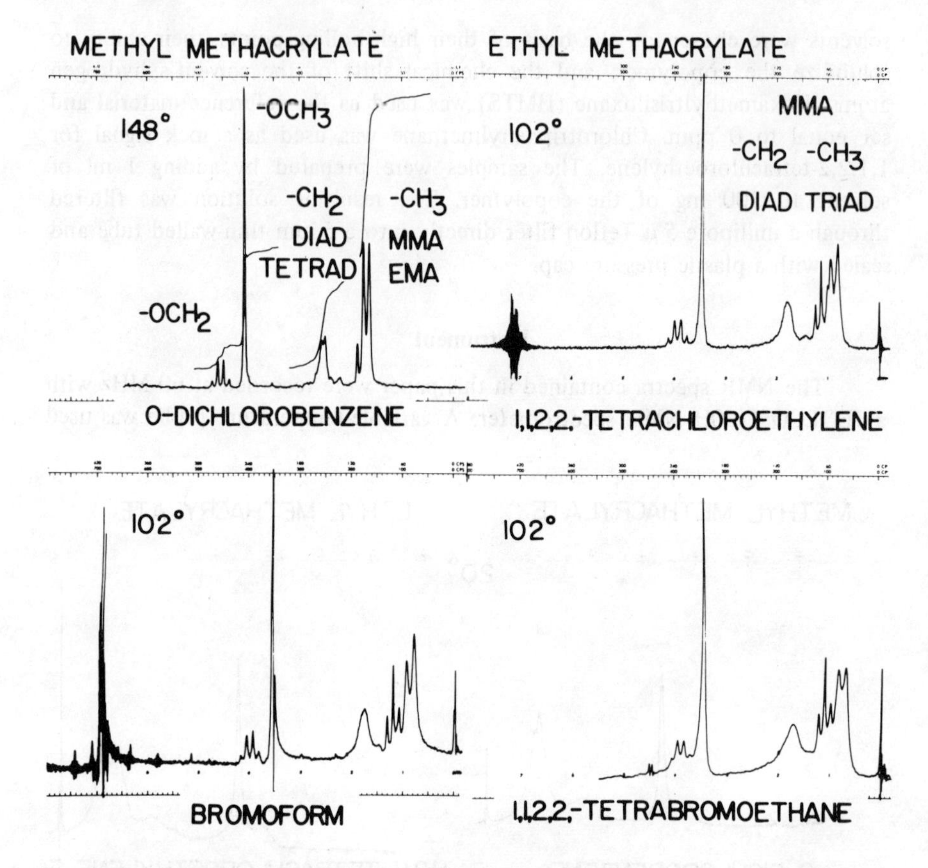

Fig. 2. NMR spectra of MMA–EMA copolymer in four solvents at elevated temperature.

to change and control the temperature. The chemical shift of ethylene glycol was used to calibrate the variable-temperature probe.

RESULTS AND DISCUSSION

Figure 1 shows the NMR spectra of 70% methyl methylacrylate (MMA) and 30% ethyl methacrylate (EMA) in four different solvents at room temperature. Figure 2 shows the NMR spectra of the MMA–EMA copolymer at elevated temperatures. Figures 3 and 4 are the NMR spectra of 70% MMA and 30% ethyl acrylate (EA) in four different solvents at room temperature and at elevated temperatures, respectively.

The room-temperature spectra of the copolymer systems have broad lines due to dipolar magnetic coupling of adjacent molecules. This effect is reduced considerably by obtaining spectra at elevated temperatures, as shown.

The bands near 4 ppm are due to the OCH_2 and OCH_3 proton resonance of the ester functionality of the copolymers, the OCH_2 resonance being downfield from the OCH_3 resonance. The resonance bands near 2 ppm are due to the *a*-H of EA and $-CH_2$ groups within the polymer backbone, while the bands in the region of 1 ppm are due to the *a*-CH_3 of MMA and the $-CH_3$ of the ethyl group. It is evident from these figures that the quality of the spectra obtained depends upon the solvent used and the temperature.

The spectra of the copolymers in *o*-dichlorobenzene at 148°C shows very clearly the quartet due to the coupling of the methylene hydrogens with the methyl hydrogens of the ethyl ester. Upfield from the OCH_2 and OCH_3

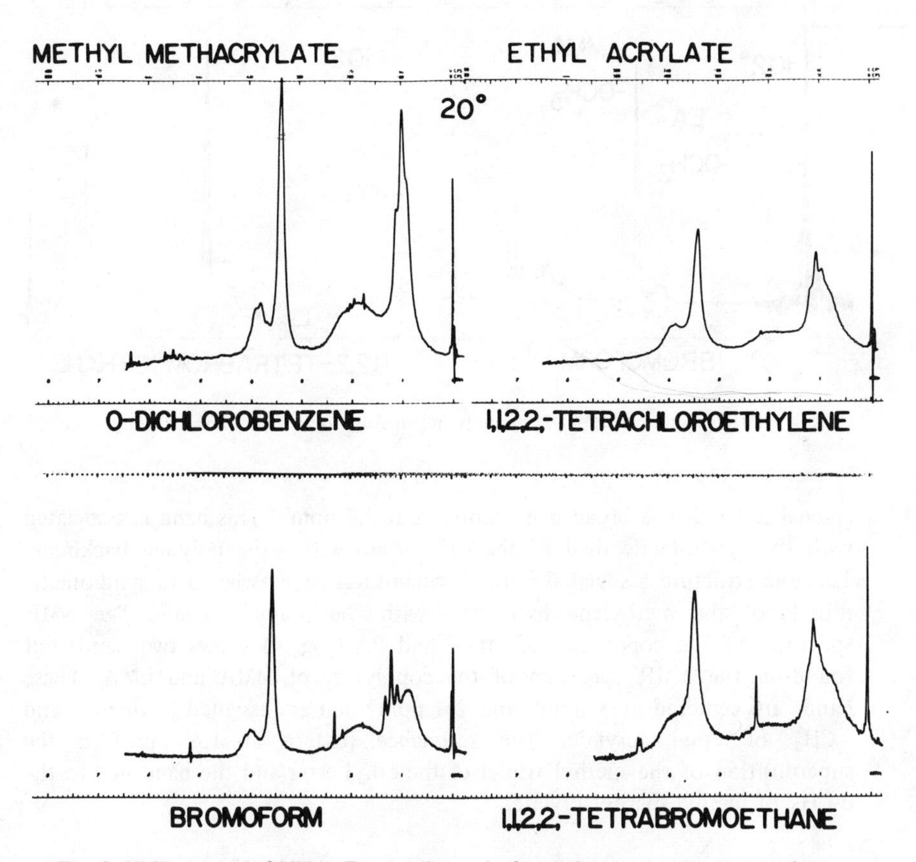

Fig. 3. NMR spectra of MMA–EA copolymer in four solvents at room temperature.

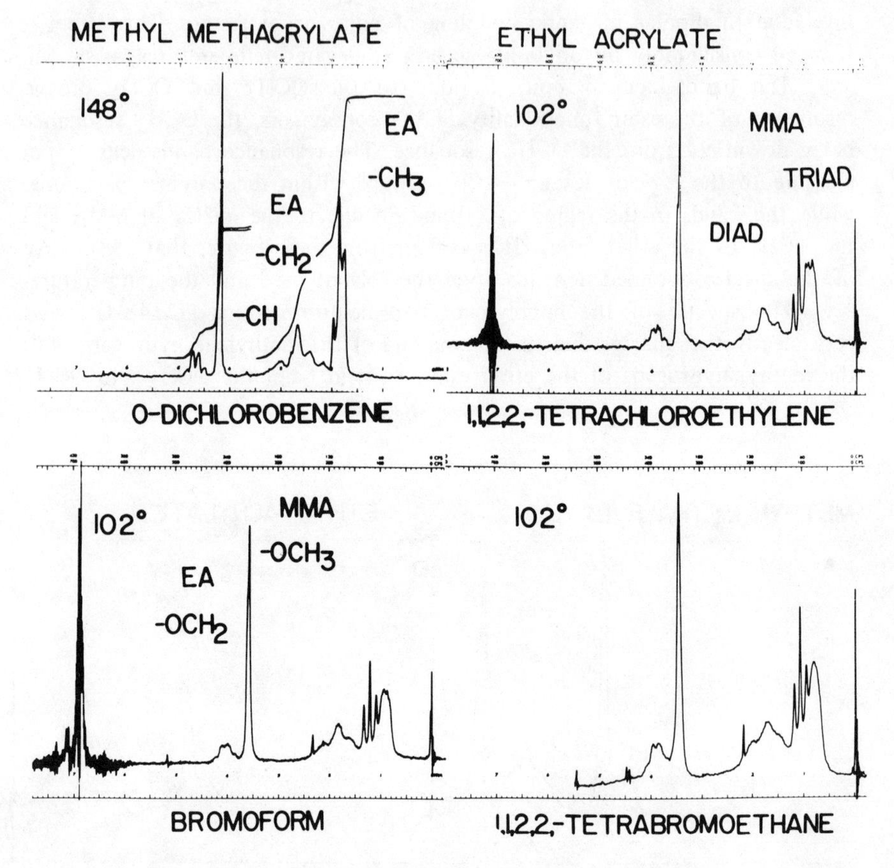

Fig. 4. NMR spectra of MMA–EA copolymer in four solvents at elevated temperature.

resonance bands is a broad band centered at 2.7 ppm.* This band is associated with the syndiotactic diad of the CH_2 group within the polymer backbone. The fine structure associated with this band has been assigned to syndiotactic tetrads of the methylene hydrogens with the polymer chain. The NMR spectrum of the copolymer of MMA and EA (Fig. 4) shows two bands not found in the NMR spectrum of the copolymer of MMA and EMA. These bands are centered at 3 ppm* and 2.4 ppm* and are assigned to the α-H and $-CH_2$ of ethyl acrylate. The resonance pattern at 1.8 ppm* is the superposition of the methyl triplet of the ethyl ester and the band due to the α-CH_3 of methyl methylacrylate.

* Spectra obtained in o-dichlorobenzene are not referenced to HMTS.

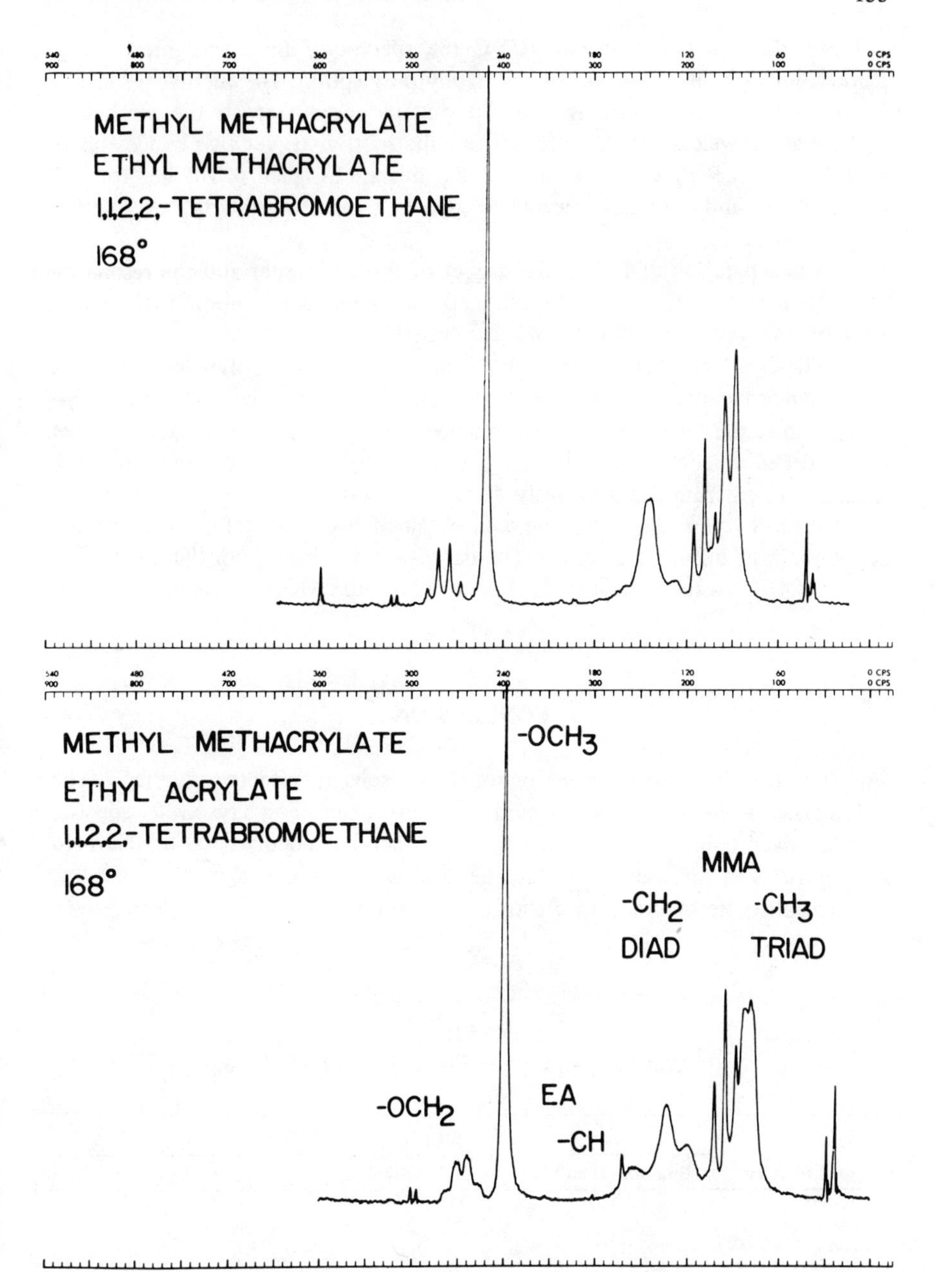

Fig. 5. Comparison of the NMR spectra of MMA–EMA and MMA–EA copolymers in 1,1,2,2-tetrabromoethane at 168°C.

In the other solvents at 102°C the spectra of the copolymers shows marked changes in the region of the methyl absorption. The methyl resonance of the ethyl ester, which is a triplet due to coupling with the methylene hydrogens, is well defined. Upfield from the methyl triplet two bands appear near 1.8 and 1.9 ppm. These resonance bands are assigned to the heterotactic (1.9 ppm) and syndiotactic (1.8 ppm) triads of α-CH_3 of methyl methacrylate.

The separation of the methyl triplet of the ethyl ester and the resonance bands due to the triads of methyl methacrylate depends upon the solvent. Bromoform appears to give the best resolution.

The NMR spectra of MMA–EMA and MMA–EA copolymers in 1,1,2,2-tetrabromoethane at 168°C are shown in Fig. 5. It is evident from these spectra that the small chemical differences of the copolymers can easily be distinguished qualitatively. The assignment of the resonance bands in these spectra follows from the previously discussed spectra.

Table I lists the quantitative data obtained from the NMR spectra of the copolymers in different solvents. The data were obtained from the integration of the OCH_2 resonance due to EA or EMA and OCH_3 resonance due to MMA.

CONCLUSION

Of the four solvents used to study solvent effects on the acrylic polymers, bromoform was found to give the best resolved spectra. Quantitative data was obtained on the acrylic copolymers at elevated temperatures in different solvents. The data was found to agree within a few per cent absolute of the known starting monomer content.

TABLE I
Acrylic Copolymer Composition Data

		Weight per cent by NMR in:		
Polymer system	Starting material (wt.%)	*o*-Dichlorobenzene	$Cl_2C{=}CCl_2$	$Br_2CHCHBr_2$
MMA	70	67	66	68
EMA	30	33	34	32
MMA	70	69	73	
EA	30	31	27	

ACKNOWLEDGMENTS

The author expresses his appreciation to C. M. Taubman for the preparation of the copolymer used in this study and to K. E. Isakson for obtaining some of the spectra.

REFERENCES

1. K. J. Liu, *J. Polymer Sci.* **Part A-2** 5, 1199 (1967).
2. K. J. Liu, J. S. Szutz, and R. Ullman, *Am. Chem. Soc., Div. Polymer Chem., Preprints* **5** (2), 761 (1964).
3. A. Nishioka, H. Watanabe, I. Yamagushi, and H. Shimizu, *J. Polymer Sci.* **45**, 323 (1960).
4. T. Yoshino, Y. Kikuchi, and J. Komiyama, *J. Phys. Chem.* **70**(4), 1059 (1966).
5. U. Johnsen, *Kolloid Z.* **178**, 161 (1961).
6. L. J. Merrill, J. A. Sauer, and A. E. Woodward, *J. Polymer Sci. Part A* **3**(12), 4243 (1965).
7. K. Matsuzaki, T. Uryn, A. Ishida, and T. Ohki, *J. Polymer Sci. Part B* **2**(12), 1139 (1964).
8. J. G. Powles, J. H. Strange, and D. J. H. Sandiford, *Polymer* **4**(3), 401 (1963).
9. N. M. Bazhenov, M. V. Vol'kenshtein, and A. S. Khachaturov, *Vysokomolekul. Soedin.* **5**(7), 1025 (1963).
10. F. A. Bovey, *J. Polymer Sci.* **46**(147), 59 (1960).
11. A. Nishioka, Y. Kato, T. Vetake, and H. Watanabe, *J. Polymer Sci.* **61**, S32–S33 (1962).
12. H. Watanabe, Y. Kato, and A. Nishioka, *Kogyo Kagaku Zasshi* **65**, 270 (1962).
13. M. L. Miller and C. E. Rauhut, *J. Polymer Sci.* **38**(133), 63 (1959).
14. K. M. Sinnott, *J. Polymer Sci.* **42**, 3 (1960).
15. N. Grassie, B. J. D. Torrance, J. D. Fortune, and J. D. Gemmell, *Polymer* **6**, 653 (1965).
16. F. A. Bovey, *J. Polymer Sci.* **62**, 197 (1962).
17. K. C. Ramey and J. Messiek, *J. Polymer Sci. Part A-2* **4**, 155 (1961).
18. C. Haney, F. A. Johnson, and M. G. Baldwin, *J. Polymer Sci. Part A* **4**(7), 1791 (1966).
19. H. A. Willis and M. E. A. Cudby, *Appl. Spectry. Rev.* **1** (2), 237 (1968).

The Use of Digital Computers for NMR Calculations

Milton I. Levenberg

Advanced Technology Department
Abbott Laboratories
North Chicago, Illinois

This paper is directed to the person unfamiliar with the details involved in using a digital computer to calculate NMR spectra. Three related topics are discussed: (1) The calculation of NMR line positions and intensities from chemical-shift and coupling-constant data, and the inverse problem of obtaining NMR parameters from line positions; (2) the calculation of peak profiles and exchange rates for nuclei exchanging between two magnetically-different sites; and (3) the adaptation of computer programs to computers other than those for which they were originally written. The manipulation of data required to use the computer programs is considered, rather than the theoretical and mathematical spects of the problem.

Spectral Analysis

An organic chemist using NMR as a tool for structure determination is often called upon to perform calculations on the NMR spectrum, either to confirm the validity of his proton assignments or to extract from the spectrum parameters of use in assigning stereochemistry or other physical-chemical properties of interest. In the simplest systems these calculations can be performed by hand, using well-documented formulas and methods (Pople *et al.*[1], Chapter 6; Wiberg and Nist[2]; Roberts[3]). As the system grows in complexity it becomes necessary to resort to more sophisticated techniques to extract the desired information, and it soon becomes apparent that the speed and precision of a modern computer are essential for the completion of an exact calculation for even a three-spin problem.

TABLE I

Some NMR Computer Programs Available

Program	Reference
FREQINT IV	(5)
LAOCOON	(5)
NMRIT/NMREN	(6)

The problem of obtaining NMR line positions and intensities from chemical-shift and coupling-constant data may be easily formulated in quantum-mechanical terms[1]; however, the mathematics involved are beyond the scope of this paper. We will instead consider the various computer programs available to perform these calculations, and how an individual naive in the ways of quantum mechanics may nevertheless use these programs in his own laboratory.

Some of the more well-known and easily-available programs to calculate NMR spectra are listed in Table I. LAOCOON is an advanced version of FREQINT IV which, after calculating line positions, will compare them with experimentally-measured line positions, and through an iterative procedure change the input parameters (chemical shifts and coupling constants) until the best possible fit of output data (line positions) is obtained. The NMRIT and NMREN programs accomplish the same iteration to a best set of input parameters using, however, different criteria to determine the next set of trial parameters.

Through a chance circumstance in availability of programs and documentation in these laboratories we are running a one-cycle version of LAOCOON in-house on an IBM 1800 with 32K of core, but when we need an iterated refined calculation the problem is sent out to an IBM 7094 computer and uses the NMRIT/NMREN programs for the calculations.

Our version of LAOCOON contains a routine which generates Gaussian curves for the output lines and plots them as a simulated spectrum on a CALCOMP plotter on line with the computer.

Typical input data for our version of LAOCOON are shown in Fig. 1. Card one contains a problem number, number of spins, and problem name. Card two contains the NMR sweep offset, sweep width, size of minimum transition to be considered, height of tallest peak on experimental spectrum, average peak width in hertz, and a parameter to indicate a plotting mode. Card three contains a set of integers which indicate when some nuclei are to have only first order (i.e., A–X type) interactions with each other. Card four contains the NMR chemical shifts; and the remaining cards contain the NMR coupling constants.

Figures 2 and 3 show the computer output for the data illustrated in Fig. 1, and Fig. 4 demonstrates the plotted output. Figure 4 also shows, for comparison, a plot of an ABX approximation of the same data. This is the result one would obtain by performing the calculation by hand and accepting the inherent errors in the long-hand approximate solution. The actual line positions are slightly different in the two spectra, but the most striking difference is in the relative intensities of peaks in the multiplets. Thus, even in systems which can be calculated by hand, if more than two nuclei are present, the equations used in the hand calculation are only approximate, and one

```
CARD 1:      1   3    NMR  3522 4-24-68

CARD 2:    250.0   250.0   0.1   150.0   0.3   -1

CARD 3:    111

CARD 4:    407.0   366.9   344.0

CARD 5:    16.75   10.25

CARD 6:    2.60
```

Fig. 1. Typical input data for computer program LAOCOON

```
LAOCOON II
_______________________________________________________________
PROBLEM NUMBER     1,    NMR 3522 4-24-68
_______________________________________________________________
FREQUENCY RANGE     250.000    500.000              3 SPINS

MINIMUM INTENSITY     0.10000
_______________________________________________________________

_______________________________________________________________
8              W(1)=   407.000
8              W(2)=   366.900
8              W(3)=   344.000
_______________________________________________________________
               A(1,2)=  16.750
               A(1,3)=  10.250
               A(2,3)=   2.600
_______________________________________________________________
```

Fig. 2. Page one of typical output from computer program LAOCOON

TABLE OF ORDERED LINES

LINE	FREQ	INT
13	337.093	0.776
5	340.149	0.909
9	346.855	0.984
3	349.911	1.319
14	355.386	0.592
4	358.442	0.633
12	372.169	1.645
2	375.224	1.136
15	395.819	1.630
7	405.581	1.142
11	412.602	0.673
1	422.364	0.543

Fig. 3. Page two of typical output from computer program LAOCOON

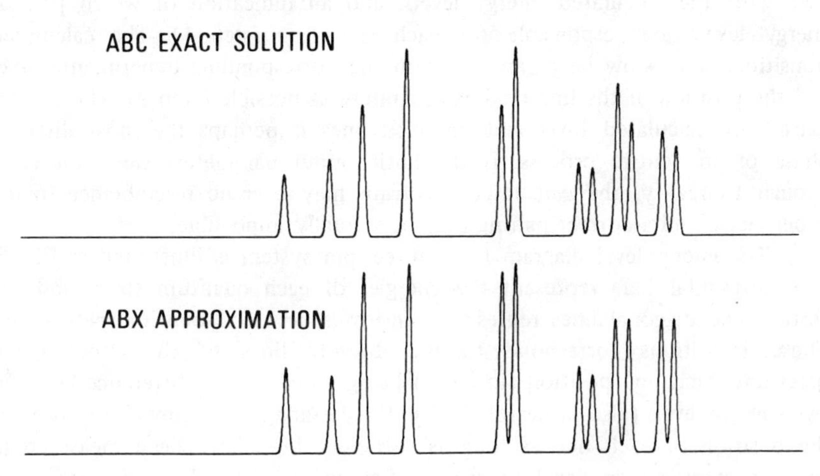

Fig. 4. Computer-generated plots of the same data treated as ABC and ABX coupled systems.

must turn to the computer for accurate calculations of line positions and intensities.

The initial input parameters may be obtained from knowledge of the structure, model compounds, measurements on the experimental spectrum, spin tickling or decoupling experiments, guesses, or ancient Egyptian hieroglyphics, whatever one's skills and inclination. If the choice of parameters is good, the computer-generated spectrum will be similar to the experimental spectrum, though probably not identical. By a comparison of the two spectra an experienced observer can often predict some of the changes necessary in the input parameters to bring the two spectra into closer agreement; however, a point is reached in complex systems where no amount of second-guessing will result in further improvements in the calculated spectrum. If a further refinement of the input data is still required, the full power of the iterative version of one of the above computer programs may be brought to bear.

We had available to us a complete punched program deck for the NMRIT/NMREN programs in a form ready to run on an IBM 7094, and we also had access to a 7094 on a cost-per-time-used basis. Rather than punch up the entire iterative LAOCOON deck or try to fit either program into the somewhat limited core size of our IBM 1800, we chose the path of least resistance – to send the available program out when we needed a full calculation. We are currently trying to fit a version of LAOCOON on our computer and will report on this at a later date.

Figure 5 shows the flow of data in the Reilly–Swalen programs. The first pass (noniterative pass) of either program will, as an option, provide the user with the calculated energy levels, and an indication of which pair of energy levels are responsible for each transition (step 1). The calculated transitions must now be paired off with the corresponding experimental line, and the position of the line read as accurately as possible (step 2). The correct pairing of calculated lines and observed lines is perhaps the most difficult phase of the entire process. If the initial input parameters were not close enough to reality, the calculated spectrum may bear no resemblance to the experiment, and a correct pairing may be virtually impossible.

The energy-level diagram for a three-spin system is illustrated in Fig. 6. The horizontal bars represent the energies of each quantum state, and the vertical and diagonal lines represent transitions between some of these states. These transitions correspond to the discrete lines of the experimental spectrum, each line position (in hertz) being equal to the difference between two energy levels (also in hertz). It is not necessary to experimentally measure the position of each line as long as sufficient lines have been measured to connect every energy level to every other energy level through some path, direct or otherwise. The one exception to this rule is that in symmetrical systems the energy levels may factor into a symmetrical and an antisymmetrical set with no transitions large enough to observe between any energy level of one set and a level of the other set.

In Fig. 6 energy levels 1 and 4 are not connected to the rest of the set, so a complete solution to the problem would not be possible. One would like to provide as many line positions to the computer program as possible to allow the program to average errors of individual measurements and provide a more accurate final solution. If lines are assigned to incorrect transitions, the errors in the final solution will be larger, and the program may even converge to an entirely false solution.

In the Reilly–Swalen programs the measured experimental line position of each assigned line is punched on a computer card along with the numbers of the two energy levels responsible for the transition (Fig. 5, step 3). These data are used by computer program NMREN (step 4) to find the set of energy levels which are the best fit for the experimental problem (step 5). These will

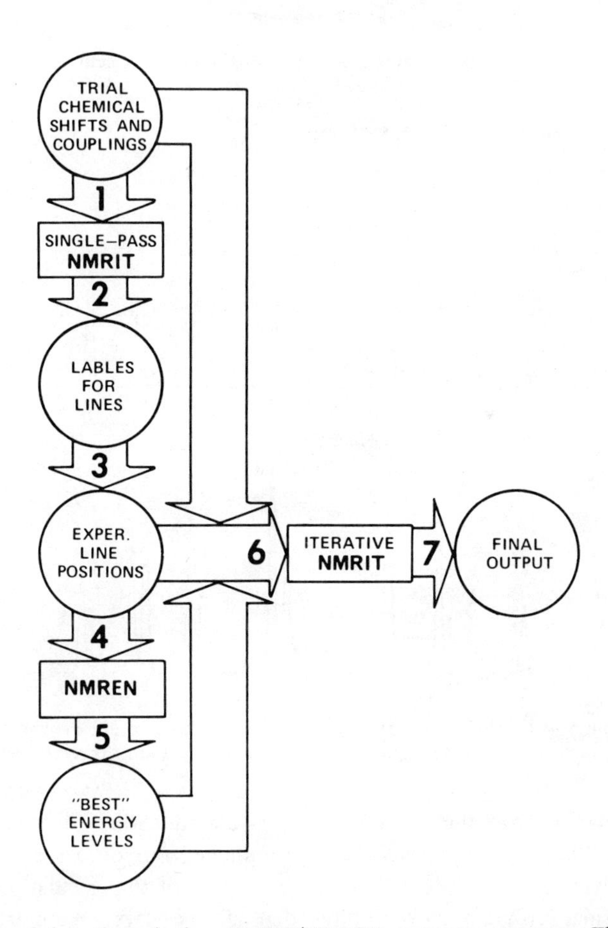

Fig. 5. Flow of data through the NMRIT/NMREN computer programs. The rectangles represent computer programs and the circles represent input and/or output data. The final output data includes adjusted chemical shifts and coupling constants, line positions and intensities, and errors between experimental and calculated line positions.

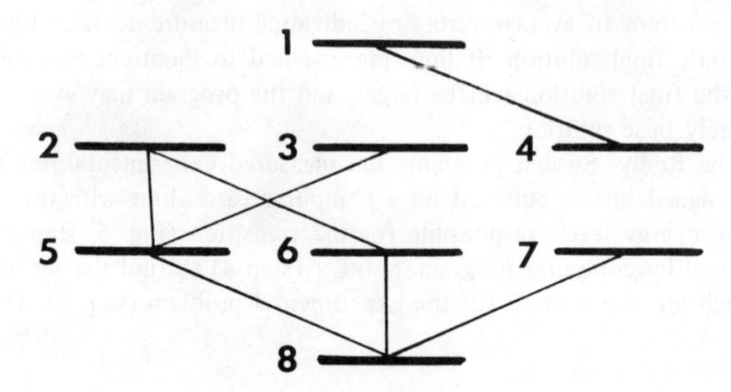

Fig. 6. Energy-level diagram for a three-spin system.

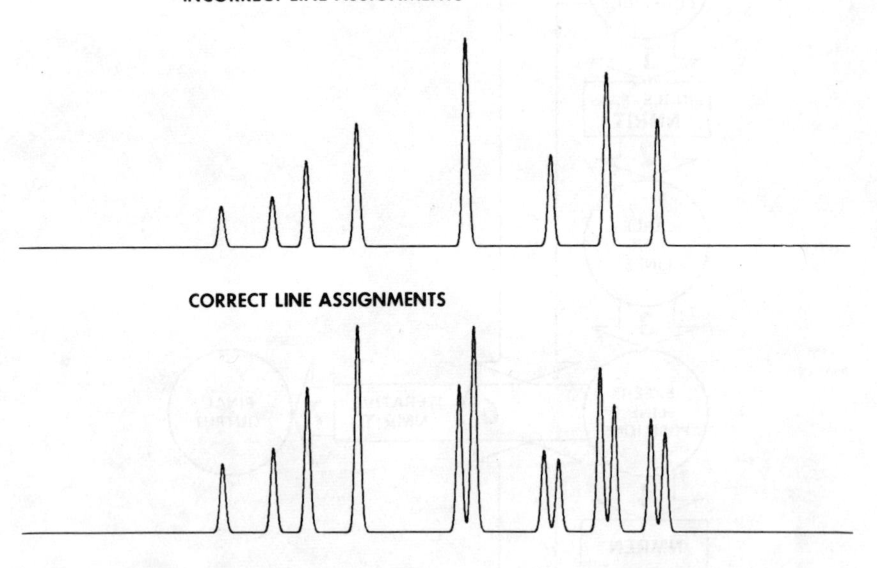

Fig. 7. Computer-generated plots of iterated solutions to a problem where several line assignments had initially been made correctly and incorrectly.

differ somewhat from the energy levels calculated earlier for the trial input data. These "best" energy levels will be punched out on a set of computer cards so they can be used directly in the next step of the calculation.

The final program pass requires that all the data accumulated in the preceding calculations be fed back into the NMRIT program (step 6). The sets of data cards – chemical shifts, coupling constants, energy levels calculated in

NMREN, and measured line positions – are stacked in the order listed, along with a few control cards to provide the program with a problem name and some other parameters it requires. The program then uses the difference of the input and newly-calculated energy levels to calculate a new set of input parameters. The new input parameters are used to calculate a new set of energy levels, and the process is repeated. Ten to twenty iterations through this cycle are usually sufficient to allow the NMR parameters to converge to a constant set (step 7).

Figure 7 shows a comparison with the results when several lines were deliberately assigned to the wrong transitions before iterating the data. Figure 8 shows a comparison of the final iterated data with the original experimental spectrum.

In doing any sort of NMR calculation, and particularly a computer fit of data, one must take particular care to avoid touting a wrong solution. A particular NMR spectrum may not have a unique solution; i.e., very different sets of input parameters may result in almost (or even exactly) identical calculated NMR spectra. Only one of the sets of input parameters can, in fact, be chemically correct, but there may be no way of distinguishing this set from the others without additional data from a spin tickling experiment, change in solvent, etc.

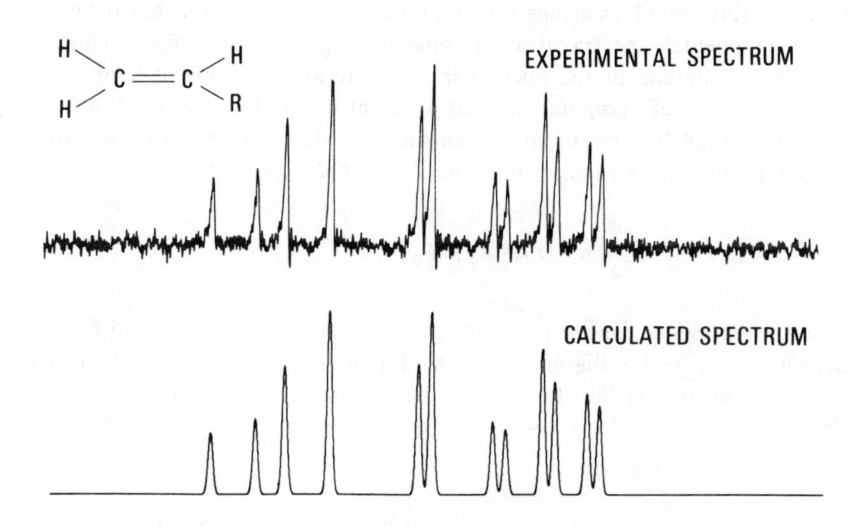

Fig. 8. Comparison of the plot of the computer solution of a problem with the original experimental spectrum.

EXCHANGING NUCLEI

Another area in which the computer can be helpful to an NMR spectroscopist is in the determination of the rate of exchange of a nucleus between two or more magnetically-dissimilar environments. The equations are given in Pople *et al.*[1] p. 222 for the intensity of radio-frequency absorption of a nucleus exchanging at an intermediate rate between two dissimilar sites. Though the equations may look complex at first glance, a computer may be readily programmed to step through the spectral range of interest and at each step use these equations to calculate an absorption intensity. The profile of this absorption as a function of frequency may then be plotted, also under computer control. In this laboratory we have developed a computer program, XCH1, to perform this calculation and plot the results, patterned after a similar program written by Lambert.[4]

To use this technique effectively, one should ideally have an NMR spectrum of the system under conditions of temperature and pH where the rate of exchange is slow on an NMR time scale. The basic data obtained from this spectrum and required in the calculation are the chemical-shift difference between the two peaks, the relative peak intensities, and the peak widths at half-height. NMR spectra are run at various temperatures* to provide spectra over a range of intermediate exchange rates. A series of peak profiles are calculated and plotted with the computer program using arbitrarily chosen lifetimes (inverse of exchange rates), and these profiles are matched up with the experimental spectra at each temperature. This matching should now provide the lifetime of the nucleus in each site as a function of temperature. A typical series of computer plots at different lifetimes is presented in Fig. 9.

The mean lifetime of the nucleus in each state is related to the absolute temperature by the Arrhenius equation (Pople *et al.*,[1] p. 367).

$$1/\tau_a = k_a = k_0 \exp(-E_a/RT)$$

where τ represents the mean lifetime, k is the rate of exchange, and E_a is the activation energy for the exchange. A plot of $\ln(1/\tau_a)$ vs $1/RT$ will yield the activation energy as the slope of the best-fit straight line drawn through the data.

* Care must be taken that the temperature of the sample is accurately known. The dial calibration on a temperature controller often drifts too much to be a reliable index of the sample temperature.

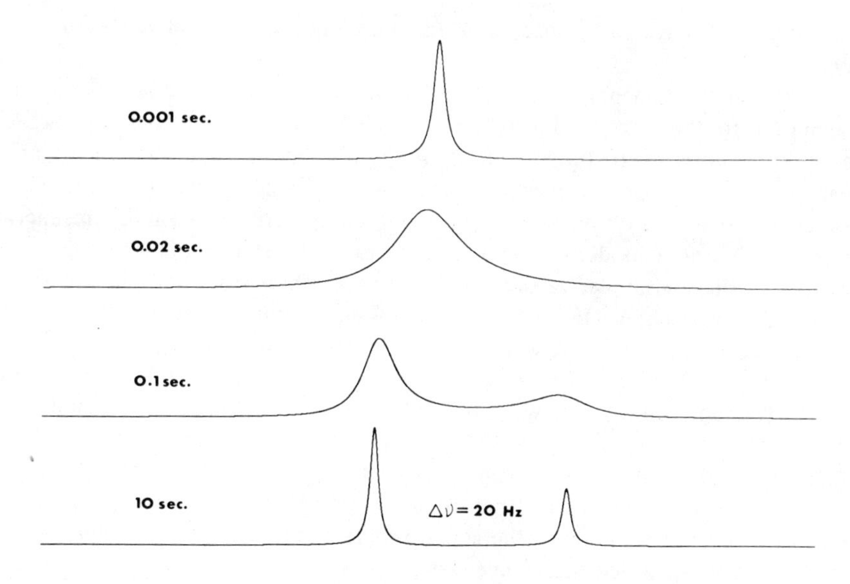

Fig. 9. Computer-generated plots of two nuclei exchanging at various rates between two magnetically different sites.

ADAPTATION OF PROGRAMS TO COMPUTERS

A problem frequently encountered by the investigator is to adapt an existing computer program so that it will run on an available computer. The most important point to stress is that it is not necessary to understand how a particular program works or how the calculation is performed to modify the program for one's needs.

By far the most popular scientific programming language in the United States today is FORTRAN. If the program has been written in FORTRAN and one's computer has a FORTRAN compiler, the modifications to transfer the program to one's own computer may be minor. One must, however, be fairly adept at FORTRAN programming.

The first step is, of course, to obtain a copy of the program of interest. A popular program may already have been fit to several different models of computers, and the correct choice of starting version may minimize the additional changes yet to be made. A second important consideration, however, is that the best write-up and discussion of the program was probably provided by the original author, although his version of the program may be for a computer substantially different than the one currently available. Each individual will have to balance these factors to decide the best route for him personally.

Certain statements in FORTRAN have evolved from one generation of computer to the next, and may have to be altered in form, though not in content or purpose, to run in a new computer. Input–output statements are perhaps the most serious offenders, and Table II illustrates the form one such statement might take on several different computers. In a similar manner, some of the more modern computers use "logic IF" statements which do not exist in earlier or simpler versions of FORTRAN These may easily be replaced with normal algebraic IF statements with equivalent arguments. The third major area of difference between various computers is in the types and forms available for plotting subroutines (assuming a plotter exists on the installation). The only convenient way to handle this incompatibility is to find the data array to be plotted in the program and then replace the existing subroutine calls with the equivalent ones on the system available.

The reader will notice that none of the above modifications require much of an understanding of the actual program, and a computer programmer is often better qualified than a chemist to perform these manipulations.

One final point of potential difficulty will occur if one's computer does not have as much memory storage available as requested by the program. There are several different solutions to this dilemma, with their respective advantages and disadvantages. A program such as the first part of LAOCOON requires approximately 8000 words of array storage to handle a seven-spin system. If the user is content to consider only problems of five spins and less, the same arrays need take only 1100 words of storage. The easiest approach, then, is often to limit the size of the problem one will handle. A second approach – which often requires the skill of an experienced programmer – is to break the program into sections, and have only one section in the computer core memory at a time. The large data arrays often put a greater strain on available core storage than actual program commands, so care must be taken that the data arrays are available when needed by the program.

With the aid of this introduction to the use of computers in NMR, the neophyte should not be reluctant to try some of these techniques for himself. Once one takes the initial step, the enthusiasm becomes contagious.

TABLE II

Variations in FORTRAN Output Statements

1620 FORTRAN	PUNCH 100, LIST
7090 FORTRAN II	WRITE OUTPUT TAPE 6, 100, LIST
7090 FORTRAN IV	WRITE (6, 100) LIST
1130/1800 FORTRAN IV	WRITE (3, 100) LIST

ACKNOWLEDGMENTS

The author wishes to thank Professor Peter Beak, Dr. Richard W. Mattoon, and Mr. Richard S. Egan for generous assistance in the preparation of this manuscript.

REFERENCES

1. J. A. Pople, W. G. Schneider, and H. J. Bernstein, *High-Resolution Nuclear Magnetic Resonance,* McGraw-Hill Book Co., New York (1959).
2. K. B. Wiberg and B. J. Nist, *The Interpretation of NMR Spectra*, W. A. Benjamin, New York (1962).
3. J. D. Roberts, *An Introduction to the Analysis of Spin–Spin Splitting in High-Resolution Nuclear Magnetic Resonance Spectra*, W. A. Benjamin, New York (1961).
4. Joseph B. Lambert, private communication, Northwestern University, Evanston, Illinois.
5. S. Castellano and A. A. Bothner-By, *J. Chem. Phys.* **41**, 3863 (1964).
6. J. D. Swalen and C. A. Reilly, *J. Chem. Phys.* **37**, 21 (1962).

The Analysis of ABX Spectra

George Slomp

Physical and Analytical Chemistry
The Upjohn Company
Kalamazoo, Michigan

Possible errors that may arise in interpreting ABX spectra are reviewed, and methods for detecting and avoiding these errors are discussed.

INTRODUCTION

ABX spectra[1] are frequently encountered in NMR studies on organic compounds. Under certain conditions these spectra may be difficult to recognize and difficult to interpret. Often several solutions are possible and the problem is to pick the correct one. The analysis of ABX spectra has already been discussed by several authors.[2-4] * Theoretical ABX spectra can be calculated by hand from the six magnetic parameters using the approximate equations of Pople *et al.*,[2] or the spectra can be exactly calculated and plotted by use of a computer.[5] The speed of the latter method makes it easy to calculate and plot large numbers of spectra in order to see how they change as the individual parameters are varied.

The objective of this investigation was to examine the errors that could occur in analyzing ABX spectra and to determine how these errors could be detected and avoided.

Many laboratories now use an iterating computer program to fit a computed spectrum to the experimental spectrum of interest. Where several solutions are possible the computer gives only the one that was specified. Thus the computed solution may be an exact frequency fit, but the param-

*Figure 6-7 of Pople *et al.*[2] and Figs. 8 and 18 of Emsley *et al.*[3] are confusing. The X part of the spectrum does not match the accompanying AB part.

eters may be incorrect because the wrong solution was requested of the computer. The investigation also showed that ABX spectra are easily misinterpreted by treating them as first order, by ignoring subtle differences attributed to the signs of the weak couplings, by assuming that a given solution is unique, or by failing to recognize the consequences of deceptive situations.

In this investigation spectra were calculated and iterated using LAOCN 3, written by S. Castellano and A. Bothner-By.[6] EXAN II written by Castellano and Waugh,[7] was occasionally used to be sure all the solutions were obtained. Spectra were plotted on an EAI plotter by a program written in these laboratories.[8] Figure 1 shows a standard ABX spectrum with chemical shifts of 50, 60, and 175 Hz for nuclei A, B, and X and coupling constants of 4.5, 3.5, and 1.5 Hz for J_{AB}, J_{AX}, and J_{BX}. An approximate spectrum is included for comparison. The spectrum is typical of ABX spectra and has twelve resolved lines. The X part is a quartet, the two satellite lines do not show in the spectrum because of the small difference in magnitude of J_{AX} and J_{BX}, and the AB part contains two quartets which are interlaced.*

FIRST-ORDER FACTORING

First-order factoring would take as J_{AX} the spacing between the lines (numbered from right to left in Fig. 1) 1 and 2, 3 and 4, 9 and 11, or 10 and 12, and would take as J_{BX} the spacing between lines 5 and 6, 7 and 8, 9 and 10, or 11 and 12. These separations are not J_{AX} or J_{BX}, although they may be related to the magnitude of J_{AX} and J_{BX}. The J_{AX} and J_{BX} do not occur in the spectrum. Only their algebraic sum appears in the X part as the separation of the outer lines or the inner lines of the quartet. Since J_{AX} and J_{BX} do not occur in the spectrum, first-order analysis is only approximate and an accurate method must be used to find the values of these parameters.

PICKING THE QUARTETS

If the X part contains six lines, the intense pair will be the sum-lines whose spacing is $J_{AX} + J_{BX}$. If only four lines are seen, the sum-lines will be the outside pair if the weak couplings have like signs and the inside pair if they have unlike signs. The analysis of the X part is difficult when only four lines are observed because they have approximately equal intensity, and without independent knowledge of the signs of the weak couplings there is no way of knowing which pair are the sum-lines.

*The term "quartet" here denotes four symmetrically-spaced lines and does not indicate they are equally spaced or have binomial intensities.

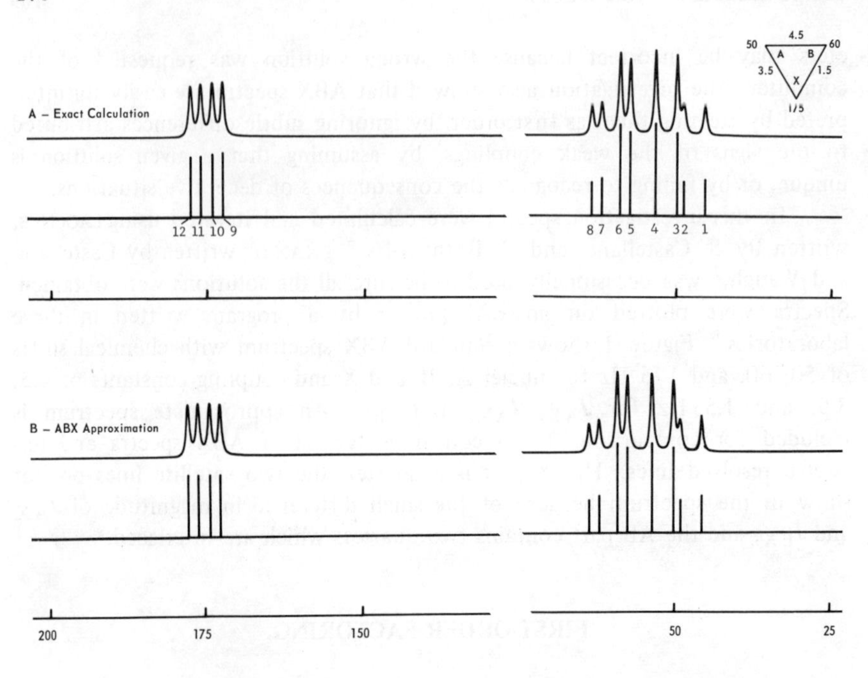

Fig. 1. Calculated spectra for very-loosely-coupled like-sign ABX system.

Therefore the first task in an accurate analysis of the ABX spectrum is to pick the two quartets in the AB part. The lines in the AB subspectrum can occur in such a way that the two *ab* quartets can be either noneclipsed, eclipsed, partly- eclipsed, or degenerate. The four situations are diagrammed in Figs. 2A and 2B.* In the diagrams the AB part is separated into two quartets, the one shown with heavy lines, designated $\delta ab+$, originates from the interaction of an A nucleus with a B nucleus on all the molecules that happen to have the X nucleus in the up orientation, and the one with the narrow lines, designated $\delta ab-$, originates from the interaction of A and B nuclei on all the molecules with the X nucleus in the downward, or lower-energy, orientation. The problem is that the noneclipsed (I, Fig. 2A), eclipsed (II, Fig. 2A), and partly-eclipsed (III, Fig. 2B) spectra all have the same line frequencies, with differences occuring only in line intensities. Thus for any experimental spectrum there may be as many as three different ways to pick the two quartets, depending on line frequencies only, all yielding different results. Crossover of the central lines of a quartet is forbidden, and this often eliminates some arrangements.

*This analysis was proposed by P. Diehl and R. J. Chuck, private communication.

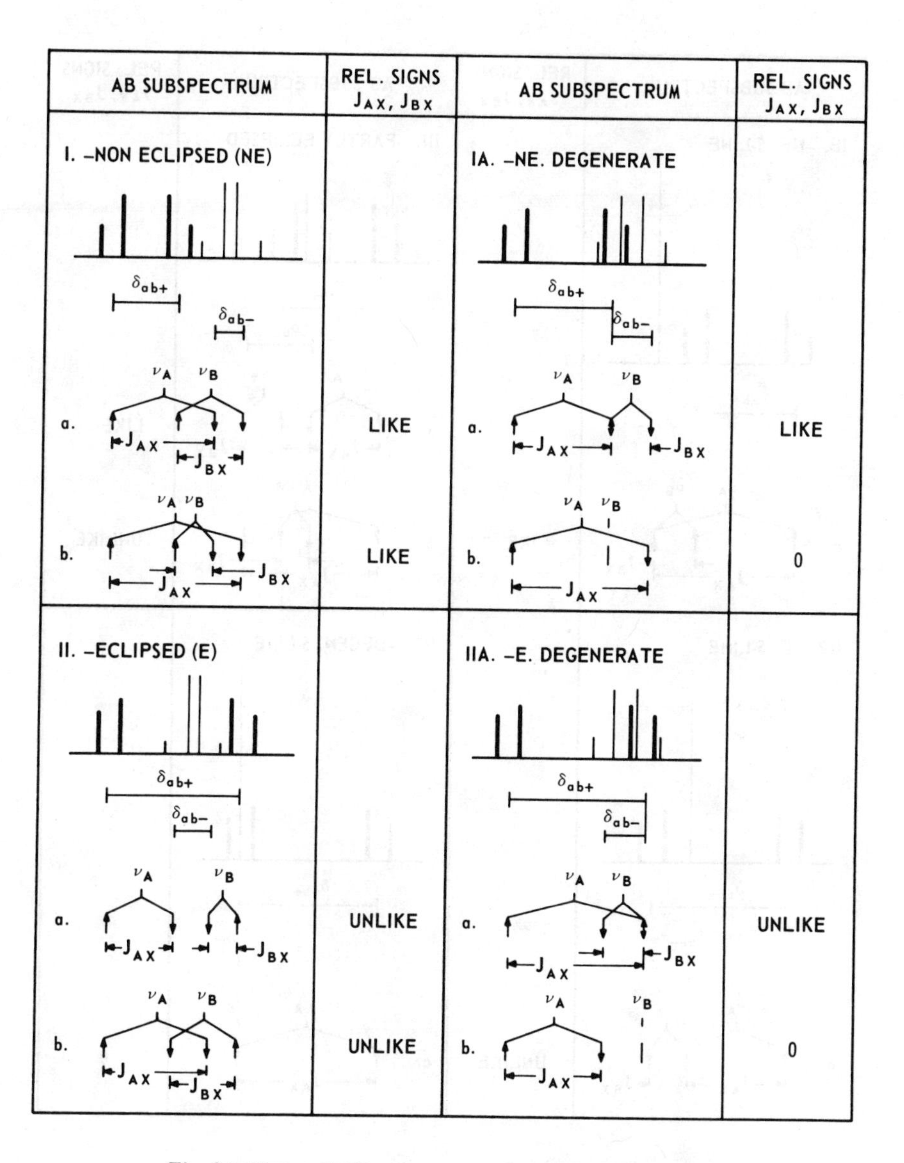

Fig. 2A. Types of AB subspectra observed in ABX systems.

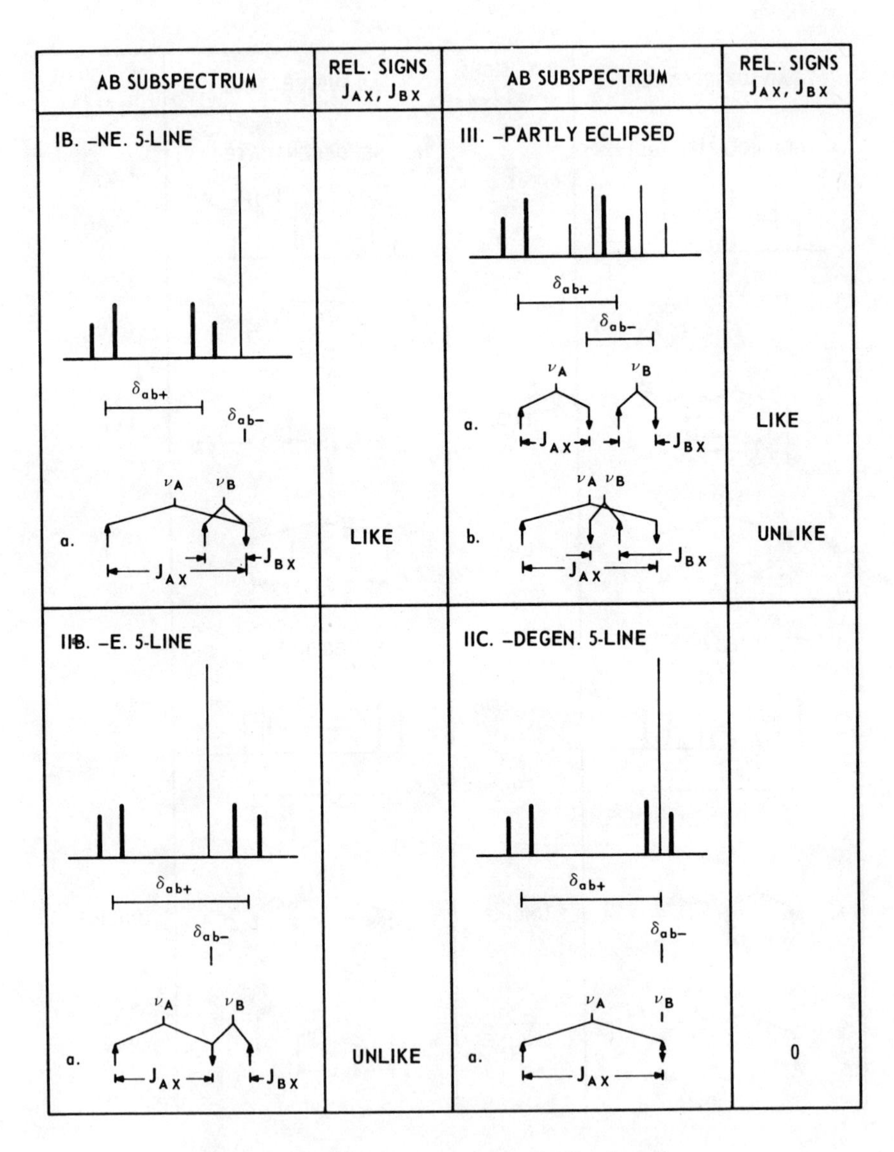

Fig. 2B. Additional types of AB subspectra.

The two quartets are then two AB subspectra and they can each be factored by arithmetic or construction methods.

PAIRING THE DOUBLETS

To find the parameters from the two factored AB subspectra, one line of the resulting *ab*+ doublet is connected with one line of the *ab*- doublet and the remaining two lines are connected accordingly. There may be two ways to do this, as shown by the diagrams labeled *a* and *b* in Figs. 2A and 2B. In the noneclipsed case (I, Fig. 2A), e.g., the first line of the *ab*+ is connected with the nearest line of the *ab*- (diagram *a*). The chemical shift ν_A is taken as the midpoint of these two lines, since the chemical shift of X is so different in magnitude, and the coupling constant J_{AX} is taken as the separation of these two lines. Similarly, the chemical shift of B is taken as the midpoint of the remaining two lines, and the coupling constant J_{BX} is taken as the separation of these two lines. In the alternate method for connecting the doublets, as diagrammed in *b* the first line of the *ab*+ is connected with the second line of the *ab*-. Since in this arrangement the first line of the *ab*- is passed over, this arrangement will be referred to as wide coupling of the doublets. Now the chemical shift of A is the midpoint of the first and fourth lines, and the coupling constant J_{AX} is the separation of the first and fourth lines. The chemical shift of B is the midpoint of the second and third lines, and the coupling constant J_{BX} is the separation of the second and third lines.

As shown in the diagrams of Figs. 2A and 2B, the two methods for connecting the doublets may give quite different parameters. In the partly-eclipsed situation (III, Fig. 2B), the narrow coupling of the doublets (diagram *a*) corresponds to J_{AX} and J_{BX} having like signs. Here nucleus A and nucleus B both have a low-field transition with X in the up orientation. Thus the signs of J_{AX} and J_{BX} are both positive. In Diagram III*b* of Fig. 2B, illustrating the wide coupling of the doublets, nucleus B has a low-field transition with X down. Therefore the sign of J_{BX} is negative and unlike the sign of J_{AX}. Since the six possible solutions for the noneclipsed, eclipsed, and partly-eclipsed arrangements of the quartets have three with like signs and three with unlike signs, a knowledge of the parity of the signs of the weak couplings gathered from selective spin-decoupling experiments may be helpful in the choice of the correct solution.

RESULTS

The effect of changing the sign of one of the weak coupling constants was investigated. Figure 3 shows a comparison of the standard ABX spectrum

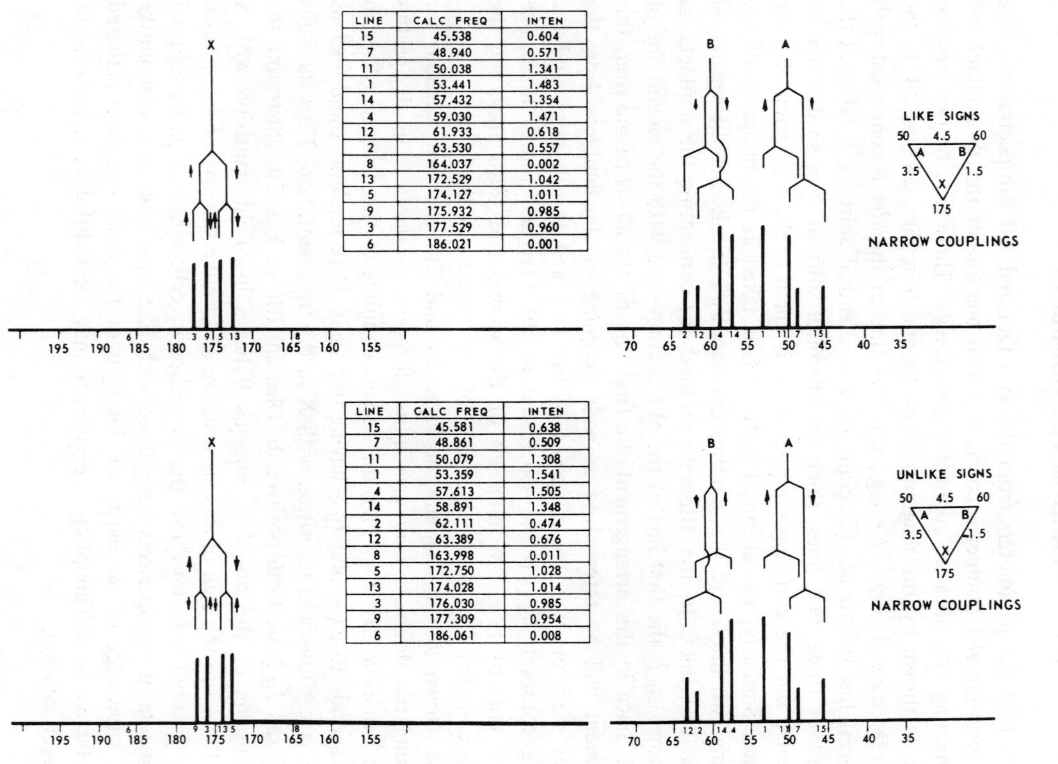

LINE	CALC FREQ	INTEN
15	45.538	0.604
7	48.940	0.571
11	50.038	1.341
1	53.441	1.483
14	57.432	1.354
4	59.030	1.471
12	61.933	0.618
2	63.530	0.557
8	164.037	0.002
13	172.529	1.042
5	174.127	1.011
9	175.932	0.985
3	177.529	0.960
6	186.021	0.001

LINE	CALC FREQ	INTEN
15	45.581	0.638
7	48.861	0.509
11	50.079	1.308
1	53.359	1.541
4	57.613	1.505
14	58.891	1.348
2	62.111	0.474
12	63.389	0.676
8	163.998	0.011
5	172.750	1.028
13	174.028	1.014
3	176.030	0.985
9	177.309	0.954
6	186.061	0.008

Fig. 3. Calculated spectra for (a) like-sign and (b) unlike-sign cases.

with the new spectrum resulting from changing the sign of J_{BX} to negative. The pitchforks have been added to aid in identifying the positions of the two *ab* quartets. In the like-sign case the two *ab* subspectra are partly eclipsed, and in the unlike-sign case the two *ab* subspectra are eclipsed. The new spectrum has three changes. The lines of the spectrum have been renumbered*. In the AB part lines 14 and 4 have been interchanged and lines 12 and 2 have been interchanged. In the X part lines 13 and 3 (which represent the sum J_{AX} + J_{BX}) have been interchanged with lines 9 and 5 to accommodate the change in the sign of J_{BX}. A second change was noted in the intensities of the lines in the AB part. Comparing the intensities of adjacent pairs of lines from left to right in the AB part of the standard like-sign case gives increasing-decreasing-decreasing-increasing intensities. In the unlike-sign case the intensities are decreasing-increasing-decreasing-increasing. Thus an intensity reversal is noted in the B lines of the spectrum. Careful examination of Fig. 3 also shows slight changes in the line frequencies. Thus the spacing between lines 4 and 14, 2 and 12, 5 and 13, and 3 and 9 in the unlike-sign spectrum is diminished compared to the corresponding spacing in the like-sign spectrum. These differences may be useful in identifying the correct solution for an experimental ABX spectrum.

Since changing the sign of the BX coupling resulted in a new arrangement of the quartets, yielding new line numbers, different line frequencies, and different line intensities, would it be possible to analyze the unlike-sign case incorrectly by the wrong choice of quartets? To answer this question, the computer was instructed to fit a spectrum with the line frequencies as shown in the unlike-sign case (Fig. 3b), but by choosing the quartets incorrectly as a partly-eclipsed like-sign case. The results are shown in Fig. 4. The computer found an exact frequency fit to the unlike-sign spectrum by the incorrect choice of quartets, but the parameters and the line intensities are now different. The changes in the parameters are compared in Table I. The chemical shifts of A and B are only slightly in error. The chemical shift of X and the AB coupling are unchanged within the round-off error of the calculation. The AX and BX couplings show the largest error. The results of first-order analysis of the spectrum of Fig. 4 are also included in Table I. The AB coupling is again correct within the round-off error of the calculation, but the AX and BX couplings are in error. Although the errors appear small in this example, they increase in magnitude when $|J_{AX}|-|J_{BX}|$ increases and when the nuclei A and B are more tightly coupled (i.e., when $\nu_A - \nu_B$ decreases and J_{AB} increases).

The second ambiguity was in the pairing of the doublets. To illustrate

*Numbering of the lines is that assigned by the LAOCN program and results from the order of treating the interactions of the various spin states.

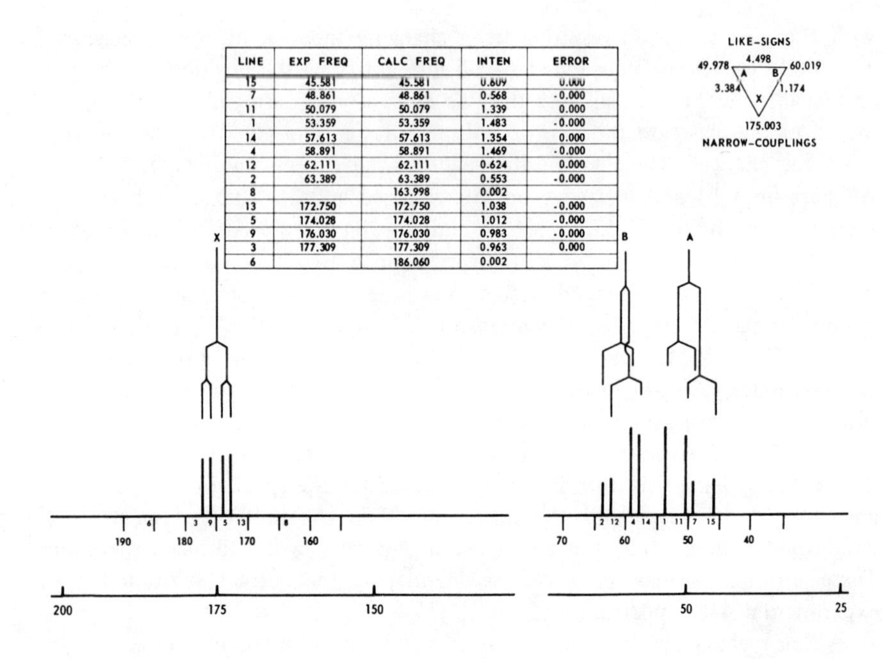

LINE	EXP FREQ	CALC FREQ	INTEN	ERROR
15	45.581	45.581	0.609	0.000
7	48.861	48.861	0.568	-0.000
11	50.079	50.079	1.339	0.000
1	53.359	53.359	1.483	-0.000
14	57.613	57.613	1.354	0.000
4	58.891	58.891	1.469	-0.000
12	62.111	62.111	0.624	0.000
2	63.389	63.389	0.553	-0.000
8		163.998	0.002	
13	172.750	172.750	1.038	-0.000
5	174.028	174.028	1.012	-0.000
9	176.030	176.030	0.983	-0.000
3	177.309	177.309	0.963	0.000
6		186.060	0.002	

Fig. 4. Incorrect analysis of unlike-sign spectrum as if it were a like-sign case.

the consequences of this, the computer was instructed to fit the standard like-sign spectrum of Fig. 3a by the correct choice of quartets (partly-eclipsed) but with wide couplings in place of narrow couplings. The results are shown in Fig. 5. The partly-eclipsed wide-coupled analysis of Fig. 5 is a perfect frequency-fit to the earlier spectrum. The lines have been renumbered and intensity changes are noted, especially in the X part. The parameters are now notably different, and the sign of the BX coupling has changed from positive to negative. The sum lines, 3 and 13, have remained the same, but the combination lines, 6 and 8, have exchanged places with lines 9 and 5 and have grown in intensity in accord with the increased difference between the magnitudes of J_{AX} and J_{BX}. The intensities in the AB part now read decreasing-decreasing-decreasing-decreasing. If the X nucleus were another isotope, or if the X part of the spectrum were otherwise buried in the experimental spectrum, it might have been difficult to make a choice between these two analyses of the spectrum.

In a similar manner, the computer was instructed to analyze the standard unlike-sign case with alternative connection of the doublets. The results are shown in Figure 6. The computer again iterated to a perfect frequency-fit, but gave differents parameters, line numberings, and intensities. The combination lines, 6 and 8, have again exchanged positions with lines 9 and 5 in the X part, and the intensities in the AB part now read decreasing-increasing-decreasing-increasing.

As an additional example of the consequences of incorrect choice of the quartets the computer was instructed to solve the previous spectrum by choosing the quartets as partly eclipsed instead of eclipsed, maintaining the wide couplings of the doublets. The results are shown in Fig. 7. Again the

TABLE I
Analysis of Unlike-Sign Case As If Signs Were Alike

Parameter	True value	Calculated results: As if signs were alike	Calculated results: First-order solution
ν_A	50.000	49.978	-
ν_B	60.000	60.019	-
ν_X	175.000	175.003	-
J_{AB}	4.500	4.498	4.498
J_{AX}	3.500	3.384	3.28
J_{BX}	−1.500	1.174	1.278

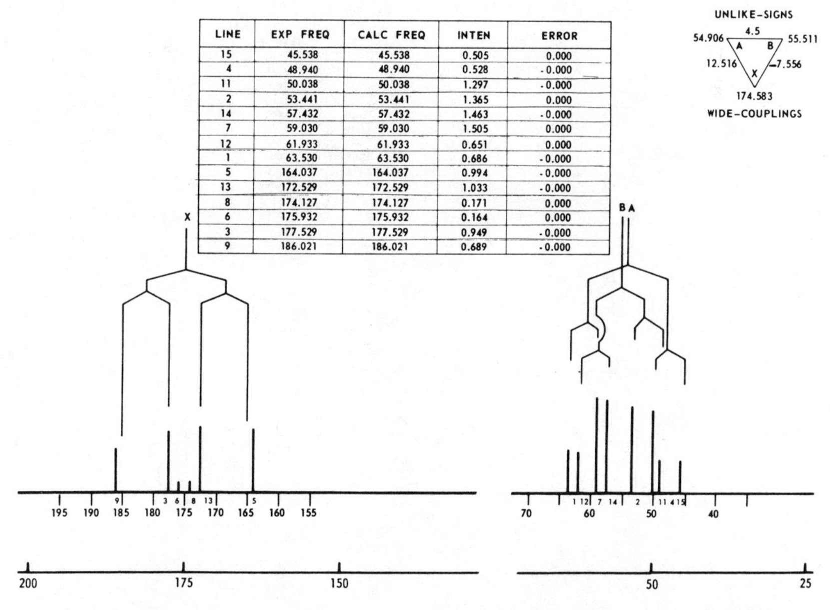

LINE	EXP FREQ	CALC FREQ	INTEN	ERROR
15	45.538	45.538	0.505	0.000
4	48.940	48.940	0.528	-0.000
11	50.038	50.038	1.297	-0.000
2	53.441	53.441	1.365	0.000
14	57.432	57.432	1.463	-0.000
7	59.030	59.030	1.505	0.000
12	61.933	61.933	0.651	0.000
1	63.530	63.530	0.686	-0.000
5	164.037	164.037	0.994	-0.000
13	172.529	172.529	1.033	-0.000
8	174.127	174.127	0.171	0.000
6	175.932	175.932	0.164	0.000
3	177.529	177.529	0.949	-0.000
9	186.021	186.021	0.689	-0.000

Fig. 5. Analysis of like-sign case with alternative connection of the doublets.

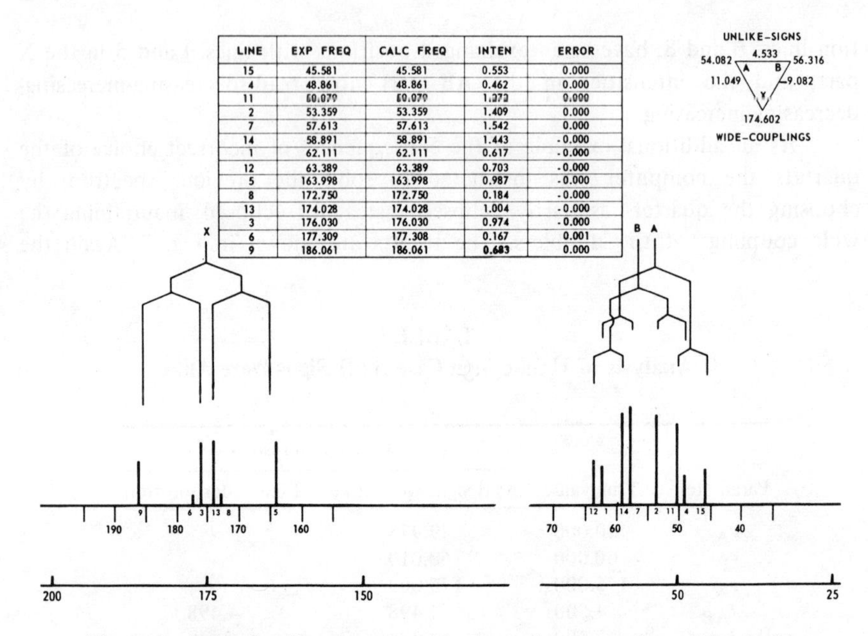

LINE	EXP FREQ	CALC FREQ	INTEN	ERROR
15	45.581	45.581	0.553	0.000
4	48.861	48.861	0.462	0.000
11	50.079	50.079	1.272	0.000
2	53.359	53.359	1.409	0.000
7	57.613	57.613	1.542	0.000
14	58.891	58.891	1.443	-0.000
1	62.111	62.111	0.617	-0.000
12	63.389	63.389	0.703	-0.000
5	163.998	163.998	0.987	-0.000
8	172.750	172.750	0.184	-0.000
13	174.028	174.028	1.004	-0.000
3	176.030	176.030	0.974	-0.000
6	177.309	177.308	0.167	0.001
9	186.061	186.061	0.683	0.000

Fig. 6. Analysis of unlike-sign case with alternative connection of the doublets.

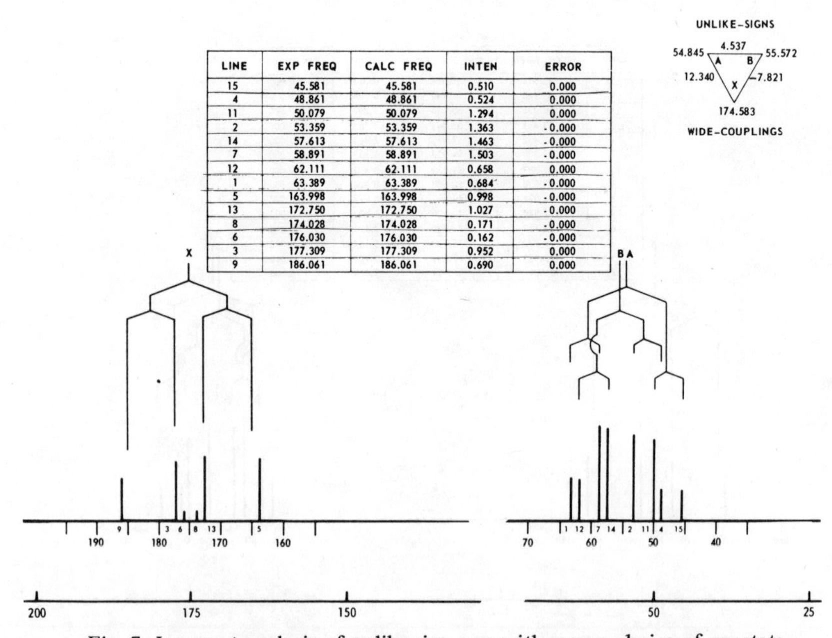

LINE	EXP FREQ	CALC FREQ	INTEN	ERROR
15	45.581	45.581	0.510	0.000
4	48.861	48.861	0.524	0.000
11	50.079	50.079	1.294	0.000
2	53.359	53.359	1.363	-0.000
14	57.613	57.613	1.463	-0.000
7	58.891	58.891	1.503	-0.000
12	62.111	62.111	0.658	0.000
1	63.389	63.389	0.684	-0.000
5	163.998	163.998	0.998	-0.000
13	172.750	172.750	1.027	-0.000
8	174.028	174.028	0.171	-0.000
6	176.030	176.030	0.162	-0.000
3	177.309	177.309	0.952	0.000
9	186.061	186.061	0.690	0.000

Fig. 7. Incorrect analysis of unlike-sign case with wrong choice of quartets.

TABLE II
Analysis of Eclipsed Wide-Coupled Case As If It Were Partly Eclipsed

Parameter	True value	Calculated results as eclipsed AB
ν_A	54.082	54.845
ν_B	56.316	55.572
ν_X	174.602	174.583
J_{AB}	4.533	4.537
J_{AX}	11.049	12.340
J_{BX}	9.082	−7.821

computer iterated to a perfect frequency-fit, but gave different parameters, line numbers, and line intensities. In the X part the sum lines, 3 and 13, have exchanged places with the lines 6 and 8. The intensities in the AB part now read decreasing-decreasing-decreasing-even. The differences in the parameters resulting from the two analyses are compared in Table II. Again the errors in the chemical shift of nucleus A and nucleus B are small. The chemical shift of X and the AB coupling are unchanged within computational error, but the largest errors appear in J_{AX} and J_{BX}.

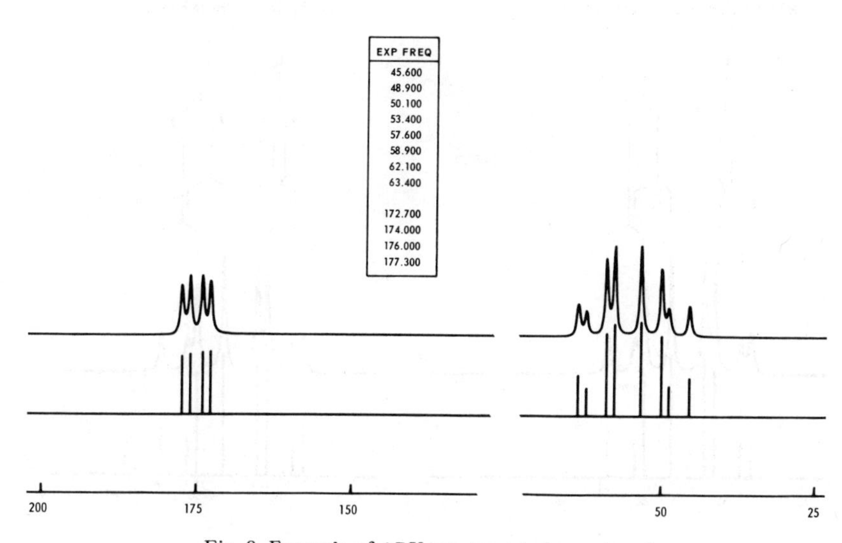

Fig. 8. Example of ABX spectrum to be analyzed.

Thus when factoring an ABX spectrum two decisions must be made for the analysis. There may be more than one way to pick the quartets, and there may be a choice of narrow *vs* wide couplings of the doublets. The choices must be made on the basis of the intensities of the lines, especially in the X part of the spectrum. If the structure of the molecule is known, the reasonableness of the parameters and the signs of the weak couplings should be considered. If a computer program is available for the analysis of the spectrum, all of the possibilities should be considered and the calculated spectra for each should be plotted and compared with the experimental spectrum to choose the correct solution.

An example is the analysis of the spectrum shown in Fig. 8. The choice of quartets can be made in two ways, as illustrated in Fig. 9, with the AB subspectra either partly eclipsed or eclipsed. The pitchforks should be constructed as shown in Fig. 9, and the line numbers for each should be assigned in accord with the above discussion. A second alternative would be to replace the narrow couplings with wide couplings of the doublets, as shown in Fig. 10. The four possible solutions and their calculated spectra are shown in Figs. 11–14. The solutions in Figs. 13 and 14 can be ruled out quickly on the basis of the intensities in the X part of the spectrum. The solutions of Figs. 11 and 12 must be examined carefully to compare the intensities in the AB part of the spectra. In case the X part of the spectrum was not available, all

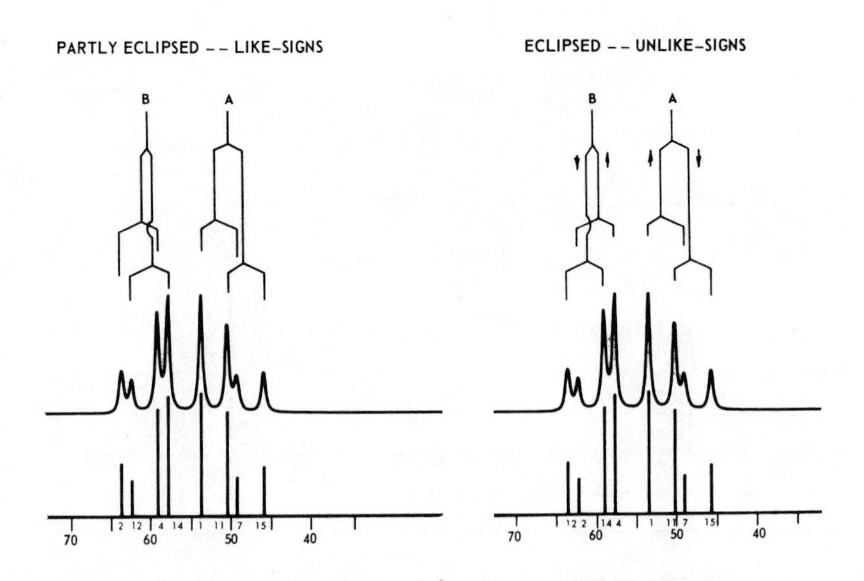

Fig. 9. Possible choice of the quartets. Narrow couplings.

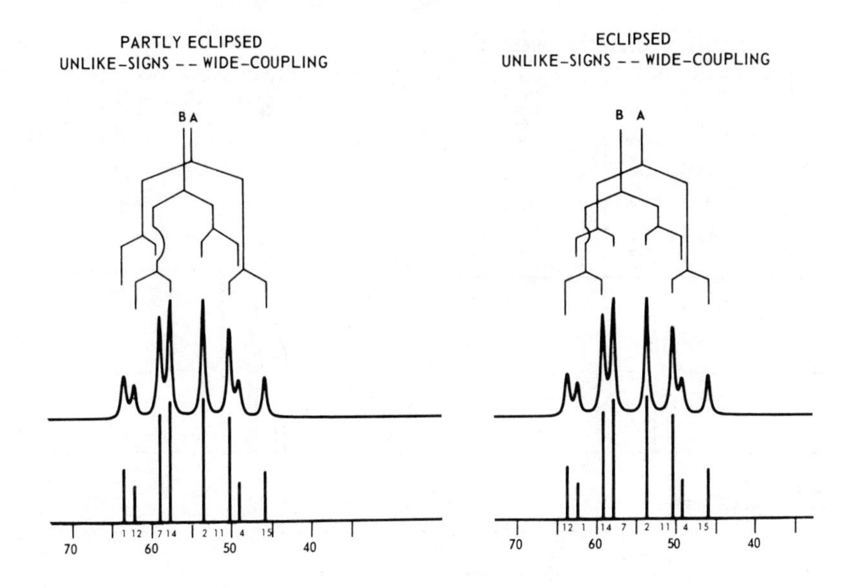

Fig. 10. Possible choice of the quartets. Wide couplings.

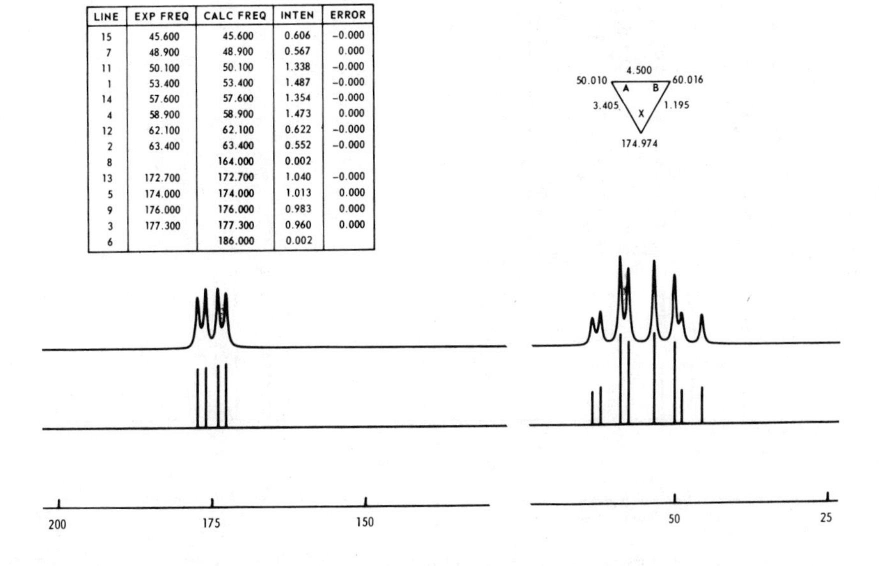

LINE	EXP FREQ	CALC FREQ	INTEN	ERROR
15	45.600	45.600	0.606	-0.000
7	48.900	48.900	0.567	0.000
11	50.100	50.100	1.338	-0.000
1	53.400	53.400	1.487	-0.000
14	57.600	57.600	1.354	-0.000
4	58.900	58.900	1.473	0.000
12	62.100	62.100	0.622	-0.000
2	63.400	63.400	0.552	-0.000
8		164.000	0.002	
13	172.700	172.700	1.040	-0.000
5	174.000	174.000	1.013	0.000
9	176.000	176.000	0.983	0.000
3	177.300	177.300	0.960	0.000
6		186.000	0.002	

Fig. 11. Iterated spectrum to fit data computed as like-sign, partly-eclipsed, narrow-coupled case.

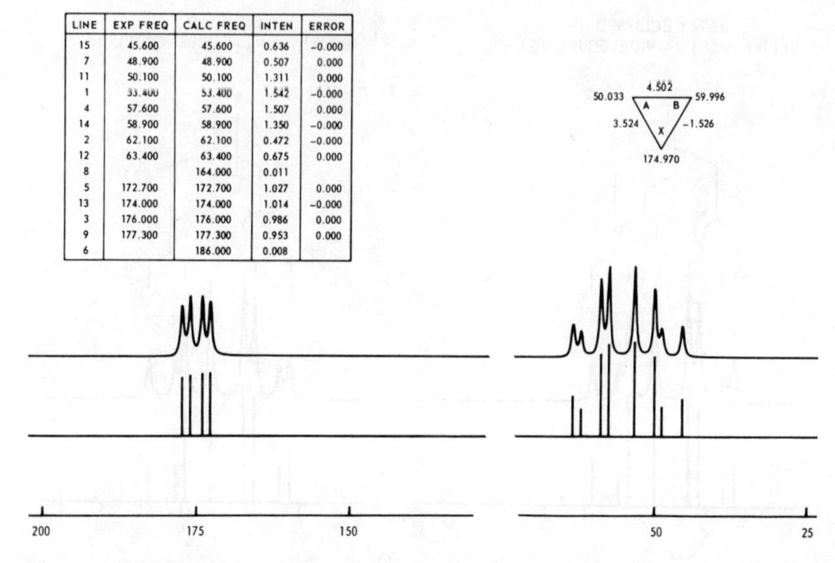

LINE	EXP FREQ	CALC FREQ	INTEN	ERROR
15	45.600	45.600	0.636	-0.000
7	48.900	48.900	0.507	0.000
11	50.100	50.100	1.311	0.000
1	53.400	53.400	1.542	-0.000
4	57.600	57.600	1.507	0.000
14	58.900	58.900	1.350	-0.000
2	62.100	62.100	0.472	-0.000
12	63.400	63.400	0.675	0.000
8		164.000	0.011	
5	172.700	172.700	1.027	0.000
13	174.000	174.000	1.014	-0.000
3	176.000	176.000	0.986	0.000
9	177.300	177.300	0.953	0.000
6		186.000	0.008	

Fig. 12. Iterated spectrum to fit data computed as unlike-sign, eclipsed, narrow- coupled case.

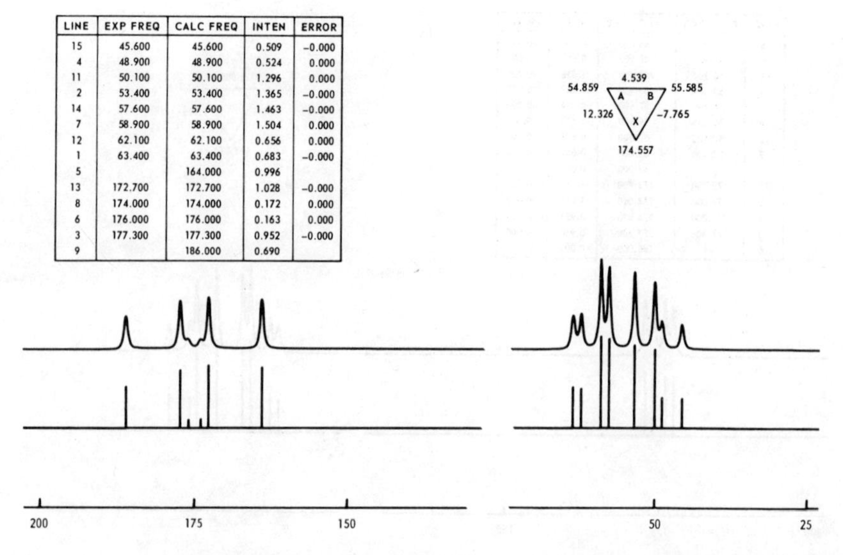

LINE	EXP FREQ	CALC FREQ	INTEN	ERROR
15	45.600	45.600	0.509	-0.000
4	48.900	48.900	0.524	0.000
11	50.100	50.100	1.296	0.000
2	53.400	53.400	1.365	-0.000
14	57.600	57.600	1.463	-0.000
7	58.900	58.900	1.504	0.000
12	62.100	62.100	0.656	0.000
1	63.400	63.400	0.683	-0.000
5		164.000	0.996	
13	172.700	172.700	1.028	-0.000
8	174.000	174.000	0.172	0.000
6	176.000	176.000	0.163	0.000
3	177.300	177.300	0.952	-0.000
9		186.000	0.690	

Fig. 13. Iterated spectrum to fit data computed as unlike-sign, partly-eclipsed, wide-coupled case.

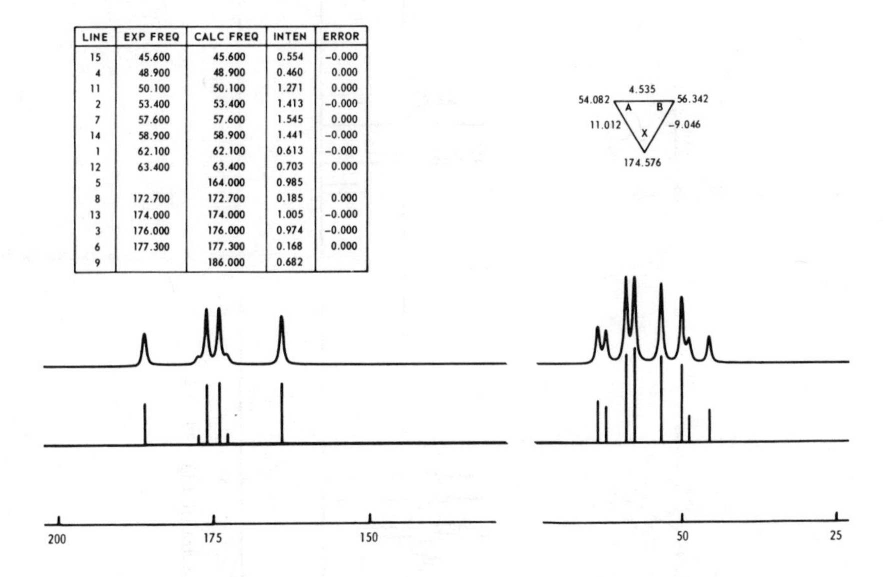

LINE	EXP FREQ	CALC FREQ	INTEN	ERROR
15	45.600	45.600	0.554	−0.000
4	48.900	48.900	0.460	0.000
11	50.100	50.100	1.271	0.000
2	53.400	53.400	1.413	−0.000
7	57.600	57.600	1.545	0.000
14	58.900	58.900	1.441	−0.000
1	62.100	62.100	0.613	−0.000
12	63.400	63.400	0.703	0.000
5		164.000	0.985	
8	172.700	172.700	0.185	0.000
13	174.000	174.000	1.005	−0.000
3	176.000	176.000	0.974	−0.000
6	177.300	177.300	0.168	0.000
9		186.000	0.682	

Fig. 14. Iterated spectrum to fit data computed as unlike-sign, eclipsed, wide-coupled case.

four solutions would have to be examined for intensity differences in the AB part. The AB parts of the four solutions are therefore compared in Fig. 15. The second solution, that of Fig. 12, is the correct one on the basis of line intensities. The differences are indeed small.

Intensity differences may be hard to detect if the lines are not well resolved or if the noise is appreciable in the spectrum. Saturation may also cause intensity errors and should be avoided.

In addition to these four solutions, there are four more solutions with all the signs of the AX and BX couplings reversed. These changes make very little difference in the intensities of the lines and in the resulting parameters. Eight more solutions having the sign of J_{AB} negative are also possible, but again these make very little difference in the line intensities or in the resulting parameters.

The aromatic portion of the spectrum of 5-hydroxynaphthoquinone serves as an additional example of an ABX spectrum which has more than one possible solution. The aromatic portion of the spectrum is shown in Fig. 16. Although the spectrum is really an ABC case, it has a strong similarity to the ABX cases discussed above, but the relative line intensities increase near the center of the spectrum. The right-hand side of the spectrum resembles a six-line X pattern with the two combination lines appearing as satellites in the spectrum. The small shoulder at 438 Hz is attributed to residual chloroform in the solvent. The AB part appears as two quartets, one of which is nearly

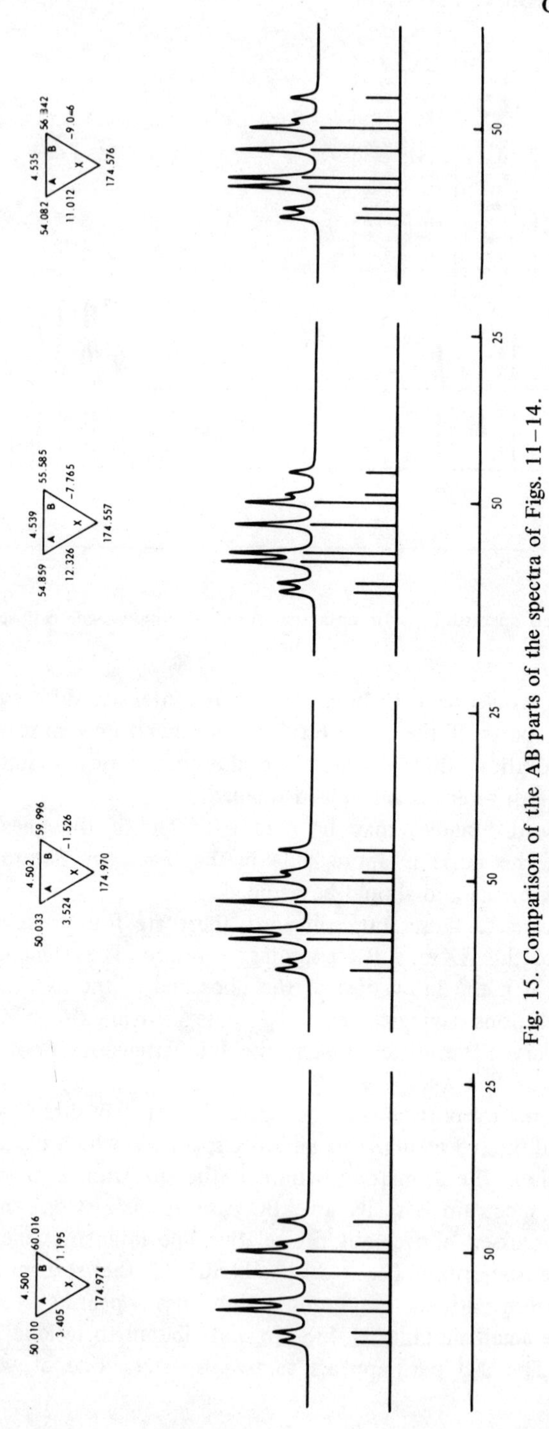

Fig. 15. Comparison of the AB parts of the spectra of Figs. 11–14.

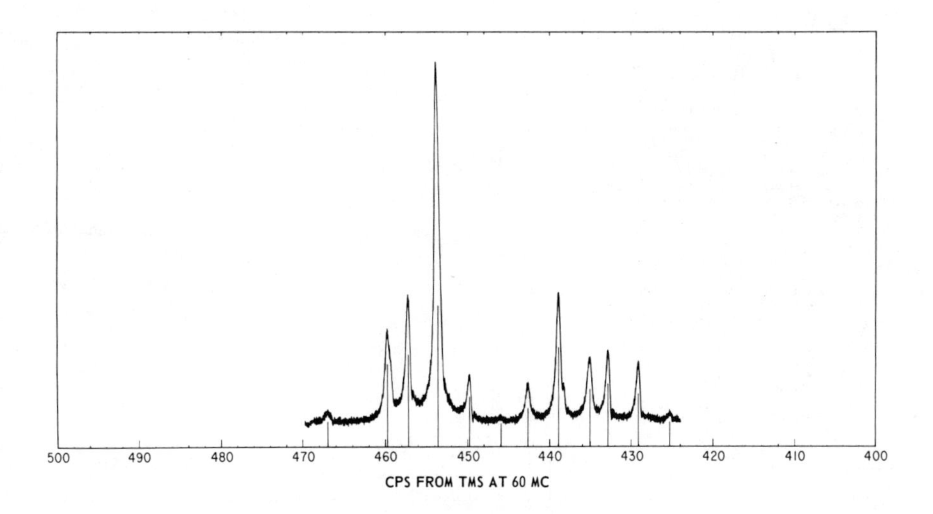

Fig. 16. Portion of the NMR spectrum of 5-hydroxynaphthoquinone.

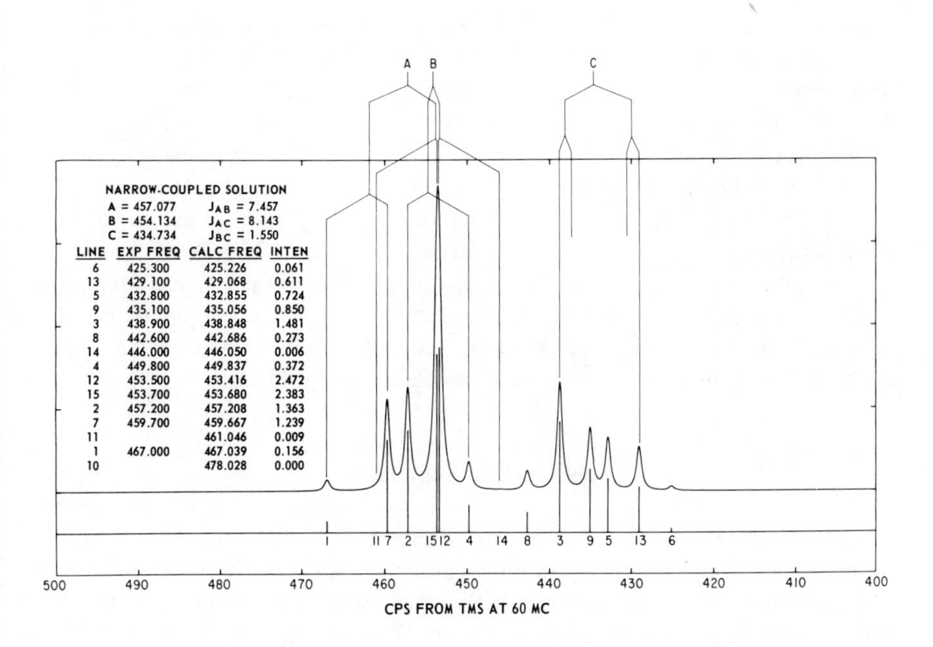

LINE	EXP FREQ	CALC FREQ	INTEN
6	425.300	425.226	0.061
13	429.100	429.068	0.611
5	432.800	432.855	0.724
9	435.100	435.056	0.850
3	438.900	438.848	1.481
8	442.600	442.686	0.273
14	446.000	446.050	0.006
4	449.800	449.837	0.372
12	453.500	453.416	2.472
15	453.700	453.680	2.383
2	457.200	457.208	1.363
7	459.700	459.667	1.239
11		461.046	0.009
1	467.000	467.039	0.156
10		478.028	0.000

Fig. 17. Calculated spectrum for 5-hydroxynaphthoquinone. Narrow-coupled solution.

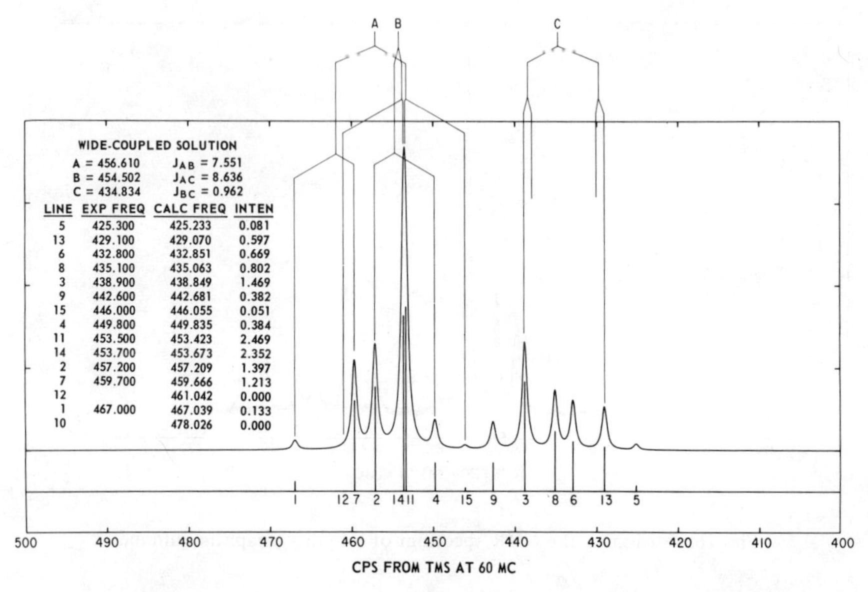

A B C
WIDE-COUPLED SOLUTION
A = 456.610 JAB = 7.551
B = 454.502 JAC = 8.636
C = 434.834 JBC = 0.962
LINE EXP FREQ CALC FREQ INTEN
5 425.300 425.233 0.081
13 429.100 429.070 0.597
6 432.800 432.851 0.669
8 435.100 435.063 0.802
3 438.900 438.849 1.469
9 442.600 442.681 0.382
15 446.000 446.055 0.051
4 449.800 449.835 0.384
11 453.500 453.423 2.469
14 453.700 453.673 2.352
2 457.200 457.209 1.397
7 459.700 459.666 1.213
12 461.042 0.000
1 467.000 467.039 0.133
10 478.026 0.000
1 12 7 2 14 11 4 15 9 3 8 6 13 5
500 490 480 470 460 450 440 430 420 410 400
CPS FROM TMS AT 60 MC

Fig. 18. Calculated spectrum for 5-hydroxynaphthoquinone. Wide-coupled solution.

degenerated to a broad, single line. In this case there is only one way to pick the quartets – noneclipsed (Fig. 17). Figure 18 shows the alternative solution with wide coupling of the doublets. The pitchforks of Figs. 17 and 18 also illustrate the large magnitude of the discrepancy between the first-order analysis for J_{AX} and J_{BX} and the exact analysis for these parameters. Thus the spacing between lines 5 and 13 and lines 3 and 9 in the X part of Fig. 17 (3.8 Hz) is more than twice that of the true J_{BX} (1.55 Hz) represented by the separation of the pitchforks immediately above it. In Fig. 18 the error in J_{BX} is approximately fourfold. The spectrum of Fig. 16 was also analyzed by EXAN, which gave the same two solutions, identical within experimental error. The solution shown in Fig. 18 was selected as the correct result on the basis of the matching of the line intensities, especially that of line 15.

DECEPTIVE SPECTRA

ABX spectra can at times be very deceptive.[9] It is important to be able to recognize these deceptive spectra so that they will not be misinterpreted. The criteria for deceptive spectra are the magnitude of the coupling of the A and B nuclei, J_{AB}, compared to the difference $\nu_A - \nu_B$, between their chemical shifts and the difference between the weak couplings, J_{AX} and J_{BX}.

We can distinguish three types of deception. The spectrum can be deceptively simple, deceptively complex, or just deceptive. ABX spectra become deceptively simple when

$$J_{AB} \gg \nu_A - \nu_B + \tfrac{1}{2} \, |J_{AX} - J_{BX}| \tag{1}$$

It is instructive to choose hypothetical parameters such that the various terms in (1) approach zero, and then compute the corresponding spectra.

If in the standard ABX spectrum J_{AX} is made about equal to J_{BX} (ignoring the signs), the right-hand term of the inequality will approach zero. Then J_{AB} is much larger than the frequency difference between the chemical shifts of nuclei A and B, and the spectra should be deceptively simple. The resulting spectra are illustrated for like signs in Fig. 19, and for unlike signs in Fig. 20. In both cases the X part of the spectrum becomes a triplet and the AB part is normal. An example of a deceptively-simple ABX spectrum of this type is found in the Varian catalog[10] (spectrum 407) (Fig. 21). In this spectrum of mercaptosuccinic acid the X part is a triplet owing to the near-equivalence of J_{AX} and J_{BX}.

If in the standard ABX spectrum the parameters are altered so that $\nu_A \approx \nu_B$ the first term on the right of inequality (1) will approach zero, and the magnitude of J_{AB} will be much greater than the sum of the terms on the right of the inequality sign. The spectrum should again be deceptively simple. The variation in the appearance of the standard like-sign ABX spectrum as ν_A

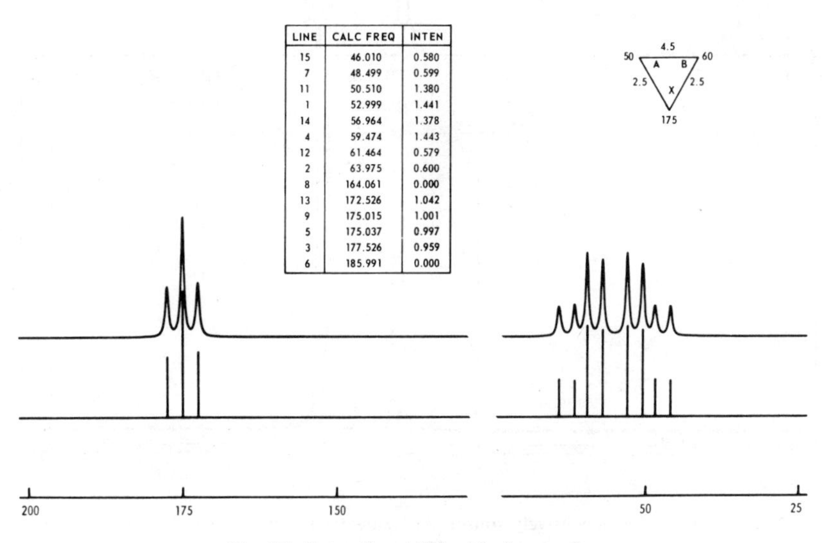

LINE	CALC FREQ	INTEN
15	46.010	0.580
7	48.499	0.599
11	50.510	1.380
1	52.999	1.441
14	56.964	1.378
4	59.474	1.443
12	61.464	0.579
2	63.975	0.600
8	164.061	0.000
13	172.526	1.042
9	175.015	1.001
5	175.037	0.997
3	177.526	0.959
6	185.991	0.000

Fig. 19. Deceptive ABX with $J_{AX} = J_{BX}$.

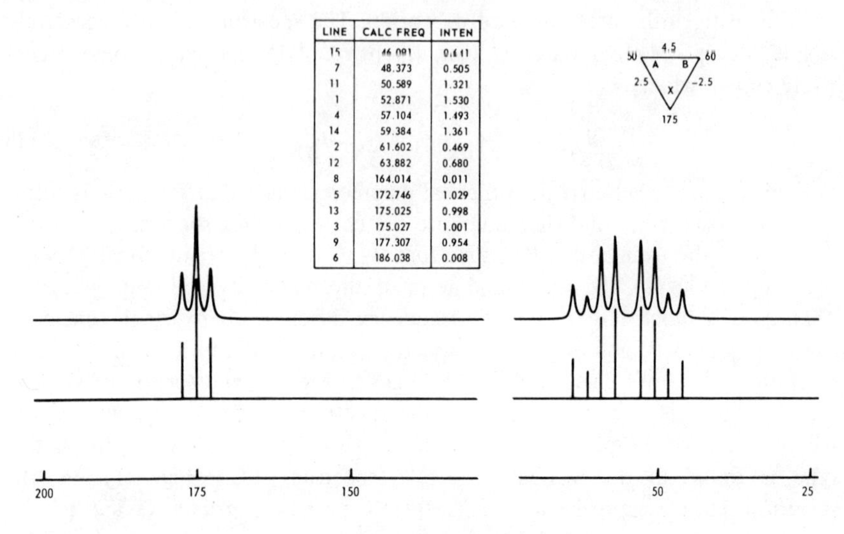

LINE	CALC FREQ	INTEN
15	46.091	0.611
7	48.373	0.505
11	50.589	1.321
1	52.871	1.530
4	57.104	1.493
14	59.384	1.361
2	61.602	0.469
12	63.882	0.680
8	164.014	0.011
5	172.746	1.029
13	175.025	0.998
3	175.027	1.001
9	177.307	0.954
6	186.038	0.008

Fig. 20. Deceptive ABX with $J_{AX} = -J_{BX}$.

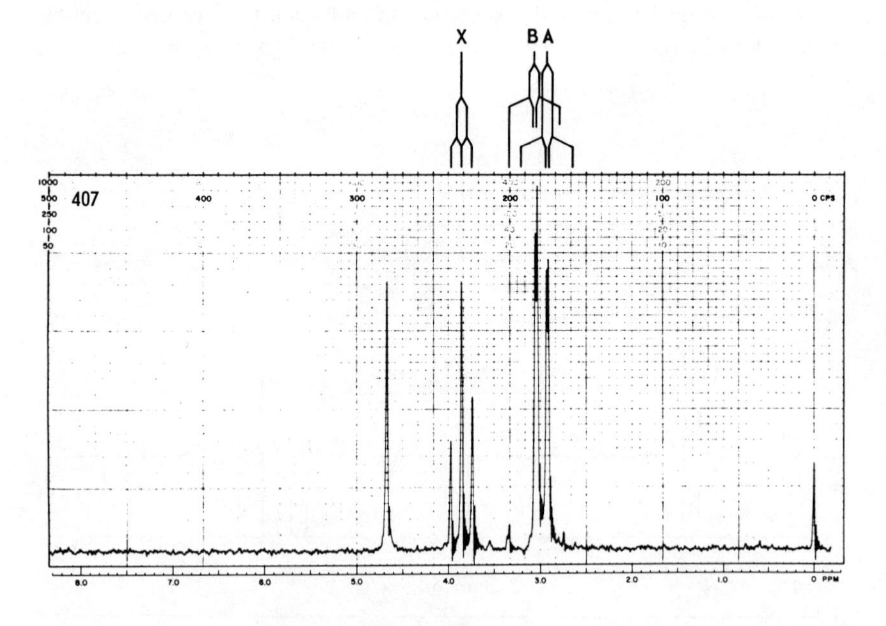

Fig. 21. Example of a deceptively-simple ABX spectrum having three-line X and nearly a two-line AB part.

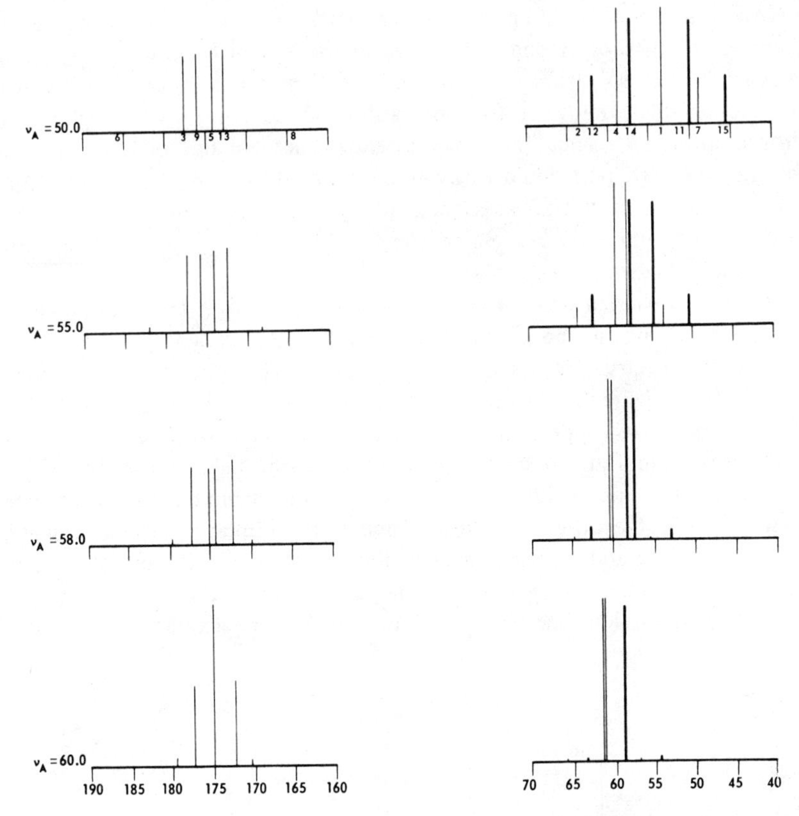

Fig. 22. Variation in like-sign ABX spectra as ν_A is shifted toward ν_B.

is moved toward ν_B until they coincide is shown in Fig. 22. The X part changes from four to six to three lines. In the X part the sum lines, 3 and 13, maintain their separation, but the other two lines, 9 and 5, are displaced inward and finally meet in the center. The corresponding spacings in the AB part track accordingly. The limiting case, where the chemical shift of A is identical to the chemical shift of B (more appropriately denoted as an $A_2'X$ spectrum) is shown in Fig. 23. For this like-sign case the X part has collapsed to a triplet and the AB part has collapsed to a doublet. The small satellite lines in the AB part and in the X part are usually lost in the noise of the spectrum. The infinitesmal spacing d in the overlapping lines, 11 and 14, and 2 and 7 of the AB part will be small if J_{AB} is equal to or greater than the absolute magnitude of $J_{AX}-J_{BX}$, according to the equation[4]

$$|J_{AX}-J_{BX}| = 2(2dJ_{AB} + d^2)^{1/2} \tag{2}$$

An example of a deceptively-simple ABX owing to the coincidence of the chemical shifts of A and B is found in the Varian Catalog[10] (No. 566) (Fig. 24). In this spectrum of thymidine the X part is a triplet and the AB part is a doublet, allowing for one additional coupling to a neighboring hydrogen at 4.49δ. Since the small satellite lines are not observed in the spectrum, the ABX part could easily be interpreted as an A_2X spectrum with an apparent J_{AX} of 7 Hz, an average of the true J_{AX} and J_{BX}. This, of course, is quite unreasonable. Since one is a *cis* coupling and one is a *trans* coupling, it is highly unlikely that $J_{AX} = J_{BX}$.

In the unlike-sign case with $\nu_A = \nu_B$ the $J_{AX} - J_{BX}$ term is not small, owing to the change in the sign of J_{BX}. Therefore the inequality (1) does not hold and the spectrum is not deceptively simple. The spectrum is, nevertheless, deceptive. Figure 25 illustrates the variations in the ABX spectra for the unlike-sign case as ν_A approaches and becomes coincident with ν_B. Here the X part changes from four to six to five lines. The sum lines, 3 and 13, which start out as the inside lines of the X quartet, maintain the same line separation, while the other two lines, 9 and 5, are displaced inward, crossing over the sum lines and meeting again in the center of the X multiplet. Also, since the magnitude of $J_{AX} - J_{BX}$ is larger, owing to the negative sign of J_{BX}, the combination lines 6 and 8 in the X part acquire a noticeable

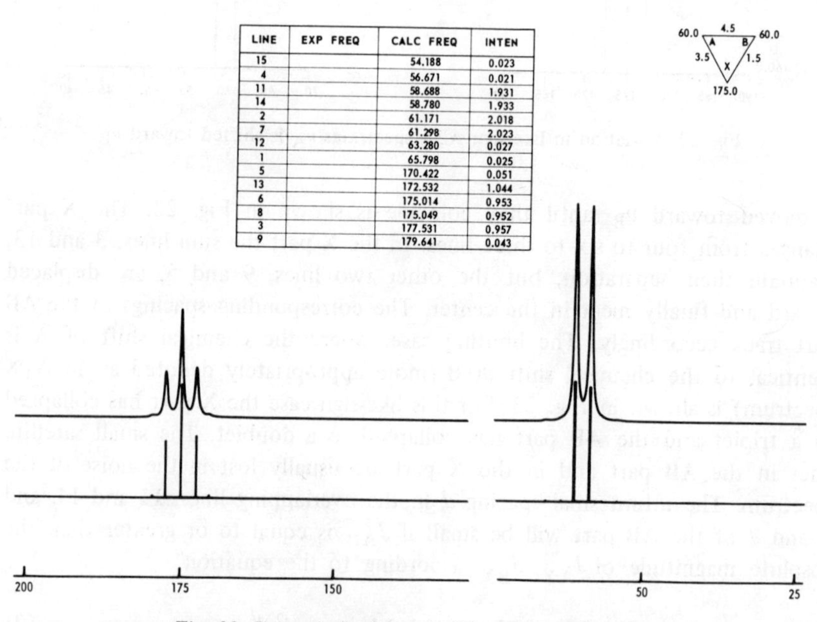

LINE	EXP FREQ	CALC FREQ	INTEN
15		54.188	0.023
4		56.671	0.021
11		58.688	1.931
14		58.780	1.933
2		61.171	2.018
7		61.298	2.023
12		63.280	0.027
1		65.798	0.025
5		170.422	0.051
13		172.532	1.044
6		175.014	0.953
8		175.049	0.952
3		177.531	0.957
9		179.641	0.043

Fig. 23. Deceptively simple ABX spectrum; like signs.

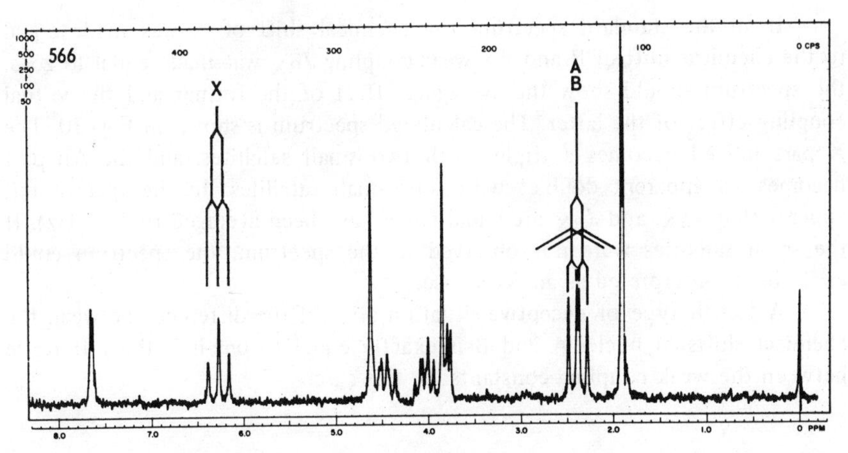

Fig. 24. Example of a doubly deceptive ABX or $A_2'X$ spectrum having $\nu_A = \nu_B$ and $J_{AX} \neq J_{BX}$.

intensity. The corresponding line separations in the AB part track accordingly. The limiting case with $\nu_A = \nu_B$ is shown in Fig. 26. The X part is a triplet plus two satellites and the AB part is a triplet plus four satellites.

A disturbing feature in both the like-sign and unlike-sign cases is that J_{AX} and J_{BX} are averaged in the spectra. From Fig. 23 it appears that $J_{AX} = J_{BX} = 2.5$, which is the average of 3.5 and 1.5 Hz. Similarly, in Fig. 26 the apparent $J_{AX} = J_{BX} = 1$ is the average of 3.5 and −1.5 Hz. Thus the analysis of such a spectrum does not give the true values for these couplings; only averaged values are obtained.

If one of the weak couplings J_{AX} or J_{BX} is equal to zero, a deceptively-complicated ABX spectrum can be observed.[11] Under certain conditions the AB part can still be eight lines and the X part can be four lines. The virtual coupling[11] S_{BX} may be large when J_{AB} is equal to or greater than $\nu_A - \nu_B$, the frequency difference between nucleus A and nucleus B. These are the degenerate types of AB subspectra that are diagrammed in Fig. 2. If in the standard ABX spectrum, J_{BX} is made equal to zero, only an eight-line ABX spectrum results (Fig. 27). Here $J = 0.45\delta$ for the A and B nuclei and the virtual coupling is not apparent in the spectrum. If the chemical shift of A is adjusted to 55 Hz, $J = 0.9\delta$ and the virtual coupling is now apparent in the AB part and in the X part (Fig. 28), and a twelve-line ABX spectrum results. If the chemical shift of A is further adjusted to 58 Hz, $J = 2.25\delta$, and in the resulting spectrum (Fig. 29) the virtual coupling is clearly seen in the separation of lines 3 and 9, 5 and 13 in the X part, and lines 1 and 15 in the AB part. Hand factoring of this spectrum would yield a J_{BX} that was too large and a J_{AX} that was too small.

If in the standard spectrum the chemical shift of A was made equal to the chemical shift of B and the weak coupling J_{BX} was made equal to zero, the spectrum should show the averaging effect of the former and the virtual coupling effect of the latter. The calculated spectrum is shown in Fig. 30. The X part indeed becomes a triplet with two small satellites, and the AB part becomes an apparent doublet with four small satellites. In the spectrum it appears that J_{AX} and J_{BX} are equal (they have been averaged to 1.75 Hz). If the small satellites were not observed in the spectrum, the spectrum could easily be misinterpreted as an A_2X case.

A fourth type of deceptive situation arises if the difference between the chemical shifts of nuclei A and B is exactly equal to one-half the difference between the weak coupling constants. In this case

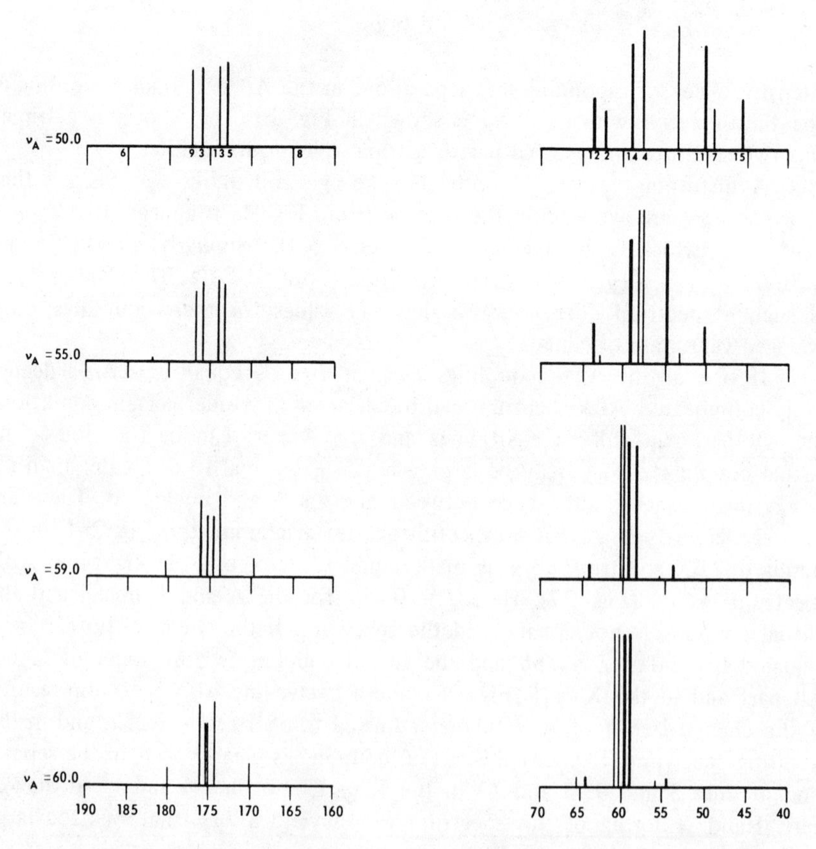

Fig. 25. Variation in unlike-sign spectra as ν_A is shifted toward ν_B.

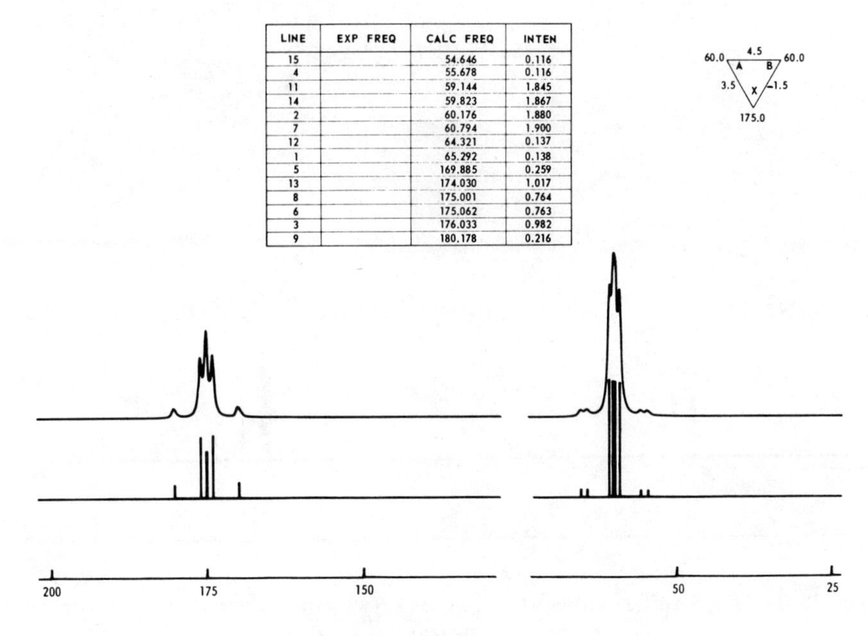

LINE	EXP FREQ	CALC FREQ	INTEN
15		54.646	0.116
4		55.678	0.116
11		59.144	1.845
14		59.823	1.867
2		60.176	1.880
7		60.794	1.900
12		64.321	0.137
1		65.292	0.138
5		169.885	0.259
13		174.030	1.017
8		175.001	0.764
6		175.062	0.763
3		176.033	0.982
9		180.178	0.216

Fig. 26. Deceptive ABX spectrum; unlike signs.

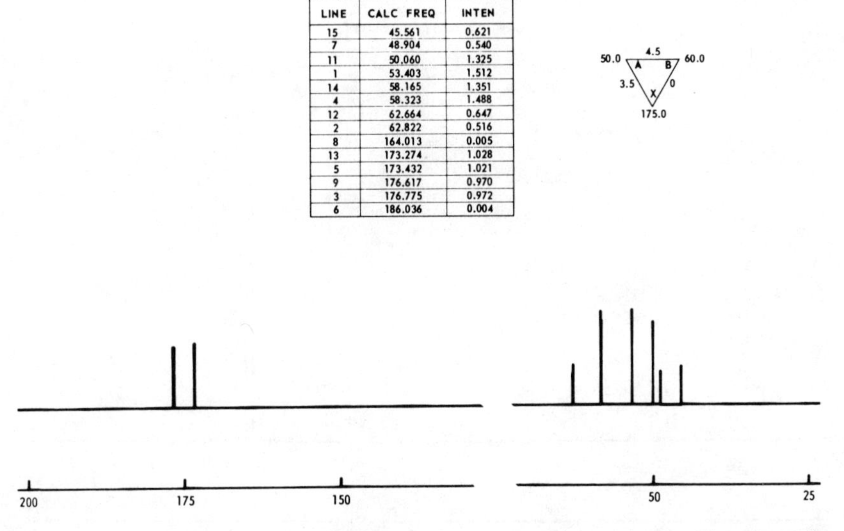

LINE	CALC FREQ	INTEN
15	45.561	0.621
7	48.904	0.540
11	50.060	1.325
1	53.403	1.512
14	58.165	1.351
4	58.323	1.488
12	62.664	0.647
2	62.822	0.516
8	164.013	0.005
13	173.274	1.028
5	173.432	1.021
9	176.617	0.970
3	176.775	0.972
6	186.036	0.004

Fig. 27. Deceptive ABX spectrum where J_{BX} = O. Eight-line spectrum. Virtual coupling not apparent.

LINE	CALC FREQ	INTEN
15	50.296	0.427
7	53.330	0.203
11	54.795	1.518
1	57.829	1.844
14	58.429	1.542
4	58.896	1.827
12	62.928	0.453
2	63.395	0.186
8	168.176	0.035
13	173.275	1.029
5	173.742	0.989
9	176.309	0.948
3	176.776	0.971
6	181.875	0.028

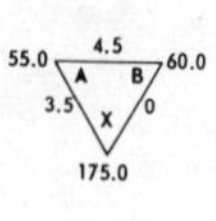

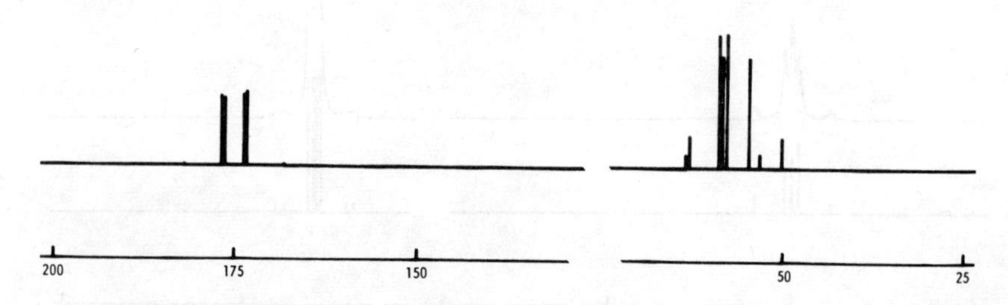

Fig. 28. Deceptive ABX spectrum where J_{BX} = O. Twelve-line spectrum. Virtual coupling apparent.

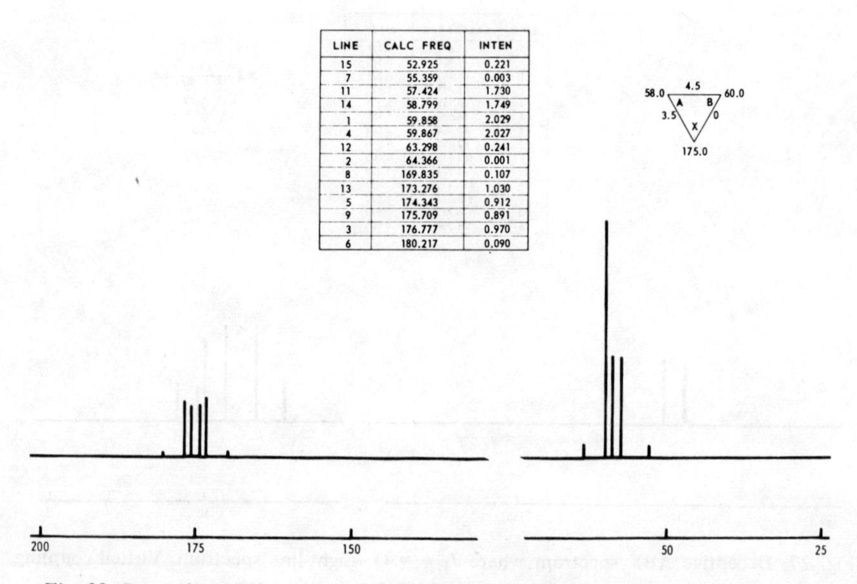

LINE	CALC FREQ	INTEN
15	52.925	0.221
7	55.359	0.003
11	57.424	1.730
14	58.799	1.749
1	59.858	2.029
4	59.867	2.027
12	63.298	0.241
2	64.366	0.001
8	169.835	0.107
13	173.276	1.030
5	174.343	0.912
9	175.709	0.891
3	176.777	0.970
6	180.217	0.090

Fig. 29. Deceptive ABX spectrum where J_{BX} = O. Virtual coupling clearly apparent.

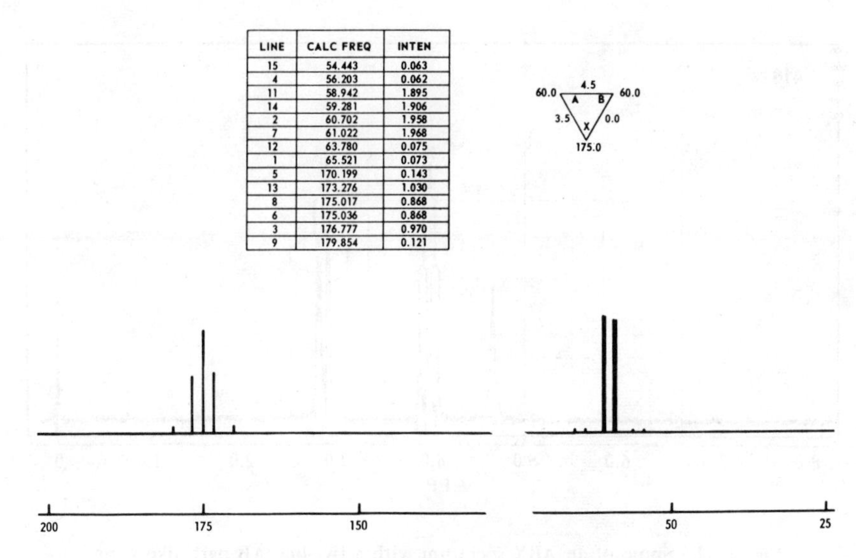

LINE	CALC FREQ	INTEN
15	54.443	0.063
4	56.203	0.062
11	58.942	1.895
14	59.281	1.906
2	60.702	1.958
7	61.022	1.968
12	63.780	0.075
1	65.521	0.073
5	170.199	0.143
13	173.276	1.030
8	175.017	0.868
6	175.036	0.868
3	176.777	0.970
9	179.854	0.121

Fig. 30. Deceptively-simple ABX spectrum. Virtual coupling. J_{BX} = O and $\delta_A = \delta_B$.

$$\nu_A - \nu_B - \tfrac{1}{2}\,|J_{AX} - J_{BX}| = 0 \tag{3}$$

Nuclei A and B now share a common energy level. One quartet degenerates to a singlet and the AB part of the spectrum will have only five lines. These situations are diagrammed in Fig. 2. If the degenerate singlet lies outside the other quartet (noneclipsed case, IB), the weak couplings will have like signs. If the degenerate singlet lies inside the other quartet (eclipsed case, IIB), the weak couplings will have unlike signs. A degenerate situation (IIC) could also arise in which the degenerate singlet exactly coincided with one end of the remaining quartet, in which case the J_{BX} coupling must be zero. It is interesting to note that in these five-line deceptive situations there is only one way to pick the quartet and only one way to pair the doublets. The solution which is obtained by analysis of the spectrum is therefore unique. Several examples of five-line AB types may be found in the Varian Catalog.[10] Spectra No. 410, 382, and 503 (Figs. 31, 32, and 33) are all examples of like-sign cases. In the 60-MHz spectrum of Fig. 32 the degenerate singlet happens to coincide with one of the outside lines of the remaining quartet, and a four-line AB part results. Spectrum No. 503 (Fig. 33) was analyzed by the computer program to give the parameters and the computed spectrum shown in Fig. 34, and Eq. (3) was approximately verified.

It should also be possible to have a five-line AB part with the weak coupling of unlike signs. Since no example of this could be found in the

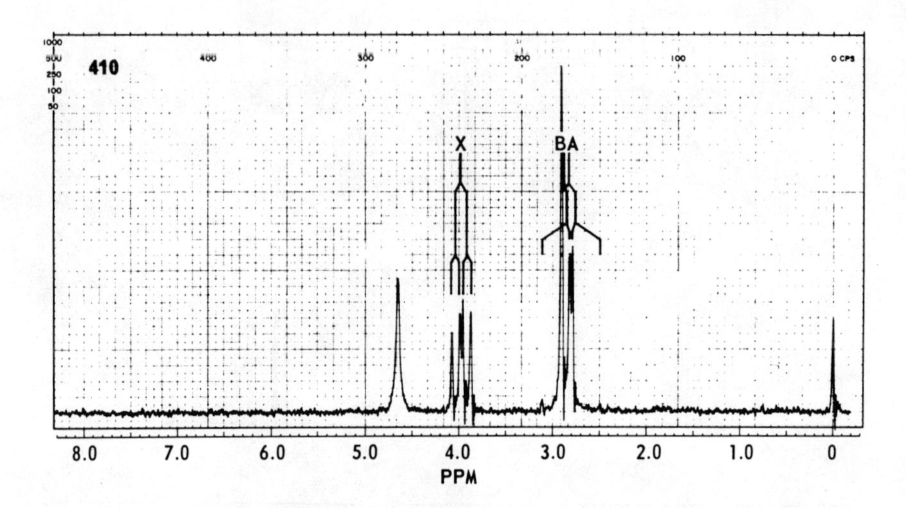

Fig. 31. Example of an ABX spectrum with a five-line AB part; like signs.

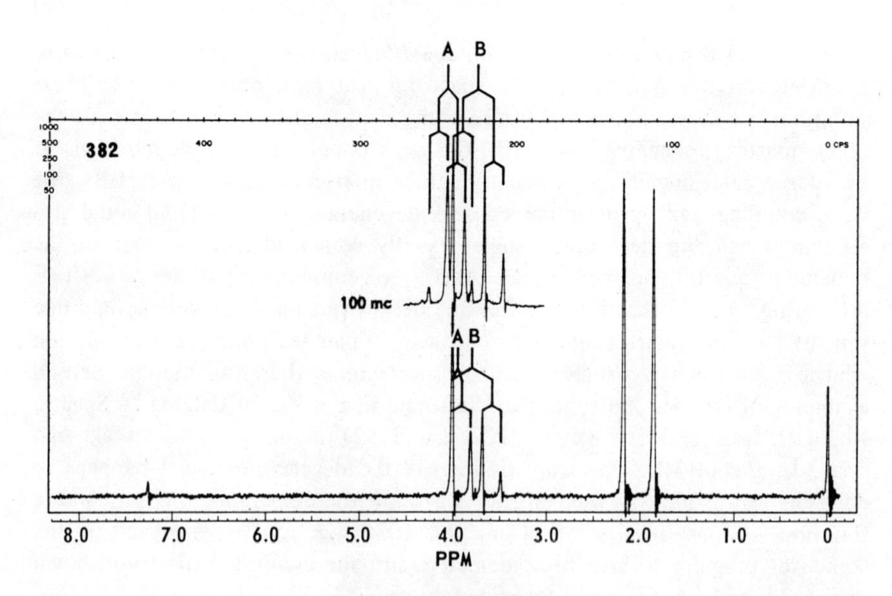

Fig. 32. Examples of a seven-line and a four-line AB part of an ABX spectrum; like signs.

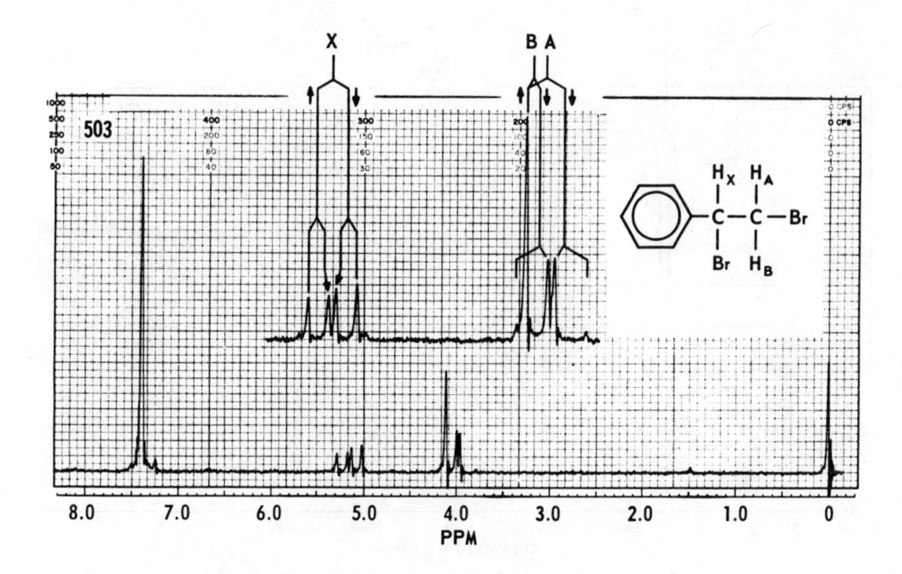

Fig. 33. Example of a five-line AB part of an ABX spectrum; like signs.

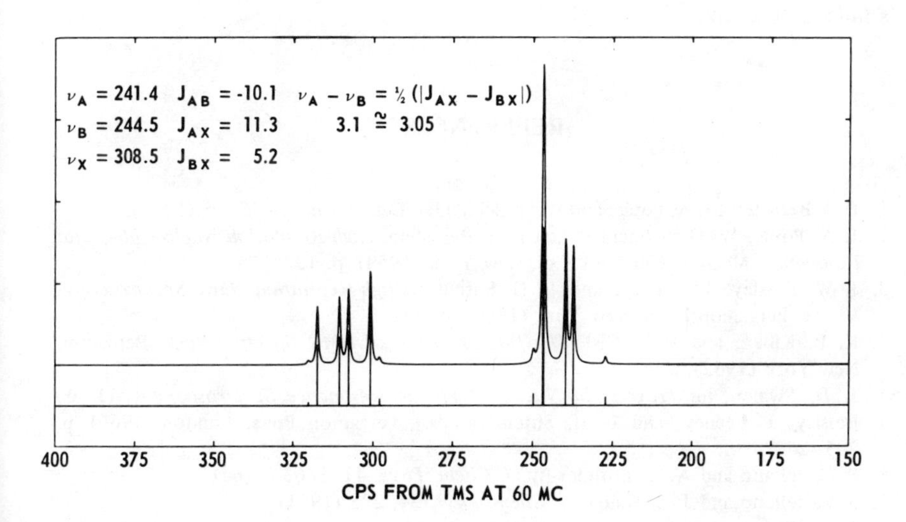

Fig. 34. Computed spectrum. ABX with five-line AB; like signs.

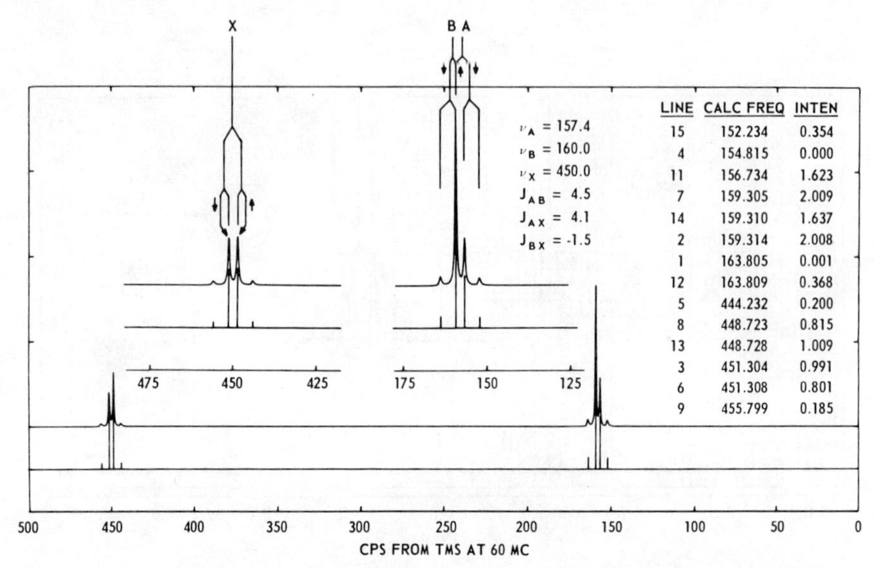

Fig. 35. Example of a four-line AB part of an ABX spectrum with unlike signs.

catalog, an example was calculated (Fig. 35). Here the parameters were carefully chosen so that the degenerate singlet overlapped one of the inside lines of the remaining quartet, and a four-line AB part resulted. This spectrum is indeed deceptive.

REFERENCES

1. H. J. Bernstein, J. A. Pople, and W. G. Schneider, *Can. J. Chem.* **35**, 65 (1957).
2. J. A. Pople, W. G. Schneider, and H. J. Bernstein, *High-Resolution Nuclear Magnetic Resonance*, McGraw-Hill Book Co., New York (1959), p. 132.
3. J. W. Emsley, J. Feeney, and L. H. Sutcliffe, *High-Resolution NMR Spectroscopy*, Vol. 1, Pergamon Press, New York (1966), p. 357.
4. K. B. Kiberg and B. J. Nist, *The Interpretation of NMR Spectra*, W. A. Benjamin, New York (1962), p. 21.
5. J. D. Swalen, in: *Progress in Nuclear Magnetic Resonance Spectroscopy I*, (J. W. Emsley, J. Feeney, and L. H. Sutcliffe, eds.), Pergamon Press, London (1966), p. 205–250.
6. S. Castellano and A. A. Bothner-By, *J. Chem. Phys.* **41**, 3863 (1964).
7. S. Castellano and J. S. Waugh, *J. Chem. Phys.* **34**, 295 (1961).
8. B. V. Cheney, private communication.
9. R. J. Abraham and H. J. Berstein, *Can. J. Chem.* **39**, 216 (1961).

10. N. S. Bhacca, D. P. Hollis, L. F. Johnson, E. A. Pier, and J. N. Shoolery, *NMR Spectra Catalog*, Vols. 1 and 2, Varian Associates, Palo Alto, California (1963).
11. J. I. Musher and E. J. Corey, *Tetrahedron* **18**, 79, (1962).

Spectrochemical Applications to Textiles and Fibers

Spectrochemical Elemental Analyses of Textiles and Textile Fibers*

Robert T. O'Connor

Southern Regional Research Laboratory
Southern Utilization Research and Development Division
USDA, ARS, New Orleans, Louisiana

Among the numerous spectroscopic techniques and procedures available for analytical purposes, some ten are methods for the qualitative identification and for the quantitative estimation of the chemical elements, both metallic and nonmetallic. These ten different spectroscopic approaches are described very briefly. Three spectroscopic procedures have been of most importance in elemental analyses of textiles and textile fibers. These three, electronic (ultraviolet or visible) emission (or so-called spectrochemical analysis), x-ray fluorescence, and atomic absorption are described in detail. Applications of these three techniques are illustrated, mainly in the analysis of cotton cellulose fibers, as these are the more familiar analysis in the author's laboratory and the chemical modification of cotton cellulose has created a special need for elemental analyses for a fairly wide number of both metals and nonmetals. However, the techniques described are applicable with little if any modification to similar analyses of any natural or synthetic fiber. They can also be applied without modification to the analysis of yarn or fabric. The three techniques are compared with reference to such factors as, speed of analysis, sensitivity, cost of original equipment, requirements for operator skill, etc.

INTRODUCTION

There is now available to the analytical spectroscopist a very large number of spectroscopic techniques. The various frequencies of the electromagnetic spectrum can be used in many different ways to develop analytical procedures tailored to meet specific requirements.[1] This paper is concerned only with

*Presented at the Symposium entitled "Spectrochemical Applications in the Analysis of Textiles and Textile Fibers" during the 7th International Meeting of the Society for Applied Spectroscopy in Chicago, Illinois, May 12–17, 1968.

procedures for the qualitative identification or quantitative determination of the chemical elements, both metallic and nonmetallic. The numerous spectroscopic techniques or procedures which are, or which could be, used for elemental analyses are listed in Table I.

The first technique listed in Table I, gamma-ray spectroscopy, is a simple and convenient method for the identification or quantitative measurement of radioactive elements. If the material to be analyzed is not radioactive, it can be made radioactive by bombardment with a stream of neutrons, protons, electrons, etc. Thus, gamma-ray spectroscopy is combined, e.g., with neutron activation. This combination of techniques provides a method which is probably the most sensitive of all spectroscopic techniques, capable under favorable conditions of measuring accurately as small a concentration as a fraction of a part per billion, and even in unfavorable circumstances of at least parts per million. Gamma-ray spectroscopy can, of course, be teamed with proton activation, electron activation, etc.

The resonance techniques, listed as III and IV in Table I, are rather unlikely procedures, although obviously the analytical spectroscopist could, by comparison with known standards, identify any element with a nucleus capable of producing a resonance signal by observing the magnetic field and radio frequency at which the specific nucleus gave a resonance signal and, by measuring the intensity of the signal, obtain some degree of quantitative measurement of the identified element. It is not likely that wide-line NMR spectroscopy nor quadrupole-moment NMR spectroscopy will find much application in this area.

X-ray emission, V in Table I, can be used, and indeed has been,

TABLE I
Spectroscopic Techniques Available for Elemental Analysis

I.	Gamma-ray spectroscopy
II.	Neutron absorption
III.	Wide-line NMR
IV.	Quadrupole-moment NMR
V.	X-ray emission
VI.	X-ray fluorescence
VII.	X-ray absorption
VIII.	Ultraviolet atomic emission (Spectrochemical analysis)
IX.	Ultraviolet atomic absorption
X.	Visible atomic emission
XI.	Visible atomic absorption

although the technique is somewhat awkward. X-rays emitted from any x-ray tube are, of course, characteristic of the metal of the target used in constructing the tube. Hence, a qualitative identification of any metal could be obtained by making the analytical sample the target of the x-ray tube. A much more convenient technique is to use the emitted beam of x-rays to produce a fluorescence or secondary emission of x-rays as they strike the analytical sample. The wavelength of the fluorescent beam readily provides a means for the identification of the element. A direct count of the intensity of the beam, compared to the count for a similar period of time from a known standard, or measurement of the time to reach a preselected count, will provide a suitable, accurate, quantitative procedure. X-ray fluorescence is one of the major spectroscopic methods of analysis for metals and, particularly, for nonmetallic elements.

While normally not much thought is given to the absorption of x-rays, this process follows the rules for the absorption of electronic or vibrational frequencies. One main difference, however, is that the absorption of x-rays appears to be independent of the chemical combinations of the elements in a compound. Use of x-ray absorption is therefore a method for elemental analysis. While the method is convenient, can be made highly accurate, and is reasonably sensitive, it has not received a great deal of attention, except in the petroleum industry. A classical example of its use is the determination of tetraethyl lead in gasoline by means of a quantitative measurement of the lead content.

Techniques VIII and X in Table I, ultraviolet and visible atomic emission, together constituting electronic atomic emission, are known more commonly as spectrochemical analysis. This is undoubtedly even today the most widely used spectroscopic method for the identification and quantitative determination of specific chemical elements.

The final two methods, IX and XI in Table I, ultraviolet and visible atomic absorption, are the most rapidly growing in popularity of all spectroscopic techniques today. It is surprising, considering that the absorption of atomic line spectra was understood by the earliest spectroscopists (as evidenced by the fact that this phenomenon was correctly used to explain the Fraunhofer lines in the solar spectrum), that an analytical technique based on atomic absorption was not introduced into the analytical laboratory until relatively recently.

This introduction to the subject has been very general, a deliberate survey of potential spectroscopic analytical techniques without the drawing of very specific boundaries. The rest of this paper will, however, be very specific, and confined to very limited boundaries. These limitations will be imposed by restricting the discussion of spectroscopic methods of analysis to those used in the author's laboratory and to the analysis of cotton cellulose fiber, yarn, or fabric, or to chemically-modified cottons.

There are two reasons for these restrictions. First, while there is considerable and growing evidence that the textile industry is interested in, and is, in fact, using spectroscopic methods for elemental determinations in various fibers and fabrics (as can be seen from activities of appropriate committees of textile societies or from visits to research centers of leading textile companies), there is almost no literature on this subject. With the exception of Russian publications, it is difficult to find any applications of any spectroscopic techniques for the determination of elements in any textile materials. Second, within our laboratory research is confined to investigations of cotton cellulose, and measurements of other natural or of any synthetic fibers or fabrics are solely for purposes of comparison. However, the techniques to be described are applicable to any textile fiber or fabric without further modification. In particular, the preparation of the sample for analysis by any of the three techniques to be discussed is identical for fibers, yarns, and fabrics from cotton or from any other natural or synthetic source.

SPECTROCHEMICAL ANALYSIS

Spectrochemical analysis, the identification of the elements by investigating the atomic lines in the electronic, i.e., ultraviolet and visible regions, of the electromagnetic spectrum produced when the analytical sample is excited as by a flame, arc, or spark, is the oldest of the spectroscopic methods of elemental analysis to attain any significant use. This technique has been used in thousands of laboratories to analyze almost every conceivable type of analytical sample. It is today probably the most widely used technique of all spectroscopic methods for the detection and determination of the various chemical elements. It has been employed in our Division of the US Department of Agriculture's research program for over a quarter of a century, or considerably longer if predecessor agencies are considered. When describing spectrochemical analysis, a distinction must be made between two almost entirely different techniques, based on the types of detectors employed: the direct-reading spectrometers (atom counters and quantometers), and the photographic spectrographs. The applications to cotton cellulose described here are limited to the photographic technique. However, regardless of the detector, the most exacting requirement for quantitative measurements by means of spectrochemical techniques is a uniform matrix. This requirement is too well known to need explanation. It is sufficient to mention that one part per million of copper, e.g., in a matrix of pure silicon dioxide will, under specifically-controlled conditions of excitation, produce atomic lines significantly more intense than the same concentration in a pure sodium chloride matrix. This observed difference is readily explained by the fact that the energy available, either as heat or electrical energy, is not equally distributed among the various elements of the analytical sample. Sodium will rob more than its share of energy from copper, while copper in turn will take from silicon its proportional share of the energy used to cause excitation and emission of atomic lines.

The requirement that the matrix for both standards and analytical samples must be kept constant is thus of primary importance in spectrochemical analyses. In major operations, as in routine analyses in a steel mill, this requirement is readily attained, as the matrix is always iron, the analyses being conducted to follow the concentration of alloying elements in this constant matrix. However, in an agricultural research laboratory successive samples may involve appreciably different matrices, necessitating different standards and numerous calibrations.

This difficulty has been overcome by taking advantage of the fact that agricultural samples are organic materials (cotton cellulose, fiber or fabric, is an ideal example) and must be ashed; otherwise, they would burn too rapidly in an arc or spark to permit measurement of any atomic line. The ashing is accomplished in the presence of magnesium nitrate. The magnesium nitrate serves four distinct purposes: (1) it acts as an ashing aid, reducing the temperature during ashing of the sample; (2) it acts as a carrier for the small quantity of ash which often results from the ashing and permits analysis of very small samples; (3) it acts as a spectroscopic buffer; it becomes the constant matrix and the analytical problem is reduced to a determination of the selected elements in a matrix of magnesium oxide–magnesium carbonate; and (4) the magnesium provides the required constant internal standard lines.

The direct-reading instruments undoubtedly permit much more rapid analyses. However, in the analysis of organic materials, such as the cotton fibers or fabrics with which we are dealing, ashing is an essential step, which so greatly overshadows the time requirements for the remainder of the analysis that the selection of the detector is no longer of primary importance. Undoubtedly, the greatest advantages of the direct-reading instruments are greater precision and accuracy. The precision or accuracy of the quantometer can be approached by the photographic techniques only if the sample is analyzed in multiple analyses. The sensitivities of the two techniques are probably quite comparable. The photographic technique may, especially in a research laboratory where requirements and types of analyses are constantly changing, have a slight edge over the direct-reading instruments, although the use of the multichannel instruments, which permit detection and determination of a sufficiently large number of elements, limits this advantage of the

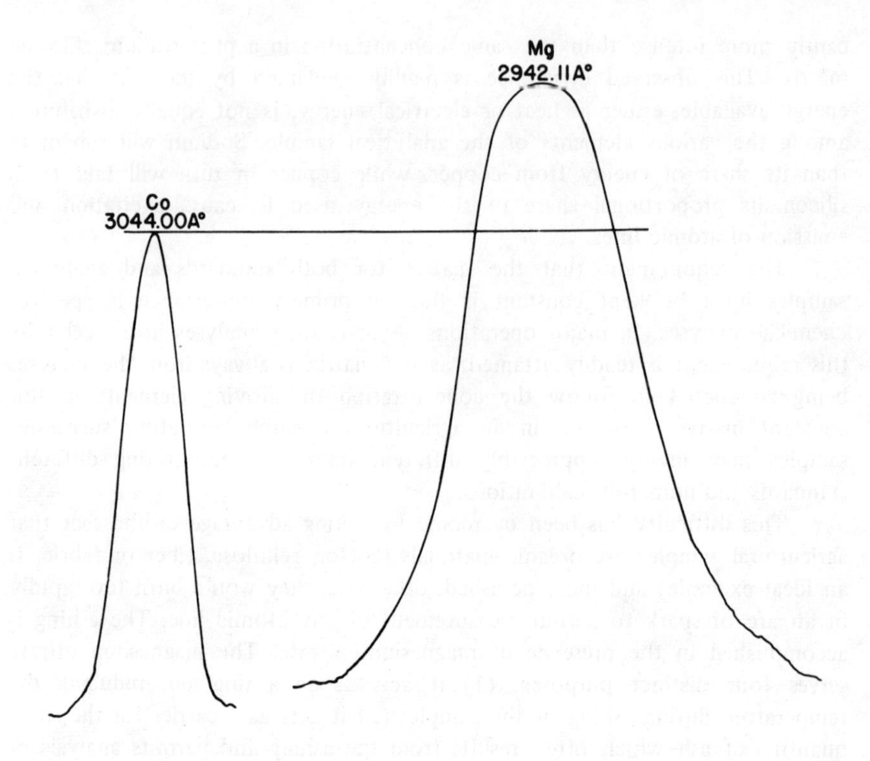

Fig. 1. Microphotometer tracing of cobalt analytical line and magnesium control line. Width of control line is measured.

photographic technique to comparisons with relatively simple or small direct-reading instruments.

The general technique of reducing the sample to a magnesium oxide–carbonate containing the metals to be determined has been combined with the little-used "linewidth" method of evaluation.[2] This method has never been widely used, probably because it was introduced at about the time most active spectrochemical laboratories were converting to direct-reading instruments. The method is based on the fact that the width of a spectral line measured through a constant external slit is directly proportional to the concentration of the element. The linewidth method is illustrated in Figs. 1 and 2. The width of the internal standard, a magnesium line relatively close to the selected analytical line, is measured at the peak of the analytical line (Fig. 1) or, if convenient, in a reverse manner (Fig. 2) the width of the analytical line is measured at the peak height of the magnesium line. A calibration or working curve is obtained by measurements from a series of magnesium nitrate standards containing known amounts of the specific elements to be determined and ashed in exactly the same manner as the cotton analytical samples. One method (Fig. 1) will produce a calibration curve with a negative slope; the second (Fig. 2), a curve with a positive slope. Typical calibration curves are illustrated in Fig. 3.

Results of these types of analysis are summarized in Tables II and III. In Table II the ash, copper, iron, and manganese content of eight varieties of cotton fiber grown in a single location (Stoneville, Mississippi) and hand-picked from bolls just prior to opening to avoid any field of dust contamination are shown. The results illustrate that even when grown under identical conditions, the genetic character has a marked effect on the metal content. In Table III, results of analyses of a single variety of cotton fiber (Coker 100 Wilt) grown in thirteen cotton-growing regions of the country are compared. Although not evident from the data shown, the copper, iron, and manganese contents of a specific variety of cotton reflect the concentrations of these metals in the soil in which it has been grown. The reproducibility of the

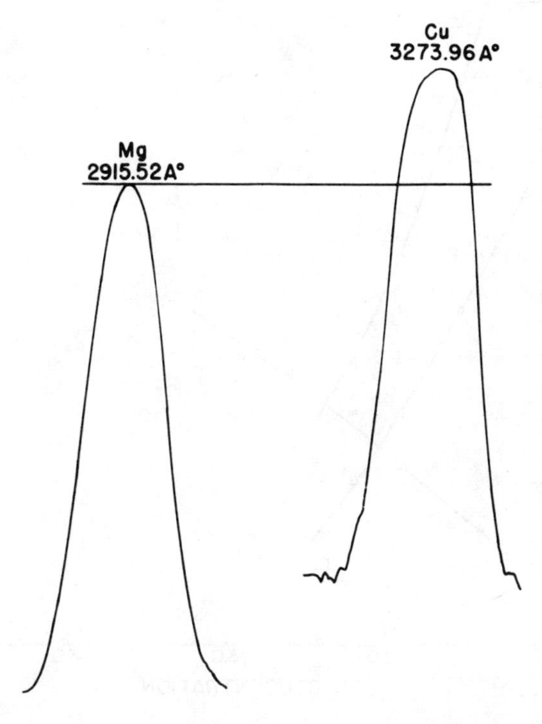

Fig. 2. Microphotometer tracing of copper analytical line and magnesium control line. Width of analytical line is measured.

spectrochemical method as outlined is shown in Table IV by the repeated analysis of a cotton fabric for its nickel content. The usual analytical recovery tests and accuracy tests based on the analysis of synthetic ash of known metal content are shown in Tables V and VI. These results are from single analyses. The precision and accuracy of spectrochemical analyses can be most easily improved by multiple analyses of the same sample. It is possible to test individual steps in the complete analytical procedure, and this has been done. First, the same spectrogram on a spectrographic plate is continually evaluated by the linewidth method. Then spectrograms from the same ashed sample on

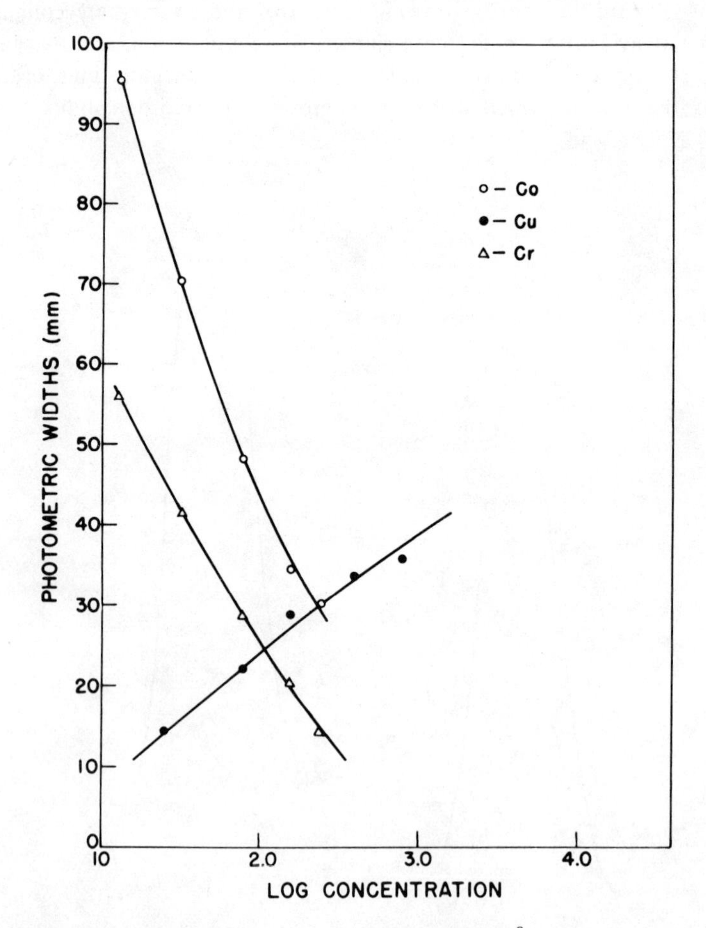

Fig. 3. Working (or analytical) curves for: o: Co (3044.00 Å). •: Cu (3273.96 Å). Δ: Cr (2835,63 Å).

TABLE II

Spectrochemical Analyses for Cu, Fe, and Mn in Varieties of Cotton Fiber from a Single Location

Location	Ash (%)	Cu (ppm)	Fe (ppm)	Mn (ppm)
Acala 4-42	1.42	1.99	11.55	7.22
Acala 1517	1.41	1.60	12.19	7.21
Coker 100 Wilt	1.13	1.59	11.14	7.46
Deltapine 15	1.09	1.40	7.90	5.90
Mebane (Watson)	1.36	1.01	5.91	3.18
Rowden 41B	1.28	1.74	10.84	5.36
Stoneville 2B	1.19	1.53	12.55	6.28
Wilds	1.26	1.00	7.97	4.91
Average	1.27	1.48	10.01	5.94

TABLE III

Spectrochemical Analyses for Cu, Fe, and Mn of a Single Variety of Cotton (Coker 100 Wilt) Grown in Various Locations

Location	Ash (%)	Cu (ppm)	Fe (ppm)	Mn (ppm)
Auburn, Ala.	1.56	0.78	9.03	8.90
Sacaton, Ariz.	1.49	2.72	12.89	12.39
Shafter, Calif.	1.35	0.93	5.60	5.81
Tifton, Ga.	1.26	1.67	7.91	9.30
St. Joseph, La.	1.33	0.68	3.74	2.79
Stoneville, Miss.	1.13	1.59	11.14	7.46
Statesville, N. C.	1.53	1.55	6.88	7.44
State College, N. M.	1.56	1.05	4.85	5.82
Chickasha, Okla.	1.37	1.13	7.25	7.87
Florence, S. C.	1.26	2.10	7.29	9.08
Jackson, Tenn.	1.22	1.45	7.25	9.20
College Station, Tex.	1.26	4.01	13.77	8.17
Greenville, Tex.	1.22	1.95	7.03	8.42
Average	1.37	1.66	8.05	7.90

TABLE IV

Precision of Spectrochemical Method Determination of Nickel in a Modified Cotton Fabric

Spectrum	% Metal		% Deviation
1	0.12		5.26
2	0.11		3.51
3	0.11		3.51
4	0.11		3.51
5	0.13		14.04
6	0.12		5.26
7	0.11		3.51
8	0.10		12.28
Average	0.114	Average deviation	6.36

Standard deviation = ± 0.009
Coefficient of variation (%) = 7.9

TABLE V

Spectrochemical Analyses for Cu, Fe, and, Mn. Recovery Tests

Metal added to cotton	Amount originally present (ppm)	Total amount present (ppm)	Amount recovered (ppm)	Deviation (%)
Cu	0.96	2.96	2.87	−3.0
	1.44	3.44	2.52	−26.7
	0.99	2.99	2.75	−8.0
Fe	5.60	9.60	9.30	−3.1
	7.47	11.47	12.52	+9.2
	6.87	10.87	13.73	+26.3
Mn	6.82	10.82	13.03	+20.4
	3.02	7.02	7.91	+12.7
	8.55	12.55	12.61	+0.5

TABLE VI

Spectrochemical Analysis for Cu, Fe, and Mn. Analysis of Known Standards

Synthetic ash	Amount in standard (ppm)	Amount Determined (ppm)	Deviation (%)
Cu	40	42.7	+6.8
	100	85.1	−14.9
	200	214	+7.0
Fe	400	468	+17.0
	1000	977	−2.3
	1600	1860	+16.3
Mn	210	214	+1.9
	841	871	+3.6
	2000	1740	−13.0

a single photographic plate are compared. The same ash is repeatedly photographed on a series of plates, and the same sample is repeatedly ashed, etc. In this manner it can be shown that each step in the analysis can probably be made precise to within about ±2 to 3%. If in a specific analysis these errors are all in the same direction, a total deviation of as much as 20% may result. However, if all analyses are conducted in triplicate or quadruplicate, the error can be reduced as the changes for the specific errors in each step will tend more and more to cancel each other. With three or four analyses of each sample the deviation from the known amount can be reduced to ±10%, an accuracy sufficient for most of the analyses for metals present in amounts of 10 ppm or less.

The types of analysis described, while they illustrate the application of the spectrochemical analysis procedure to the determination of elements found in native cotton cellulose, do not, of course, constitute any considerable activity in textile research laboratories. The chemical modification of cotton, its treatment with various reagents to impart properties not native to cellulose, has, however, created an additional need, or opportunity, for applications of spectroscopic techniques for elemental analyses. It is probably well known to most spectroscopists that infrared absorption spectroscopy has been of considerable assistance to the cotton cellulose chemist in attempts either to identify the manner in which a specific sample of cotton fiber or fabric has been modified or to determine the extent of the chemical modification. A review of this subject is given in a subsequent paper in this symposium.

The sensitivity of infrared absorption spectroscopy is, however, often

TABLE VII

Element Determined by Spectrochemical Analysis of Chemically-Modified Cotton

Element	Analytical line wavelength (Å)	Magnesium control-line wavelength (Å)	Range of concentration from typical analyses (%)	Use
Al	3082.16	2942.11	0.018 – 0.038	Oxide: Soil resistance Hydroxide: Water repellent 8-quinolinolate: Resistance to microorganisms
Bi	3067.72	3073.99	0.51 – 0.56	Oxide: Flameproofing Hydroxide: Water repellent, weatherproofing
Cd	3261.06 3261.06	2915.52 2942.11	0.037 – 0.36	Chloride: Resistance to termites and microorganisms Pentachlorophenate: Rotproofing Hydroxide: Flame and weatherproofing
Cr	2835.63 3197.08	2942.11 2942.11	0.003 – 2.61	Cr complex perfluoroctanoic acid; water and oil repellency Oxide: Rot- and weatherproofing, fungicide and bactericide protection against light Lead Chromate: Weatherproofing, flame retardant Potassium dichromate: Resistance to termites and microorganisms
Co	3044.00	2942.11	0.10 – 0.17	Hydroxide: Mildew-, rot-, and general weatherproofing Metaborate: Mildew-, rot-, and general weatherproofing

Cu	2961.16	2942.11	0.00005 – 2.02	Naphthenate: Rotproofing, fungicide, mildew-proofing
	3273.96	2915.52		8-quinolinolate: Water-, rot-, mildew-, and weatherproofing
	3273.96	2942.11		Phosphate: Waterproofing
				Carbonate: Fungicide and bactericide
Fe	3017.63	2942.11	0.17 – 4.34	Hydroxide: Fungicide and bactericide
	3020.64	2942.11		Oxide: Rot- and weatherproofing, protection against light
	3057.45	2942.11		Phosphate: Waterproofing
Mn	2576.10	2942.11	0.030 – 8.04	Oxide: Protection against actinic degradation and waterproofing
	3054.36	2942.11		Fluosilicate: Catalyst in stabilization against shrinkage
Ni	2943.91	2942.11	0.13 – 3.16	Acetate: Water repellency
	3002.49	2942.11		Oxide: Water repellency
				Phosphate: Weatherproofing and protection against actinic degradation.
Sn	2571.59	2942.11	0.002 – 2.14	Oxide: Weatherproofing and protection against action of light; flameproofing and mildewproofing
	3175.02	2942.11		Phosphate: Weatherproofing, protection against actinic degradation; fire resistance
				Bis (tri-n-butyl tin) oxide: Antibacterial agent
Zn	3345.02	2942.11	0.030 – 0.24	Oxide: Flame retardant
	4722.16	5528.46		Chloride: Flame retardant
				Phosphate: Fire and weather resistance
				Nitrate: Crease resistance, wrinkle resistance (catalyst)

not sufficient for adequate quantitative determination of the small amounts of reagents used in many of these chemical modifications to produce cotton resistant to mildew, rot, soiling, actinic degradation, and flame, or to impart by crosslinking the cellulose molecules such properties as permanent creasing and wrinkle resistance. Many of these reagents contain elements which lend themselves to spectrochemical analysis. Several of them are organometallic compounds, and many others are inorganic salts and oxides. Once a modification of this type has been identified, as by infrared absorption spectroscopy, it is a relatively simple procedure to obtain a reasonably accurate determination of the elemental content, and to calculate the concentration of the organometallic compound or the extent of modification with the inorganic salt or oxide.

In Table VII, eleven elements commonly found in chemically-modified cottons are listed, together with the spectral lines used for their determination, the concentration range in which they are most commonly found in chemically-modified cotton, and, particularly, the form in which these elements are used and the properties imparted to the cellulose by their presence. The data in Table VII illustrate the type of modifications that have created a considerable interest in spectroscopic methods of elemental analysis by the cellulose chemist. They have created the need for spectroscopic methods, of one type or another, by the textile research laboratories which deal with cotton fabrics either as pure cotton fibers or in blends with the numerous synthetic fibers.

X-RAY FLUORESCENCE

The second spectroscopic technique used in our laboratory is x-ray fluorescence, or secondary emission, for the detection and quantitative determination of elements, both metals and nonmetals. Use of this technique, compared to spectrochemical analysis, is relatively recent. While we have applied emission methods to the analysis of cotton cellulose for over 25 years, x-ray fluorescence has been used for only about five.

The x-ray instrumentation used in all the experiments to be described is a General Electric XRD-5 diffractometer modified for fluorescence analysis.* Radiation was obtained from tungsten, molybdenum, or chromium tubes operated at 50k. For detection or determination of the heavier elements emitted radiation of short wavelengths was used with a LiF crystal and xenon-filled proportional counter. For lighter elements long wavelengths were

*Use of a company or product named by the Department does not imply approval or recommendation of the product to the exclusion of others which may also be suitable.

used with an ethylenediamine tartrate (EDT) or with a potassium acid phthalate (KAP) analyzing crystal. Both of these crystals were used with gas-flow (90% argon, 10% methane) proportional counters and with a helium path.

X-ray fluorescence spectra to permit qualitative identification of any element can be obtained from the cotton cellulose in almost any physical form. Fiber can be measured as a fiber bundle, fabric as a small square of the cloth. However, for best quantitative results standardization of the particle size of the textile material is highly desirable. The textile material, in the form of fiber, yarn, or fabric is ground to pass a 20-mesh screen in an intermediate Wiley mill. A disk of 318 mg of the ground material is pressed in a 1-in. die under a pressure of 25 tons for 3 min. The resulting disks have good mechanical stability, giving reproducible counts over a period of several months. Working or calibration curves are prepared from disks made by mixing solutions containing appropriate known quantities of the desired element with the ground cotton, drying, agitating any loose material to ensure homogeneous distribution, and pressing in the die. If the standard material cannot be readily dissolved, it is finely ground and intimately mixed with the ground cotton.

Count rates of equivalent disks usually showed a maximum difference of 2%. The relationship of percentage composition to counting rate was found to be linear over a range of 0.01 to 3% for most of the elements examined. In all cases a loss in counting rate was noted at concentrations above 3%, indicating self-absorption. This loss of counts became marked at concentrations of about 10%, and dilution with pure cellulose should be employed.

Conditions used for the x-ray fluorescence of seventeen of the heavier elements, both metallic and nonmetallic, are summarized in Table VIII. In all of these measurements maximum efficiency was obtained with the LiF analyzing crystal. Tungsten radiation was used for examination of all except two of the elements listed. Chromium radiation was used for tungsten excitation and molybdenum radiation for detection or determination of selenium because tungsten radiation specifically interferes at the Se K a line region.

In Table VIII the concentration range for the quantitative determination is given as from 0.02%, about the lower level of detection, to 3.0%, the level at which self-absorption becomes apparent. In this table the objective to be obtained by treatment of the cotton with an organometallic compound or with an inorganic salt or oxide containing the determined element is indicated. Typical analyses are illustrated in Table IX, where values obtained by the x-ray fluorescent method for zirconium, copper, and silver are tabulated.

Conditions selected for detection or quantitative determination of elements lighter than titanium (atomic-number 22) are tabulated in Table X. The range of concentration for most determinations is between 0.03 and

TABLE VIII
Conditions for X-Ray Fluorescence Measurement of Heavier Elements in Modified Cottons

Element	Function	Analytical wavelength Line 2θ	°(LiF crystal)	Concentration range (%)
Ag	Fungicide	*Ka*	16.1	0.03–3
Br	Flame resistance	*Ka*	30.0	0.03–3
Cd	Fungicide	*Ka*	15.3	0.03–3
Co	Fungicide	*Ka*	52.8	0.03–3
Cr	Pigment	*Ka*	69.4	0.03
Cu	Fungicide	*Ka*	45.0	0.03–3
Fe	Pigment	*Ka*	57.5	0.03–3
Hg	Fungicide	*La*	35.9	0.03–3
Pb	Outdoor protection	*Lβ*	28.2	0.02–3
Sb	Flame resistance	*Ka*	13.5	0.05–3
Se	Fungicide	*Ka* (Mo)	31.9	0.1–10
Sn	Flame resistance	*Ka*	14.0	0.05–3
Te	Fungicide	*La*	109.5	0.05–3
Ti	Pigment	*Ka*	86.1	0.05–3
W	Fungicide	*La* (Cr)	43.0	0.03–3
Zn	Catalyst	*Ka*	41.8	0.03–3
Zr	Water repellency	*Ka*	22.6	0.05–3

TABLE IX
Heavy-Element Content of Chemically-Modified Cottons by X-Ray Fluorescence

Sample	Zirconium (%)	Copper (%)	Silver (%)
B-1	2.06	0.35	0.14
B-2	1.13	0.19	0.08
B-3	0.73	0.12	0.05
B-4	0.36	0.06	0.03

TABLE X
Summary of Conditions of X-Ray Fluorescence Analysis of Lighter Elements in Modified Cottons

Element	Function	Analytical wavelength		Concentration Range (%)
		Line 2θ	°(EDT crystal)	
Al	–	*Ka*	36.8*	0.1–3
Ca	–	*Ka*	44.9	0.03–3
Cl	Flame resistance	*Ka*	65.0	0.03–3
K	–	*Ka*	50.3	0.03–3
Mg	–	*Ka*	43.9	0.1–3
S	Crosslink	*Ka*	75.2	0.03–3
Si	Water repellency	*Ka*	108.1	0.1–3
P	Flame resistance	*Ka*	88.7	0.03–3

*KAP crystal.

3.0%. The lower limit for three lightest elements determined, aluminum, magnesium, and silicon, is 0.1%. Without a vacuum spectrometer and a special analyzing crystal such as lead stearate, magnesium is the lightest element which can be determined at these concentration levels. Again in Table X the purpose for treatment of the cotton with an organic complex or an inorganic compound is indicated. Typical analyses for three elements, phosphorus, sulfur, and chlorine, in chemically-modified cotton fabrics are given in Table XI. In this table results from the x-ray fluorescent technique are compared with the more or less classical chemical methods. X-ray results for phosphorus agree quite well with values from chemical analyses by the reduced molybdate colorimetric method. Sulfur contents by x-ray analyses are uniformly higher than values obtained by the Schoniger oxygen flash technique, and chlorine values by chemical and x-ray determination are in good agreement.

Cadmium and selenium in the form of cadmium selenide (CdSe) or cadmium sulfoselenide (CdSSe) have been used in the treatment of cotton to produce resistance to weather exposure and mildew. Obviously, the useful life of a fabric treated to impart these special properties will depend primarily upon the retention of the cadmium and/or selenium during normal use. The analytical problem is merely the determination of cadmium and selenium in decreasingly smaller amounts as a function of exposure time. In Table XII the cadmium and selenium content is charted as a per cent of these two elements in the originally-treated cotton material over a period of two years. These data reveal a more or less steady loss as a function of exposure time. The

TABLE XI
Light-Element Content of Chemically-Modified Cottons

Phosphorus (%)		Sulfur (%)		Chlorine (%)	
Chem.	X-ray	Chem.	X-ray	Chem.	X-ray
Nil	0.03	Nil	0.02	1.15	1.36
2.37	2.35	0.62	0.74	1.42	1.22
2.86	2.90	0.72	0.90	–	–
2.62	2.60	0.72	0.90	1.31	1.17
2.53	2.50	–	–	1.38	1.30

TABLE XII
Relative Cadmium and Selenium Contents of Weather-Exposed Treated Cotton Fabrics

	CdSe* % of original			CdSSe† % of original		
Months of exposure	Cd	Se	Breaking strength	Cd	Se	Breaking strength
0	100	100	100	100	100	100
4	83	88	87	87	86	93
8	64	80	77	74	73	84
12	39	66	54	21	58	67
16	30	52	38	8	40	50
20	23	49	32	–	31	45
24	17	47	23	–	24	32

*Original cadmium and selenium contents 0.63% and 0.37%, respectively.

†Original cadmium and selenium contents 0.50% and 0.25%, respectively.

usefulness of the fabric can be repeatedly checked by the simple textile test for breaking strength using an Instron tester. From the data in Table XII decrease in breaking strength, i.e., deterioration of the cotton fabric, is shown to follow very closely the per cent loss of cadmium and selenium as determined by the x-ray fluorescent technique.

ATOMIC ABSORPTION

We have only recently initiated investigations of the identification and quantitative determination of elements by means of atomic absorption. These investigations have been instituted because of the continually growing interest in this technique by the textile, and, incidentally, by other, industries. The Technical Association of the Pulp and Paper Industry has been investigating atomic absorption methods for the determination of metals in textile materials in combination with the American Society for Testing and Materials, Committee D-23 on Cellulose and Cellulose Derivatives. The American Society for Testing and Materials Committee D-13 on Textiles and Textile Materials has been cooperating with ASTM Committee E-2 on Emission Spectroscopy, and these two groups may be working with other textile groups.

In our laboratory first interest in atomic absorption was in connection with the byproduct of cotton, cottonseed oil. The chemists dealing with lipid and fatty acid in the vegetable and animal oil industry have evidenced considerable interest in the potentials of this technique as one which can be brought into the processing plants.

Our experiments with atomic absorption in the analyses of cotton cellulose have been carried out to a sufficient degree to assure us that this method can produce satisfactory results, but are as yet insufficient to completely or properly evaluate specific techniques. We have leaned strongly on experience gained in the above-mentioned collaborative studies.

For atomic absorption techniques the sample must be in solution, and cotton is insoluble in any suitable solvent. Thus recourse must be had to ashing and, like emission analysis, atomic absorption suffers from the fact that the preparation of sample, due to this requirement for the preparation of a suitable ash, requires an appreciable amount of time.

A sulfate ashing is used. Five grams accurately weighed into a platinum dish are wetted with 1 *M* sulfuric acid and ashed in a muffle furnace at 575°C until all carbonaceous material is destroyed. The ashing time is usually about 3 hr. The analytical sample is removed from the furnace, cooled, and dissolved in 10 ml of 2 *M* hydrochloric acid, warmed slightly if necessary, and brought to volume in a 50-ml volumetric flask with distilled water. If undissolved matter is present, this solution is filtered before analysis.

TABLE XIII
Determination of Trace Metals in Cellulose by Atomic Absorption

Sample	Trace metal (ppm)								
	Ca	Cu	Fe	K	Mg	Mn	Na	Pb	Zn
1	310	2.35	6.5	10.5	219	34.4	27.0	11.7	1.70
2	17.4	0.32	1.6	11.0	5.1	0.15	222	13.0	0.53
3	413	1.11	3.6	10.5	100	0.53	59.0	12.4	0.85
4	384	1.48	2.6	14.0	106	0.53	163	13.0	0.74

In the experiments described in this paper a Perkin-Elmer atomic absorption instrument model 303 equipped with a Digital Concentration Readout Accessory (DCR-I) was used. The line sources of the emission spectrum were hollow-cathode lamps of the specific element being analyzed, operating at the specific current for which each lamp is rated. The sample was atomized and burned in a Boiling Burner Head using the air–acetylene oxidizer fuel system. The results obtained were printed out on tape by the DCR-I.

In Table XIII some random analyses for nine different metals in cellulose by means of atomic absorption are shown. These data merely illustrate that the technique appears to be satisfactory. We have relied heavily on direct comparison of metal content by other spectroscopic techniques, especially x-ray fluorescence, in the limited evaluation of atomic absorption methods which have been made.

COMPARISON OF THREE SPECTROSCOPIC METHODS

In Table XIV repeated determinations of zinc content by atomic absorption spectrochemical emission and x-ray fluorescence techniques are compared. The sample was a cotton fabric treated with dimethylol ethyleneurea and varying amounts of zinc salts as catalyst to form intermolecular crosslinks between cellulose chains to produce a wrinkle-resistant or durable-press fabric. These types of comparison show that each of the three spectroscopic techniques are capable of giving, within relatively small differences, the same values. It is our conclusion that any laboratory working with one of the three techniques exclusively could sharpen up its technique to a point where results would be as reliable as could be obtained by either of the other two techniques. In other words, selection of the technique need not, probably cannot, be based on demonstrated superior results.

There are, however, several factors which can, and indeed do, influence the selection of a specific technique. These factors, which will have considerable influence in the selection of techniques during the presently expanding use of spectroscopic methods for the detection or quantitative determination

TABLE XIV
Zinc Content of Chemically-Modified Cotton by Three Analytical Methods

	Zn content (%)		
Sample no.	Atomic absorption	Emission	X-ray fluorescence*
1	0.06	0.04	0.07
2	0.06	0.06	0.08
3	0.06	0.04	0.07
4	0.18	0.08	0.22
5	0.03	0.07	0.04
6	0.09	0.08	0.10
7	0.03	0.06	0.04
8	0.58	0.88	0.72
9	0.87	0.89	0.80

*ZN Kα line, LiF crystal, tungsten target, 1 mA current.

TABLE XV
Relative Criteria for Selection of Spectroscopic Techniques for Elemental Analyses of Cotton Cellulose

	Emission analysis* "Spectrochemical analysis"	X-ray fluorescence	Atomic absorption
Sample preparation	Slow–involves ashing	Vest fast–negligible	Slow–involves ashing
Instrumental analysis	Fast	Very fast	Reasonably fast
Operator requirements	Highly skilled	Skilled	Moderately skilled
Initial cost	Expensive	Very expensive	Low
Other factors	Exhaust system, darkroom required; Controlled atmosphere highly desirable	Controlled atmosphere desirable	Exhaust system

*Based on photographic technique.

of elements in fibers, fabrics, and textile materials, are summarized in Table XV.

Spectrochemical analysis has the disadvantage that the analytical sample must be ashed (remember we are here referring only to the analysis of fibers, fabrics, and textile materials). This requirement for ashing is an essential part of a spectrochemical technique whether we are considering the photographic technique as it has been described, or whether the direct-reading instrumentation is used. In either case the ashing step dominates the time required for the complete analysis. The overall analysis must therefore be rated as relatively slow. The instrumentation for spectrochemical analysis must be considered as somewhat sophisticated and relatively expensive, although the actual cost of initial equipment varies over a considerable range. The technique requires a somewhat highly-skilled operator, and an appreciable period of training is required before one is sufficiently skilled in identifying specific atomic lines, particularly in materials which may contain a myriad of different metals. Another negative factor in the selection of spectrochemical analysis is that, unless the much more highly sophisticated vacuum instruments are considered, it is limited principally to metals. This is an important consideration to the textile chemist, for as illustrated in the example cited above, very often he is mainly interested in the nonmetallic elements chlorine, bromine, sulfur, phosphorus, tellurium, or the amphoteric elements selenium, antimony, etc. These are the elements which either cannot be determined by emission spectroscopy, or whose analysis is limited to relatively high quantities, as the sensitivity of the method for such elements is not very great.

The advantages of spectrochemical analysis are, primarily, its ability to permit a single scan of a considerable portion of the entire spectrum in a single spectrogram. If the photographic technique is being employed, such a spectrogram permits the operator to investigate in detail the presence or absence of specific metals. We have confined our use of spectrochemical analysis in the last few years primarily to such qualitative surveys, using x-ray fluorescence, or more recently atomic absorption, as the quantitative tools. The photographic spectrochemical-analysis technique is probably the simplest and most direct answer to the question as it is often submitted to the analytical spectroscopists, not "Does the analytical sample, a chemically-treated cotton, contain Cu or Cr," but, "What elements are present in the sample?" For this latter task the emission spectrogram is ideal, although the limitation that nonmetallic elements will probably not be included in such a qualitative analysis must be kept in mind. The spectrochemical analysis is probably the most sensitive for the determination of basic elements and the true metals. High sensitivity can be obtained by taking advantage of the integration, i.e., longer exposures with photographic instruments or longer counts with direct reading instruments, by analyzing larger samples. In this respect spectrochemical analysis has superior sensitivity to x-ray fluorescence methods for the basic elements and true metals. Whether atomic absorption can be made as sensitive has not, as far as can be ascertained, been clearly established for these types of analytical samples. We do know, however, that solutions used in this latter technique cannot be made too concentrated without incurring problems during atomization into the flame.

X-ray fluorescence is the only one of the three techniques considered that does not require a preliminary ashing of the analytical sample. It is therefore superior in overall time of analysis. For qualitative identification, particularly for a specific element, a piece of fabric can be analyzed while a muffle furnace is being brought to temperature for the preliminary ashing required by either spectrochemical analysis or atomic absorption. Even if a precise quantitative analysis is required, the little time needed to reduce the fiber or fabric sample to the necessary constant particle size by grinding is negligible compared to the ashing techniques of the other two procedures. This overall superior time for analysis is the outstanding advantage of the x-ray fluorescence method. However, it does have other advantages. It has been, in our experience, the most reproducible and very probably the most accurate method when compared with photographic spectrochemical analysis or with atomic absorption. Very probably, this superiority would not continue if comparisons were made with direct-reading spectrographs. Another outstanding advantage of the x-ray fluorescence technique is the ability to detect and determine nonmetallic elements of considerable interest, in several applications, to the textile chemist, including, particularly, the acid-forming elements, such as chlorine, sulfur, and bromine.

A limitation of x-ray fluorescence analysis is that light elements may escape detection or accurate measurement. However, techniques and selection of such accessories as x-ray tube, analyzing crystal, and detector have pushed these analyses down to magnesium (atomic number 12), and, as mentioned earlier, with special vacuum spectrophotometers and specially-designed analyzing crystals, such as the lead stearate, analyses are being made down to boron (atomic number 5), leaving very few elements which cannot be detected and determined by this technique. These latter determinations between magnesium and boron can be made, however, only by a significant increase in instrumental sophistication and expense.

The disadvantages of x-ray fluorescence are the sophistication and particularly the cost of the initial equipment. An adequate x-ray spectrometer is expensive, relative to the equipment used in other techniques, and to properly operate an x-ray spectrometer and adequately interpret and analyze the data obtained from this instrument, a technician with above-average training is required.

Atomic absorption requires both ashing of the sample and solution of the resulting ash. In overall time of analysis it, therefore, cannot compete with x ray fluorescence. There is some concern as to whether in all analyses, techniques used to obtain the complete solution, which is required for satisfactory aspiration into the flame, can be made without losing some of the elements to be determined. Tests made primarily by comparisons with the other techniques indicated that losses of this sort, if they occur, must be negligible; but for any specific material the question as to whether or not a significant amount of a specific element is being removed during filtration of the solution by adsorption onto the insoluble material must be raised.

The sensitivity of atomic absorption should approach that of spectrochemical analysis, as, in principle, it involves measurement of the absorption of the same atomic lines that are measured in spectrochemical analyses by direct emission. However, the limitation on the concentration of the solutions, without causing problems in atomization due to high viscosity, may mean that it will be somewhat intermediate between spectrochemical analysis and x-ray fluorescence techniques. The range of elements which can be detected by atomic absorption might again, and for the same reasons, be considered as equal to spectrochemical analysis. However, if comparisons are limited to atomic absorption and the photographic techniques of spectrochemical analysis, additional elements can probably be handled satisfactorily by the atomic absorption technique, as those elements whose strongest lines lie just below the limits of detection by photographic plates can readily be measured by direct counters. The range of elements which can be satisfactorily analyzed by atomic absorption is very probably about the same as can be handled by direct-reading spectrometers. X-ray fluorescence, therefore, has a decided advantage in the determination of acid-forming elements, e.g., chlorine, bromine, sulfur, etc.

The outstanding advantage of atomic absorption is that the instrumentation is relatively inexpensive and unsophisticated. A laboratory can get into operation for analysis by atomic absorption with an initial cost of probably less than 10% of that required for either of the other two techniques. Furthermore, a qualified technician can be trained to operate an atomic absorption spectrophotometer in a sufficiently skilled manner with only minimum training. These two advantages undoubtedly overshadow the advantages of either spectrochemical analysis or x-ray fluorescence techniques or the disadvantages of atomic absorption.

In Table XV the relative criteria for the selection of spectroscopic techniques for elemental analyses of cotton cellulose (and these comparisons could readily be extended to textile materials in general) are compared. It has not been the aim in this comparison to arrive at a firm recommendation of a technique which should be selected by any research laboratory within the textile industry. Three conclusions can, however, probably be made with a fair degree of certainty:

1. If in a textile laboratory all three techniques are available, the personnel within that laboratory will find occasions where, for a specific analysis, one of the three will be preferred over the other two, and other cases where another technique will be the preferred one.

2. Selection of the technique to be used, particularly if selection is to be limited to a single spectroscopic technique, will depend upon the operations of the specific laboratory. If, e.g., the analytical spectroscopist is being called upon to identify the elements present in a specific fabric, atomic absorption would be an awkward manner to continually perform such qualitative analyses. Emission analysis, particularly with the use of photographic plates, would be much more preferred. If, however, the laboratory is continually asked to follow the concentration of a half dozen, dozen, or dozen and a half specific elements by quantitative analyses, atomic absorption might be the choice.

3. The relative simplicity of atomic absorption, the feeling that many laboratories have that it does not require as sophisticated instrumentation or personnel, and the relatively inexpensive requirements for initial installation are going to result in considerable examination of this type of analysis in textile laboratories (and indeed in other laboratories, of which the chemist dealing with lipids or fatty acids in vegetable or animal oil processing laboratories has been cited as an example). In the immediate future (in fact, this trend is already underway) we are going to see atomic absorption spectrometers in many laboratories where spectroscopic techniques, especially spectroscopic techniques involving atomic spectroscopy, have heretofore been unknown.

REFERENCES

1. R.T. O'Connor, *J. Assoc. Offic. Anal. Chem.* **51**: 233-259 (1968).
2. R.T. O'Connor, D.C. Heinzelman, and M.E. Jefferson, *J. Am. Oil. Chem. Soc.* **25**(11): 408-414 (1948).

The Applications of Infrared Spectroscopy in the Investigations of Cotton*

Elizabeth R. McCall and Nancy M. Morris

Southern Regional Research Laboratory
Southern Utilization Research and Development Division
ARS, USDA, New Orleans, Louisiana

With the development of the potassium bromide disk technique infrared absorption spectroscopy has become an increasingly important analytical tool for the cellulose chemist. No simple chemical method exists for evaluating the changes in molecular structure which occur with the chemical treatment of cotton. Infrared spectroscopy offers a means for obtaining such information through the identification of new functional groups which are present after modification, and provides information concerning changes in crystallinity and polymorphic form. Infrared data may also be applied to the detection and quantitative determination of cotton in blends with other fibres. Methods of analysis such as hydrolysis, differential spectroscopy, and multiple internal reflectance have added to the usefulness of this tool in the examination of cotton and modified cotton. Qualitative and quantitative analytical procedures using the infrared absorption spectra of cellulose are described and examples are presented of the applications of these techniques.

INTRODUCTION

Infrared absorption spectroscopy is widely used in the textile industry today to identify fibers and to ascertain the type, and in some instances the extent,

*Presented as an invited paper at a Symposium entitled "Spectrochemical Applications in the Analysis of Textiles and Textile Fibers" during the Seventh National Meeting of the Society for Applied Spectroscopy, Chicago, Illinois, May 12–17, 1968.

of chemical modification of the fiber or fabric. Application of the infrared method of analysis to textile chemistry began about 1950, and between 1950 and 1960 several techniques were introduced for obtaining infrared spectra of cotton and other fibers. In general, these involved adaptations of existing methods to meet the specific requirements for obtaining spectra of fibers, yarns, and fabrics. Table I lists chronologically the principal methods for obtaining the spectra of textile materials.

Because of the problem of solubility, there was no satisfactory procedure for obtaining infrared spectra of fibers, yarns, and fabrics prior to the adaptation of the KBr disk technique.[1] With the application of this innovation infrared spectroscopy became an increasingly important tool for the cellulose chemist. The data obtained by this method provided a means for reliable identification and quantitative estimation of various fibers in blends and of chemically-modified fibers,[2,3] and the technique is still the most convenient and widely used. No simple chemical method exists for evaluating the changes in molecular structure which occur in the finishing of cotton and which are of considerable interest to the chemist whose goal is to improve the fabric performance. Information concerning these structural changes may be obtained from infrared absorption spectra by the identification of new functional groups which are present after modification.[4] In addition, spectral data may be used to evaluate changes in crystallinity and polymorphic form. Development and application of analytical methods of analysis such as hydrolysis, deuteration, differential spectroscopy, linear scale expansion, and multiple internal reflectance have added to the usefulness of infrared spectroscopy in the examination of cotton and its various modifications.

The applications of infrared absorption spectroscopy to cellulose and cellulose derivatives can be divided into two sections: (1) Examination of

TABLE I
Methods of Obtaining Infrared Absorption Spectra of Cellulose and Cellulose Derivatives

Method	Reference
Films from deacetylated cellulose	(5)
Mineral oil mulls of finely-ground samples	(6)
Films cast on glass plates	(7)
Potassium bromide disks	(1)
Multiple internal reflectance	(8)
Pressed fiber specimen	(9)

unmodified cellulosic fibers (either alone or blended with other fibers), and (2) examination of chemically-treated cellulose and identification of the chemical treatment.

The techniques described in this paper and the interpretation of the data are illustrated with applications to cotton cellulose, but they are suitable for use with any natural or synthetic textile material and are applicable without modification to fiber, yarn, or fabric.

UNMODIFIED COTTON

One of the primary applications of infrared data to unmodified cellulose is for the estimation of crystallinity. The loss in crystallinity which occurs in grinding cotton in a vibratory ball mill is proportional to the time of grinding. This was demonstrated by Forziati *et al.*[6] with x-ray diffraction measurements. O'Connor *et al.*[11] investigated the use of infrared to follow the extent of decrystallization by mechanical grinding. They observed that a band at 1449 cm^{-1} (6.9 μ), arising from a C–H deformation, disappears, while a very weak band at 909 cm^{-1} (11.0 μ) increases in intensity with decrease in crystallinity. This is illustrated in Fig. 1, which shows spectra of cotton ground in a ball mill for increasing lengths of time. The ratio of the "crystalline" band (1449 cm^{-1}) to the "amorphous" band (909 cm^{-1}), referred to as crystallinity index, provides a method for estimating crystallinity which correlates well with values obtained by x-ray diffraction. However, as pointed out by the authors, these same bands change during conversion of cellulose I to cellulose II as a result of mercerization, thereby limiting the ratio strictly to measurement of the crystallinity of cellulose I.

Since cellulose frequently occurs as a mixture of these polymorphic forms, further investigations were carried out by Nelson and O'Connor[12] to determine whether infrared could be used to evaluate separately the changes due to crystallinity and those due to change in lattice form. They reported some strong similarities and marked differences in the infrared spectra of highly crystalline samples of cellulose I, II, and III and of amorphous cellulose. They proposed a new ratio using bands at 1372 and 2900 cm^{-1}. Crystallinity values obtained from the ratio of these two bands correlate fairly well with x-ray values, and the ratio is independent of the lattice type of the sample. Figure 2 illustrates the method for obtaining this ratio from the infrared spectra.

A probable cause for the crystalline properties of cellulose is the hydrogen-bonding capacity of its many hydroxyl groups. O'Connor *et al.*[11] observed that chemical modification involving reaction with OH groups of the cellulose caused the hydroxyl stretching band at 3333 cm^{-1} to shift to higher

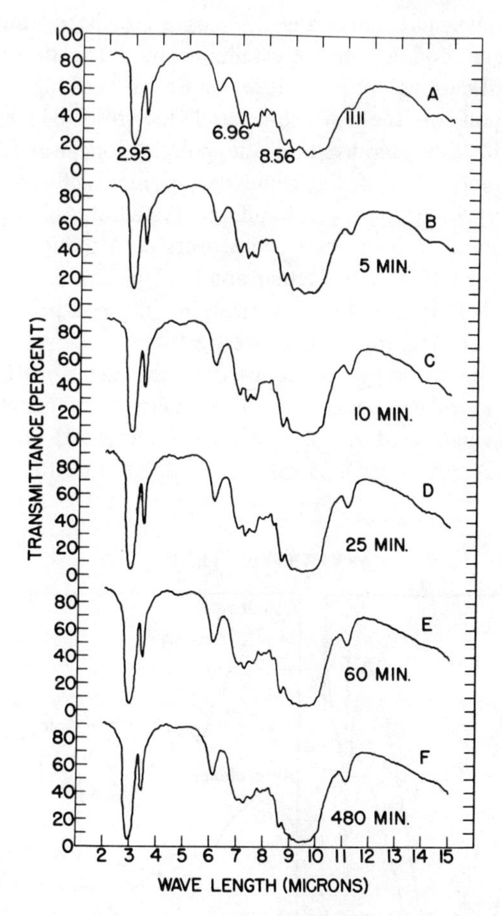

Fig. 1. Infrared spectra of Deltapine cotton: (A) ground to pass 2 mm mesh in Wiley mill; (B), (C), (D), (E), and (F) ground in vibratory ball mill. From O'Connor *et al.*[11]

frequencies. For example, the O–H bands of methyl and ethyl cellulose exhibit shifts to higher frequencies of 68 and 91 cm^{-1}, respectively, from that of native cellulose. The O–H band of cellulose acetate is shifted from 3355 to 3508 cm^{-1}, a shift of 153 cm^{-1}. These changes were interpreted as resulting from a decrease in the amount of accessible hydrogen bonding caused by the occurrence of the chemical reaction. These reactions are believed to occur in the amorphous or accessible regions of the cellulose and have the effect of causing the O–H band component from the crystalline region to become relatively more prominent.

Deuteration experiments by Mann and Marrinan[10] and polarization studies by Marchessault and Liang[13,14] have also been used in the examination of hydrogen bonding and crystallinity by infrared. In addition, certain chemical modifications of cellulose cause a striking resolution in the 3330 cm^{-1} region of the spectrum, and the resolved bands are in close agreement with those reported in the polarization and deuteration studies. Due to the complexity of the cellulose polymer, interpretation of infrared spectra with respect to hydrogen bonding, crystallinity, and polymorphic form is complicated, and in some cases controversial. Although some of the bands have been assigned to intermolecular and intramolecular hydrogen bonding on the basis of deuteration and polarization, these aspects of the cellulose molecular structure require further investigation.

Infrared spectroscopy is also used in the identification of natural and synthetic fibers and in some cases the major components in a mixture of fibers. This is illustrated by the multiple-internal-reflectance (MIR) spectra shown in Fig. 3. Although it is easy to distinguish cotton from the wool and

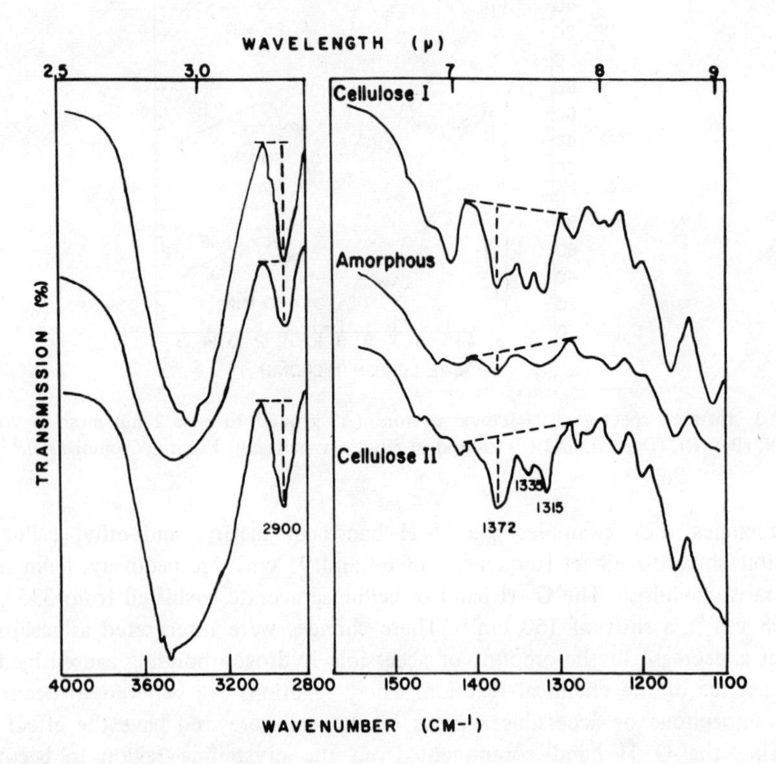

Fig. 2. Illustration of portions of infrared spectra from which crystallinity ratio A_{1372}/A_{2900} is obtained. From Nelson and O'Connor.[12]

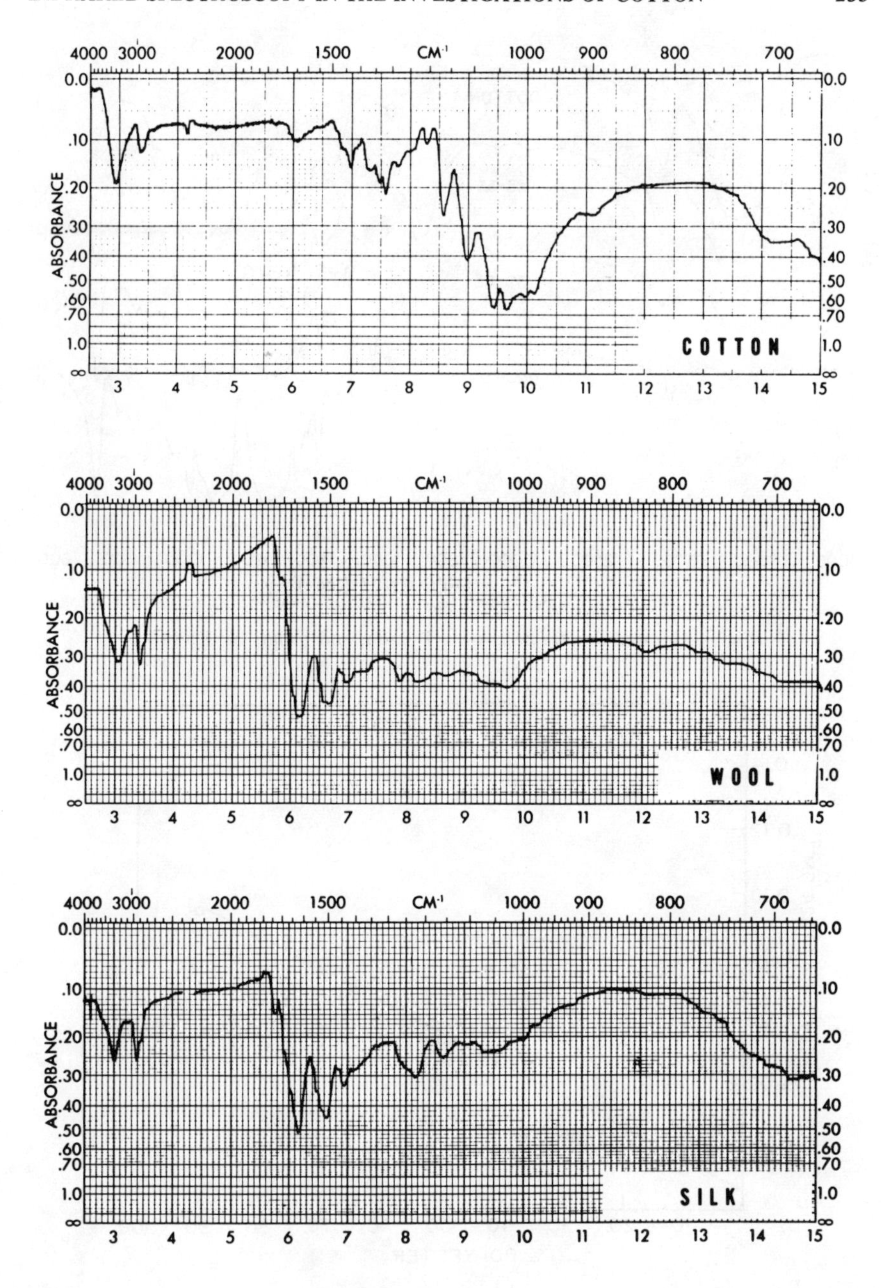

Fig. 3. Comparison of infrared spectra of cotton cellulose with spectra of natural protein fibers. From Wilks and Iszard.[20]

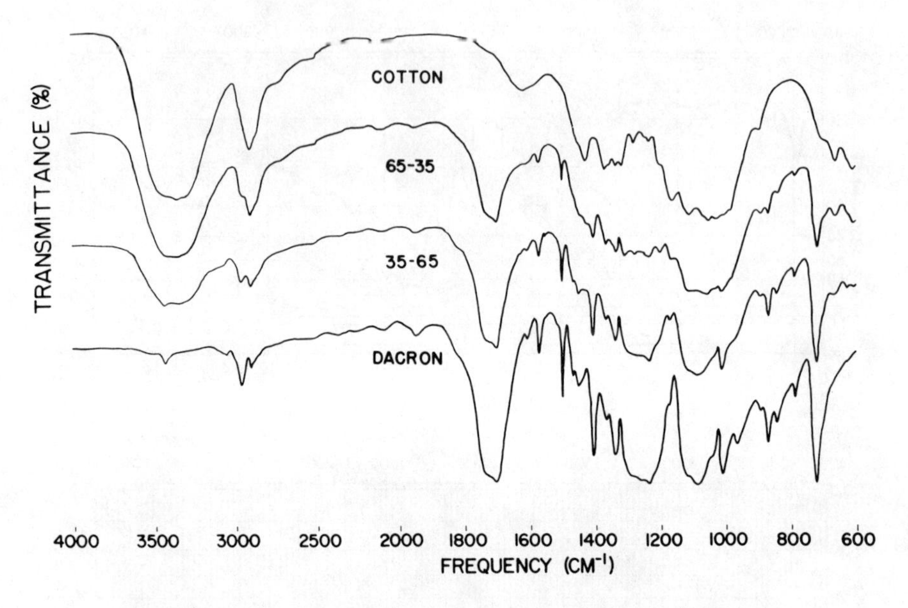

Fig. 4. Infrared spectra of cotton, Dacron, and blends of the two.

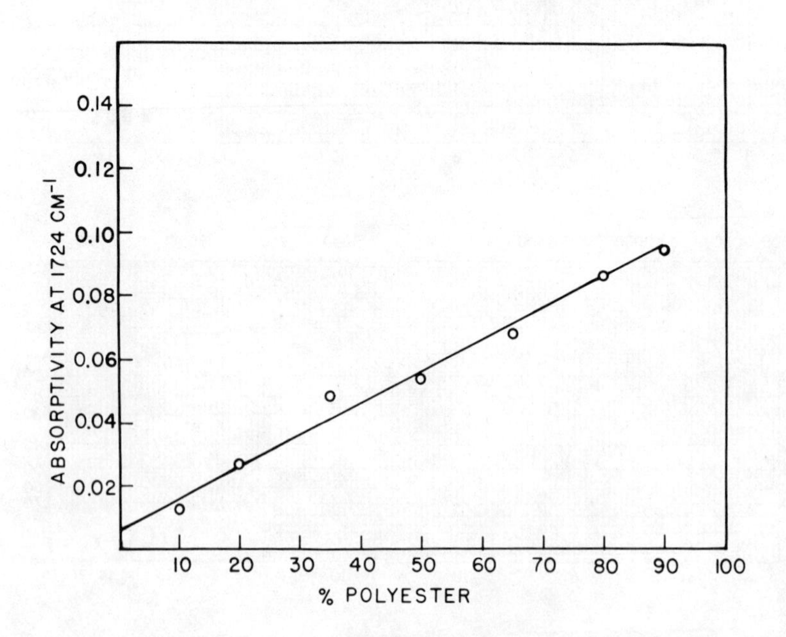

Fig. 5. Increase in intensity of the 1724 cm^{-1} band with increase in per cent of Dacron in the blend.

silk, the small differences between the latter two make differentiation considerably more difficult. However, the characteristic amide I and amide II bands in the protein fibers would be useful in detecting these fibers in blends with cotton. Hemp, jute, flax, ramie, and many synthetic fibers have also been examined by infrared spectroscopy and can be distinguished from one another by their "fingerprint" spectra.

In order to successfully distinguish fibers in blends, it is necessary that the spectra contain bands not common to both components of the blend. An estimation may be made of the amount of certain fibers blended with cotton, as illustrated in Fig. 4, which shows spectra of cotton and Dacron individually and blended with each other. Bands at 1724, 1613, 1538, and 725 cm^{-1} in the blends are characteristic of the polyester and identify this as the fiber blended with the cotton. The intensity of the 1724 cm^{-1} band when plotted against the known Dacron concentration for a series of cotton–Dacron blends can be used to determine the amount of polyester present in the blend at a concentration as low as 10%. A typical relationship between the intensity of the 1724 cm^{-1} band and the per cent polyester in the blend is shown in Fig. 5. A similar situation exists in the case of nylon and cotton. Figure 6 shows

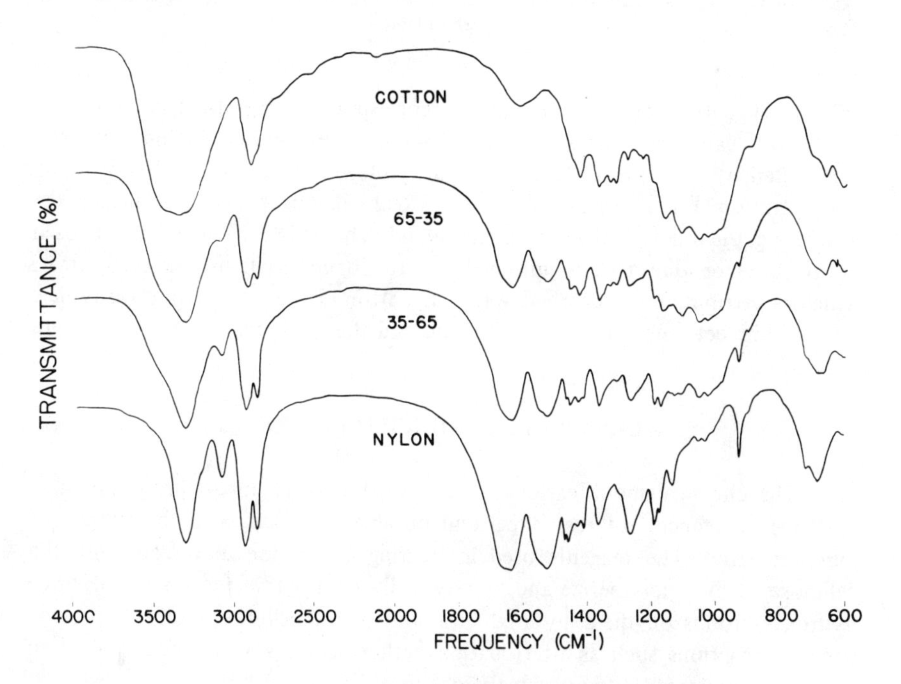

Fig. 6. Infrared spectra of cotton, nylon, and blends of the two.

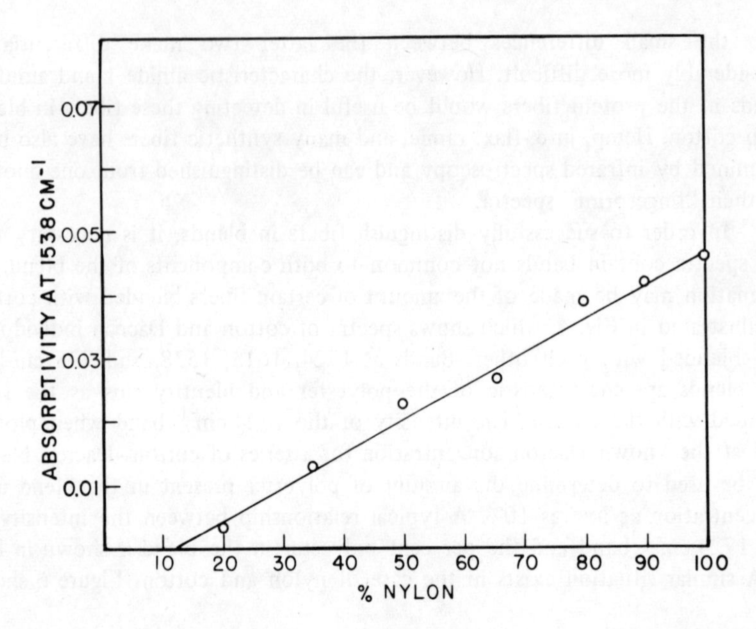

Fig. 7. Increase in intensity of the 1538 cm^{-1} band with increase in the per cent of nylon in the blend.

spectra of cotton and nylon and the corresponding blends. The 1658 and 1538 cm^{-1} bands are characteristic of nylon in the nylon–cotton mixtures. The intensity of the 1538 cm^{-1} band plotted against the known nylon concentration for a series of cotton–nylon blends provides a means for estimating the amount of nylon in the blend. The 1538 cm^{-1} band was chosen to avoid error due to any influence of the band occurring at 1639 cm^{-1}, which is attributed to adsorbed water in cotton. Figure 7 illustrates a typical relationship between the 1538 cm^{-1} band and the per cent nylon in the blend.

CHEMICALLY MODIFIED COTTON

The chemical modification of cotton cellulose to impart properties such as flame resistance, rot resistance, and durable press has been the subject of much research. The reagents used in treating the cotton may react with the cellulose or may polymerize and be physically held within the fiber. The three hydroxyl groups on the anhydroglucose unit of the cellulose molecule provide sites for reactions such as esterifications, etherifications, or introduction of an intermolecular crosslink between the chains.

Acetylation is one of the oldest methods of modifying cotton, and serves to illustrate the type of spectrum obtained from an esterification reaction (Fig. 8). The strong band at about 5.7 μ (1754 cm^{-1}) indicates the presence of the C=O of the ester and is a useful band for quantitatively estimating the degree of acetylation. The 8.1 μ (1235 cm^{-1}) band due to the C–O stretch of the ester is another characteristic of this modification. Esters of cotton have also been prepared from saturated and unsaturated acids and from aromatic acids, and many of these products can be differentiated by means of their infrared spectra.

Use of infrared spectroscopy in the investigation of etherification reactions is demonstrated in Fig. 9, which presents a comparison of the spectrum of an etherified cotton with that of an unmodified cotton. The characteristic C≡N band at 4.45 μ (2247 cm^{-1}) identifies the treatment as cyanoethylation and is useful for quantitatively estimating the extent of reaction. Infrared analysis has the advantage in this case of being specific for the content of nitrogen due to the C≡N group even in the presence of other modifying reagents containing nitrogen. Some other cellulose ethers that can be identified by their infrared spectra are methyl, ethyl, carboxymethyl, and carboxyethylcellulose and those produced by reaction with benzyl alcohol, triphenylmethanol, and various carboxylic acids.

More recent modifications of cellulose have been concerned with treatments which result in the formation of an intermolecular crosslink between

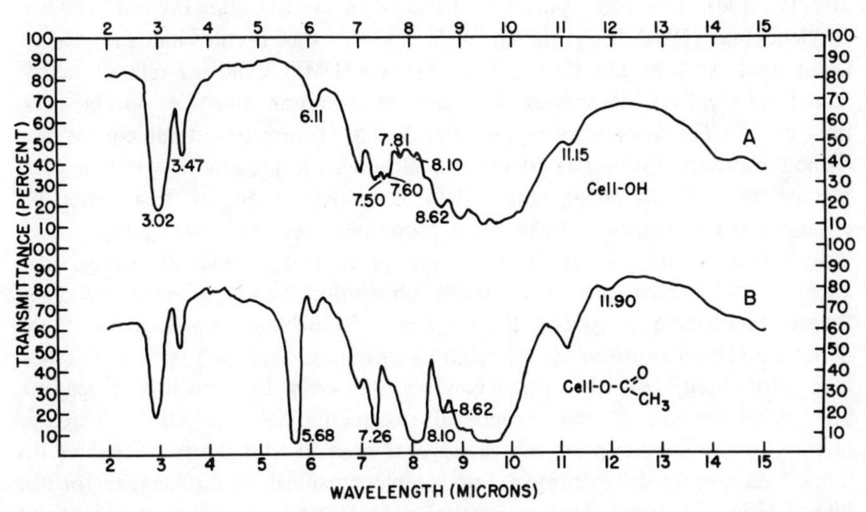

Fig. 8. Infrared spectra of (A) native cotton and (B) acetylated cotton. From O'Connor *et al.*[3]

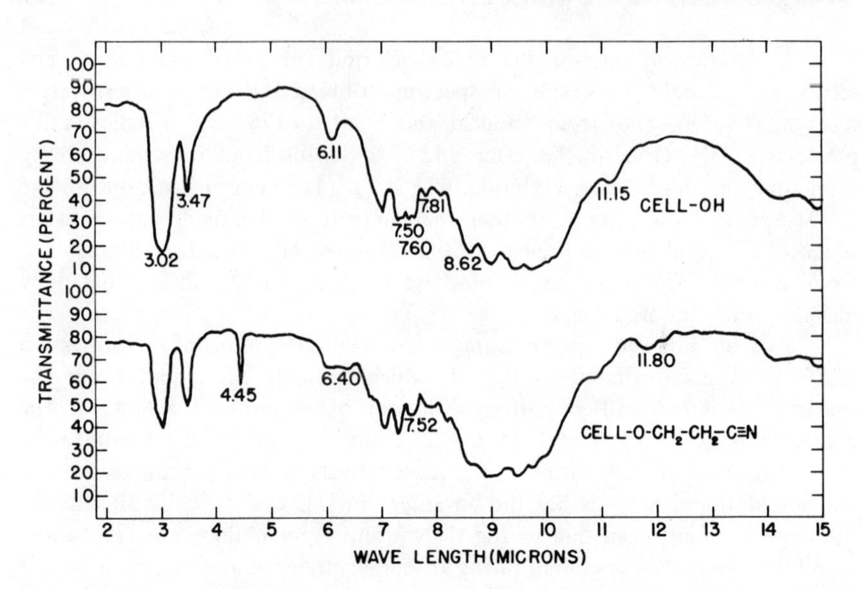

Fig. 9. Infrared spectra of (A) native cotton cellulose and (B) cyanoethylated cellulose. From O.Connor *et al.*[3]

the cellulose chains. Reagents used in these treatments are frequently difunctional or polyfunctional compounds of the *N*-methylol type as shown in Fig. 10. Many of these compounds are urea derivatives, such as dimethylolethyleneurea (DMEU). The ring hydroxyl derivative of DMEU, dimethyloldihydroxyethyleneurea (DMDHEU), is in wide use in the textile industry today. Compounds such as dimethylolethylcarbamate (DMEC) and the related methyl and hydroxyethyl derivatives are also in common use for durable-press finishes. *Tris*(1-aziridinyl)phosphine oxide (APO) imparts durable-press and flame resistance properties to cotton, although at present it is not in commercial use. Methylolmelamine (MM) is mainly used in treatments for obtaining rot resistance or wash–wear properties.

Cottons modified in this way are often treated with relatively small amounts of reagent, and the strong absorption bands of cellulose itself frequently obscure or conceal the presence of the bands characteristic of the derivative. Other finishing agents, such as softeners, sizes, and optical bleaches, may also obscure the absorption bands arising from the crosslinking reagent. The extensive use of these chemical treatments has resulted in increased interest in the identification of the reagents used to alter the properties of the fabric. As previously mentioned, no simple chemical method exists for the identification of these finishing agents, and identification through the use of infrared spectroscopy has required a new approach. A scheme of analysis has been developed[15] which has been extended when necessary to include new finishing agents.

A combination of infrared techniques is used in the analytical method for the identification of nitrogenous crosslinking reagents and certain types of softeners applied in the finishing of cotton fabric. The five basic steps in the scheme of analysis are the conventional potassium bromide disk technique, solvent extraction, acid hydrolysis, multiple internal reflectance, and the differential disk procedure. In an actual analysis it will probably not be desirable to use all of these techniques with each sample. The following discussion illustrates the application of these techniques to a number of samples which have been submitted to this laboratory for analysis.

The first sample is a durable-press fabric which was known to have been treated with a crosslinking reagent and a softener. Figure 11 illustrates the identification of finishing agents used on the durable-press fabric. The conventional potassium bromide disk spectrum of the treated fabric (Fig. 11B) reveals bands at 1712 and 1484 cm^{-1} which do not appear in the spectrum of the untreated cotton (Fig. 11A). Table II shows the bands which appear in the spectra of various chemically treated cottons. The bands marked by an asterisk are bands which have been found to be suitable for quantitative analysis. A comparison of the bands in the spectrum of the treated cotton with those of the table indicates that the resin treatment may be dimethyloldihydroxyethyleneurea, but further investigation of the sample is indicated.

The next step consists of a solvent-extraction procedure. The fabric is extracted a minimum of four times with hot trichloroethylene. The solvent is filtered into a casserole and concentrated on a steam bath. A few drops of the concentrated solution are transferred to a clean potassium bromide plate and

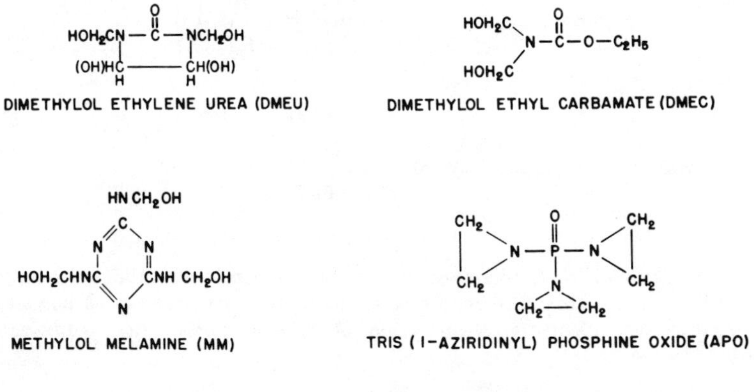

DIMETHYLOL ETHYLENE UREA (DMEU)

DIMETHYLOL ETHYL CARBAMATE (DMEC)

METHYLOL MELAMINE (MM)

TRIS (1-AZIRIDINYL) PHOSPHINE OXIDE (APO)

Fig. 10. Reagents used to crosslink cotton.

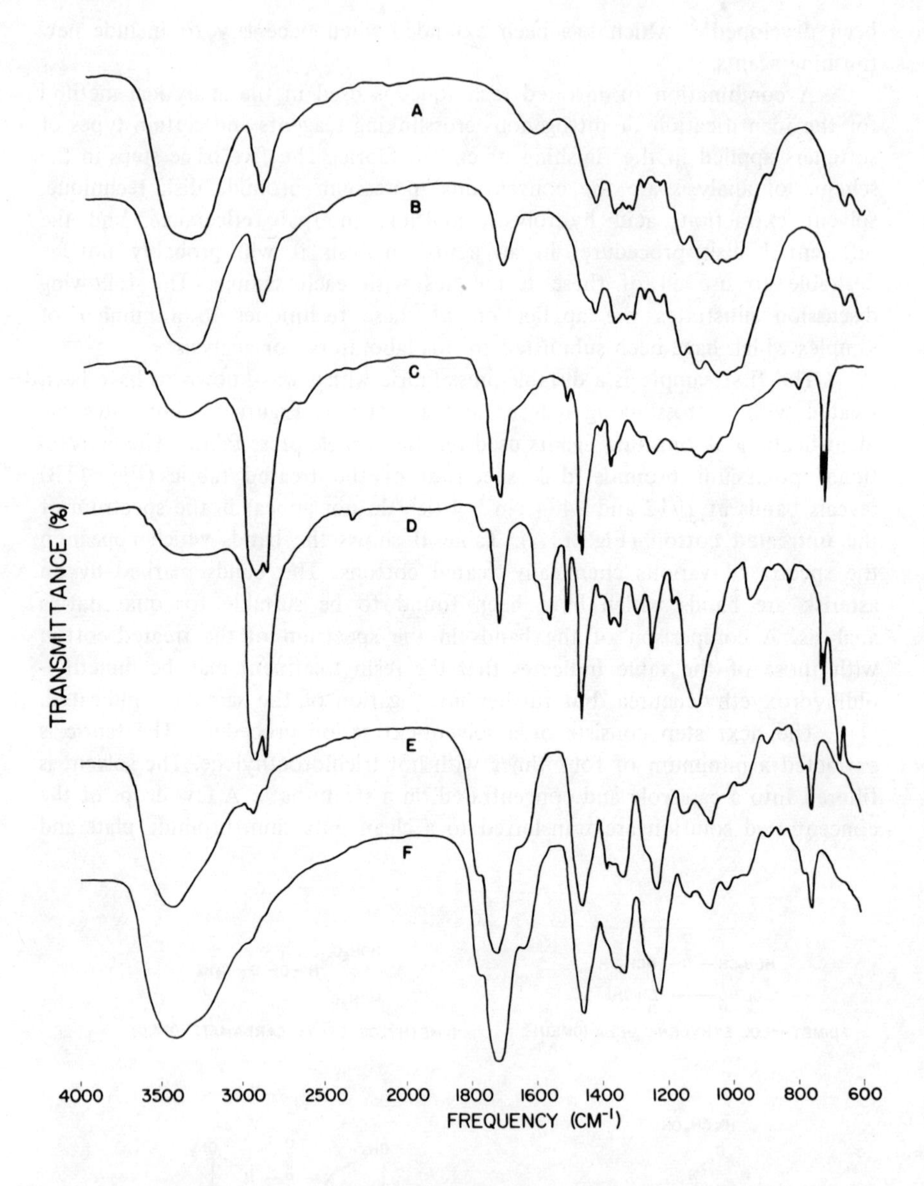

Fig. 11. Identification of finishing agents used on a durable-press fabric: (A) untreated cotton; (B) treated cotton; (C) extract of treated cotton; (D) commercial polyethylene softener (Cellusoft PXB); (E) hydrolyzate of treated cotton; (F) hydrolyzate of DMDHEU.

TABLE II
Significant Bands in the Infrared Spectra of Some Chemically Treated Cottons

Reagent	Absorption bands (cm^{-1})				
Urea–formaldehyde (UF)	1665	1545*	-	-	-
Dimethylolethyleneurea (DMEU)	1680*	1480	760*	-	-
Dimethyloldihydroxyethyleneurea (DMDHEU)	1712*	1484	-	-	-
Dimethylolethylcarbamate (DMEC)	1700*	1520	775	-	-
Methylolmelamine (MM)	1560*	1460	810*	-	-
Tris(1-aziridinyl)phosphine oxide (APO)	-	-	925	815	-
Dimethylolethyltriazone (DMET)	1640*	1500	830*	780*	750

*Bands suitable for quantitative analysis.

the solvent is allowed to evaporate. The transmission spectrum of the material which is soluble in trichloroethylene is obtained.

The spectrum of the extract of the durable-press fabric is shown in Fig. 11C. The doublet which appears at 720 and 730 cm^{-1} is characteristic of a polyethylene. Figure 11D is the spectrum of a commercial polyethylene softener. Except for the band which appears at 1560 cm^{-1} in the known polyethylene, the two spectra are identical. The 1560 cm^{-1} band is thought to be due to an emulsifying agent in the commercial softener.

Many of the nitrogenous agents in use are applied to the cotton with acid catalysts, in most cases with good efficiency. The reverse reaction, in which the crosslinking agent or some modification of it is liberated, can yield products suitable for analysis by infrared absorption techniques.[16] However, it is necessary that the reagents themselves and known reaction products be available for the purpose of comparison. Spectra of the reagent hydrolyzate and the hydrolyzate of the cotton treated with the reagent may be compared with the spectrum of the hydrolyzate of the unknown derivative. A "fingerprint" identification may thus be made.

Curve E of Fig. 11 is the spectrum of the acid hydrolyzate of the durable-press fabric being examined. Curve F is the spectrum of the acid hydrolyzate of DMDHEU. A comparison of the two spectra reveals that they are identical. Thus it could be concluded that this fabric was treated with DMDHEU crosslinking reagent and a polyethylene softener.

Multiple internal reflectance spectroscopy[8] provides a technique for obtaining spectra of cotton fabrics. No sample preparation other than cutting a piece of fabric to the appropriate size or wrapping fibers around the reflectance plate is necessary. An outstanding advantage of this technique is the ability to obtain separate spectra of the coated and uncoated sides of a fabric. This is illustrated in Fig. 12, which shows multiple internal reflectance spectra of a fabric coated on only one side with methylolmelamine. The spectrum of the coated side (Fig. 12A) shows absorption bands at 1560, 1460, and 810 cm^{-1}. These bands are characteristic of cotton treated with methylolmelamine and are not present in the spectrum of the uncoated side (Fig. 12B). Multiple internal reflectance is a reliable means for establishing that a fabric has been treated on only one side as well as for identifying the treatment.

The sensitivity of the multiple internal reflectance technique is illustrated by the spectra shown in Fig. 13. These spectra are of the two sides of a sateen fabric woven with mercerized filling threads and unmercerized warp threads. The 3300 cm^{-1} region of curve A (back of fabric) resembles cellulose I, while the small bands developing at 3440 and 3480 cm^{-1} in curve B (fabric face) are characteristic of mercerized cellulose. These bands have been reported in the literature in studies of mercerized cellulose by use of polarized radiation.[14] They are not apparent in a disk spectrum. Further indications of

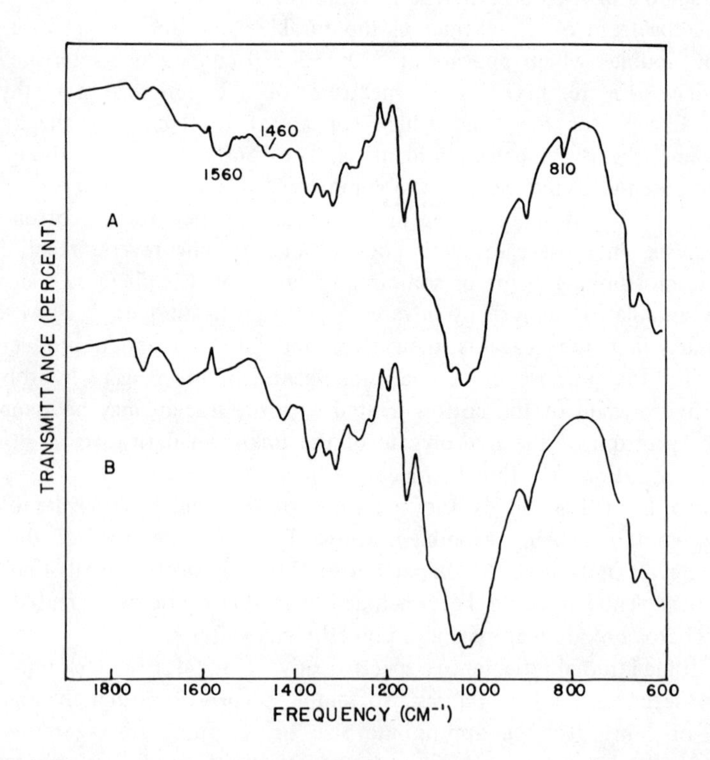

Fig. 12. Infrared spectra of a cotton fabric treated on one side with a 5% solution of methylolmelamine: (A) treated side; (B) untreated side. From McCall *et al.*[8]

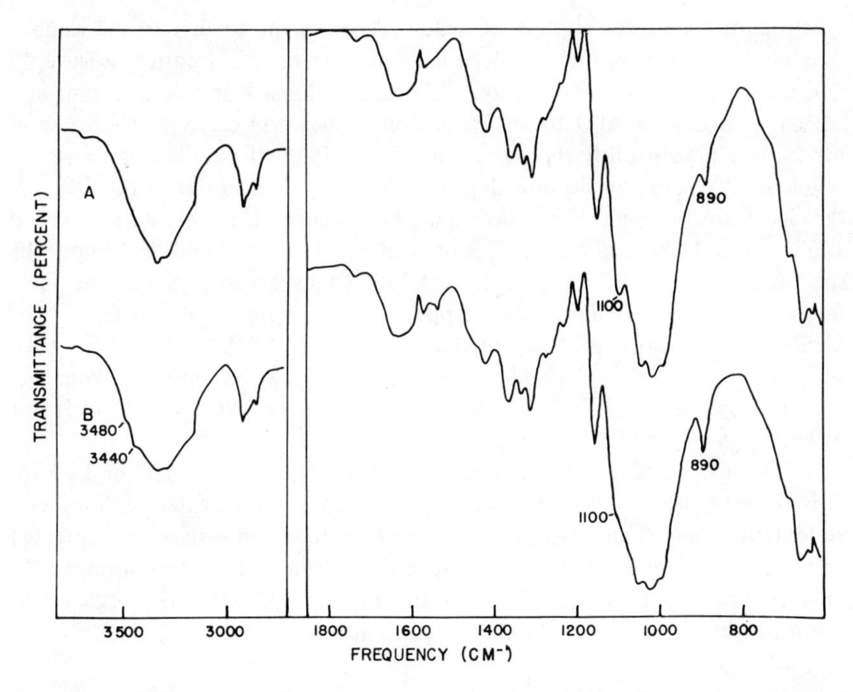

Fig. 13. Multiple internal reflectance spectra of a cotton filling faced sateen: (A) Back; (B) face. From McCall *et al.*[8]

mercerization are the disappearance of the band at 1100 cm^{-1} and the sharpening of the band at 890 cm^{-1} in curve B. It would be impossible to obtain such information from the spectrum of this fabric by the conventional KBr disk method.

The reflectance technique may also be applied in lieu of the potassium bromide disk technique. This is illustrated in Fig. 14, which compares the disk and internal reflectance spectra of the DMDHEU fabric previously identified. The bands at 1712 and 1484 cm^{-1}, which do not appear in the untreated cotton, are present in both spectra. The greatest advantage of the internal-reflectance technique is the ease and rapidity of sample preparation and the fact that the sample is not destroyed. This is somewhat offset by the time required to adjust the reflectance attachment and the rather fragile nature of the KRS-5 reflectance plates. However, if the attachment is already adjusted and suitable plates are available, spectra may be obtained rapidly by this technique.

A differential spectrum may be useful in the identification of a treatment which is impossible to detect on cotton except at a very high add-on.[17] The spectrum is obtained by placing in the reference beam of a double-beam spectrophotometer an amount of cellulose equivalent to that in the sample, thus compensating for the cellulose bands in the modified cotton. Figure 15 illustrates the use of this technique. The conventional KBr disk spectrum of a cotton treated with APO to an 8% add-on is shown in curve A. There are no bands which distinguish the spectrum of this fabric from that of untreated cellulose. When the absorption due to cellulose is blocked out in the differential spectrum (curve C) the spectrum of the same fabric exhibits bands at 1468, 1260, 1190, and 925 cm^{-1} not seen in the conventional spectrum. The spectrum of pure APO (curve B) exhibits bands at very nearly the same frequencies as the bands which appear in the differential spectrum. The 810 cm^{-1} band which is strong in the spectrum of the APO reagent has been assigned to the aziridinyl ring.[18] The opening of the ring during polymerization and reaction with cellulose accounts for the fact that this band is not present even in the differential spectrum.

The use of the differential technique is further illustrated in Fig. 16. This compares the conventional disk spectrum of cotton treated with dimethylolethyltriazone (curve A) with the conventional spectrum of untreated cotton (curve B) and the differential spectrum of the treated cotton (curve C). In this case the presence of the bands in the 700–900 cm^{-1} region was confirmed by the use of linear scale expansion (curve D).

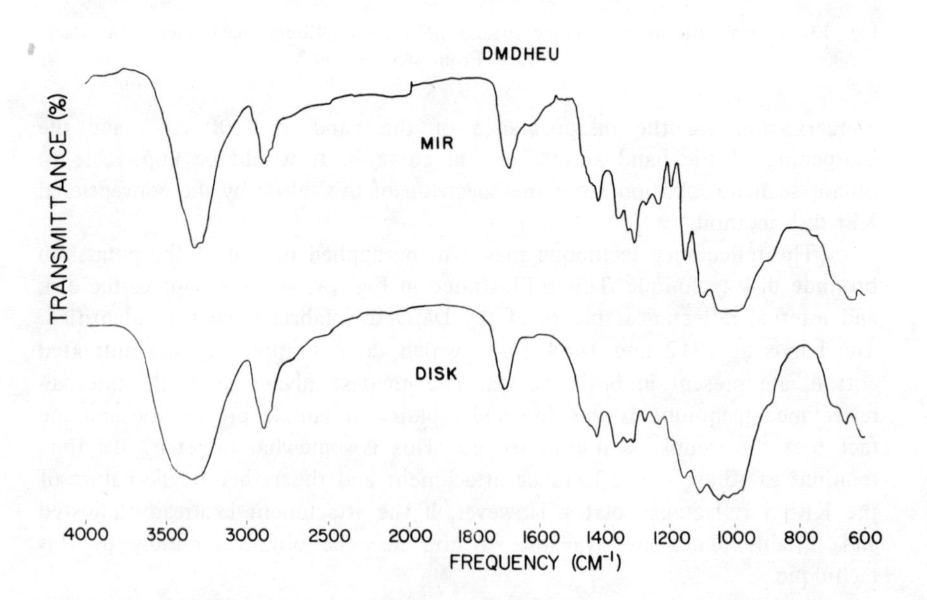

Fig. 14. Comparison of internal reflectance and disk spectra of cotton treated with DMDHEU: (A) MIR; (B) disk.

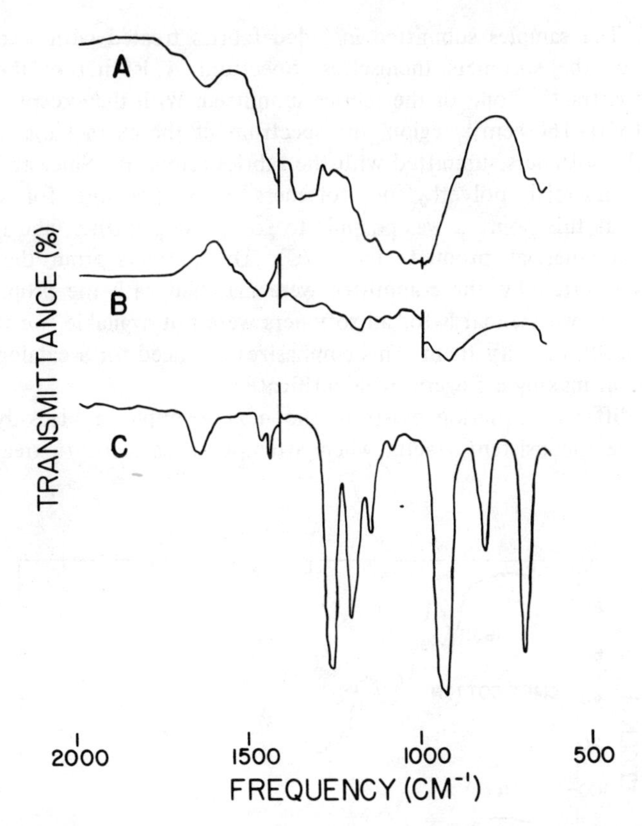

Fig. 15. Infrared spectra of (A) cotton treated with APO (8% add-on) – conventional disk spectrum; (B) cotton treated with APO (8% add-on) – differential spectrum; (C) APO reagent.

Another example of the use of different spectroscopy is a cellulose modified with a fluorocarbon ester. The conventional spectrum showed no evidence for the presence of fluorine, even at a level of nearly 5%. The use of the differential disk procedure permitted the detection of fluorine by the presence of bands in the 1140–1240 cm^{-1} region corresponding to those which appeared in the spectrum of the modifying reagent.[19]

The samples discussed thus far illustrate each of the five basic techniques. The following is a discussion of some applications of these techniques.

The application of the method for the identification of softeners is shown by the spectra in Figs. 17A, B, and C. These are spectra of samples submitted for collaborative testing by the American Association of Textile Chemists and Colorists Committee RA-45: Identification of Finishes on

Textiles. The samples submitted included fabrics treated with a softener and samples of the softeners themselves. Spectrum A is that of the trichloroethylene extract of one of the fabrics submitted. With the exception of bands in the 1560–1600 cm^{-1} region, the spectrum of the extract matches that of one of the softeners submitted with the fabrics (curve B). Since the spectra of some commercial polyethylene softeners were available for comparison purposes, at this point it was possible to go one step further, and identify the specific commercial product (Fig. 17C). The extracts from the remaining fabrics submitted by the committee were matched with the proper softener, but since known standards for all softeners were not available, we were unable to specifically identify them. This emphasizes the need for a catalog of known standards in making a fingerprint identification.

A different situation exists in the next example, a black-dyed cotton which gave inconsistent results when attempts were made to treat it with a

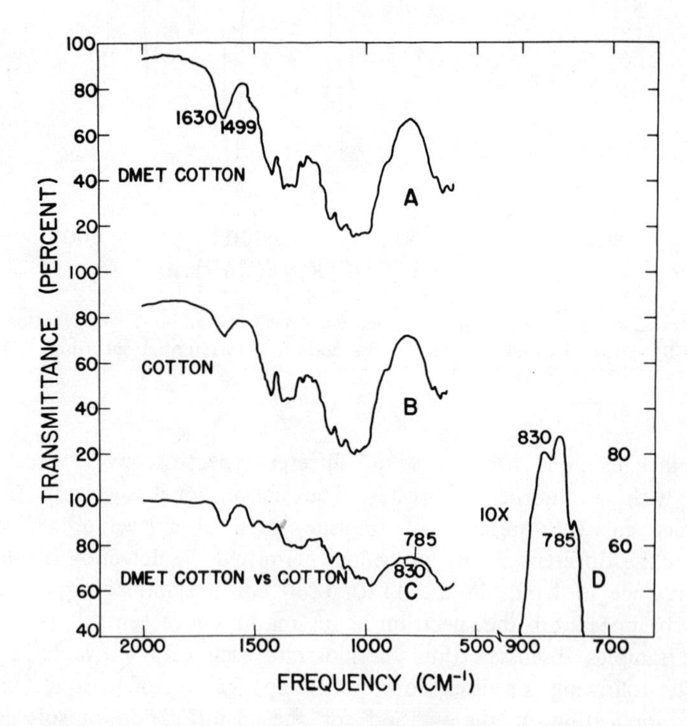

Fig. 16. Spectra of unmodified and dimethylolethyltriazone-treated cottons. (A) DMET cotton, conventional spectrum; (B) unmodified cotton, conventional spectrum; (C) differential spectrum of (A) and (B); (D) scale expansion of (C) in 700–900 cm^{-1} region. From Tripp *et al.*[4]

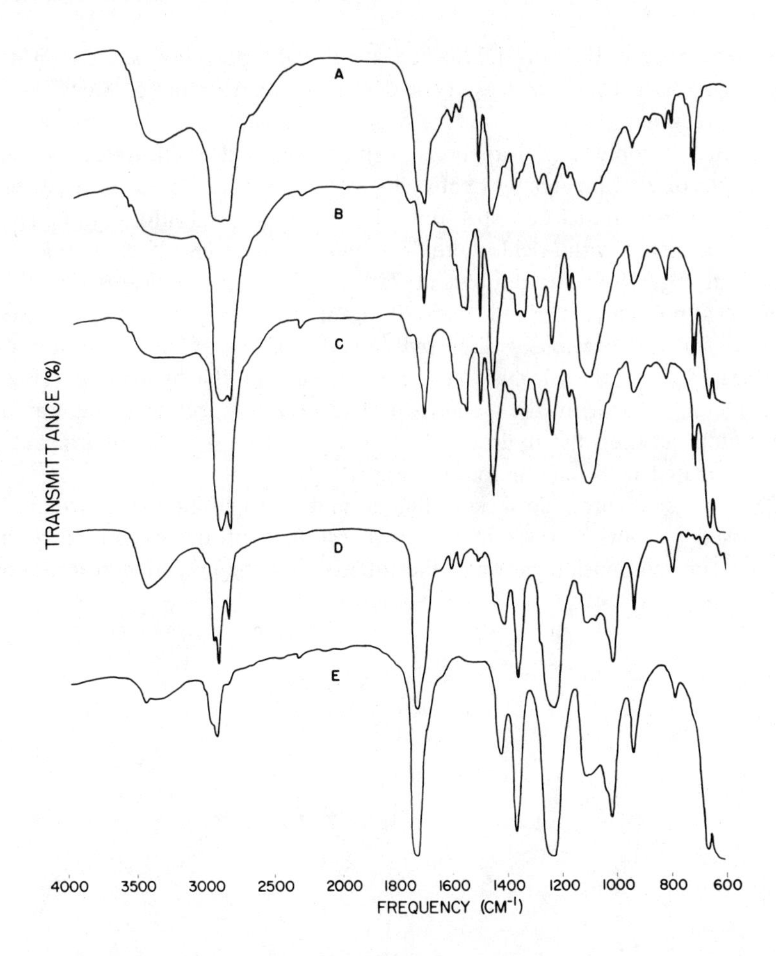

Fig. 17. Identification of trichloroethylene-extractable softeners and sizes: (A) Extract of fabric submitted by AATCC Committee RA-45; (B) unknown softener submitted with fabric; (C) commercial polyethylene softener – Cellusoft PXB; (D) extract of black-dyed cotton; (E) polyvinyl acetate.

crosslinking reagent. Although dyes do not normally interfere with resin treatment, it was thought that dye might be the cause of the problem, since this fabric allegedly had no other treatment. The sample was subjected to the scheme of analysis even though information on dyes is not included in the method. The only technique that yielded any positive results was the trichloroethylene extraction. Figure 17D is the spectrum of the extract from the fabric. This compares almost perfectly with the spectrum of a polyvinyl acetate emulsion shown in Fig. 17E. By weighing the fabric and the extract it was possible to show that the fabric had about 1.9% polyvinyl acetate on it. This may well have been the reason for the inconsistent results obtained with crosslinking reagents.

Another example of the application of the method is illustrated in Fig. 18. A sample of a commercially treated durable-press fabric was submitted for analysis. The conventional disk spectrum yielded little useful information. The fabric was extracted with trichloroethylene and the spectrum of the extract is illustrated in Fig. 18A. The doublet at 720 and 730 cm^{-1} indicates that the softener is a polyethylene, and there is again a striking similarity to the commercial softener previously discussed (Fig. 17C). The extracted fabric was then subjected to acid hydrolysis and the spectrum of the hydrolyzate (Fig. 18B) was compared with the catalog of hydrolyzate spectra. There was a perfect match between the hydrolyzate of the unknown and the hydrolyzate of a fabric treated with dimethylolurea (Fig. 18C).

The potassium bromide disk technique and multiple internal reflectance may be used to obtain a semiquantitative estimate of the extent of resin treatment. The relationship between the intensity of the measured absorption

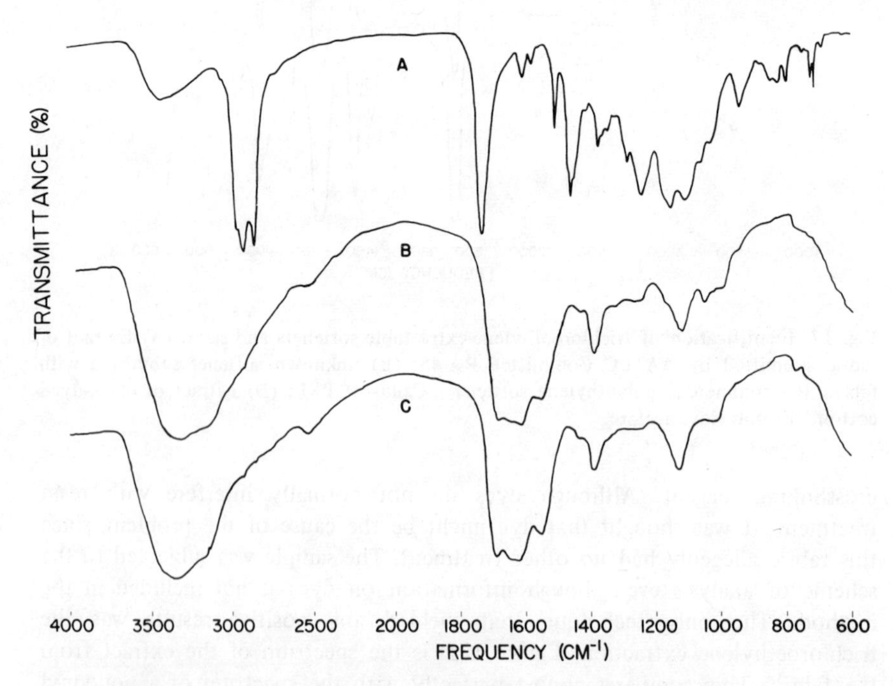

Fig. 18. Identification of finishing agents used on a commercial durable-press fabric: (A) Extract of fabric; (B) hydrolyzate of fabric; (C) hydrolyzate of DMU-treated fabric.

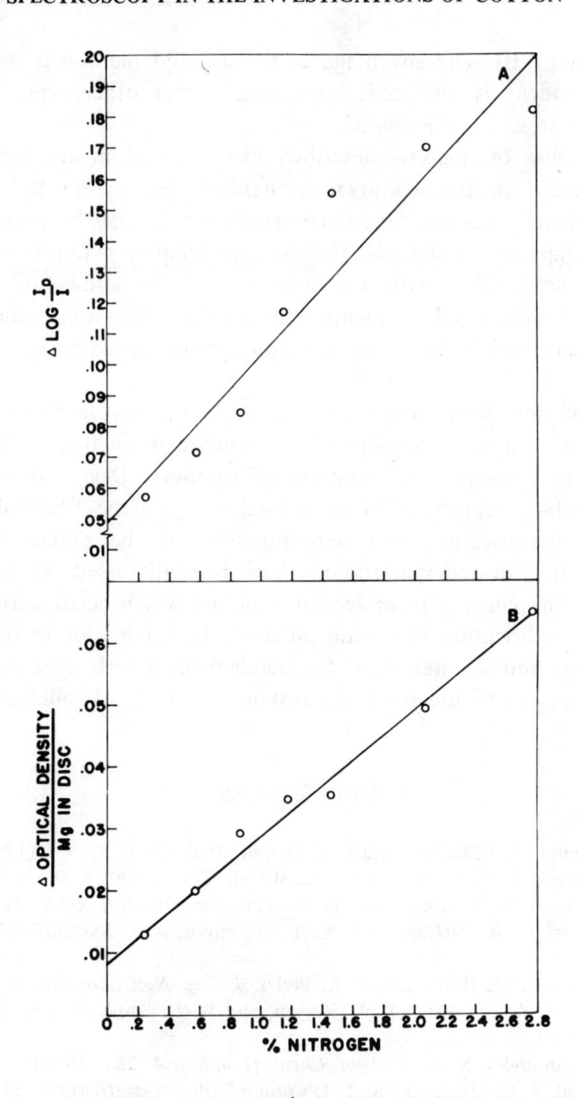

Fig. 19. Increase in intensity of 1675 cm^{-1} band of DMEU-treated cotton with nitrogen content: (A) MIR technique; (B) disk technique. From McCall *et al.*[8]

of the C=O stretching vibration and the increasing amount of nitrogen resulting from treatment with DMEU is shown in Fig. 19. The nitrogen content was determined chemically by the Kjeldahl method. The relationship may be demonstrated with multiple internal reflectance by applying an equal (measured) amount of pressure to each sample (Fig. 19A). The results show the samples to be ranked in the same order as the results obtained by the disk technique (Fig. 19B). The advantage of the infrared method is that only the crosslinking reagent is analyzed; interference from other types of nitrogen which may be present is eliminated.

The scheme of analysis described can be used in any laboratory by obtaining suitable spectra of known standards by each of the techniques. It is possible to identify certain functional groups which may be present without the use of comparison standards. However, to identify a finishing agent, it is necessary to be familiar with the spectrum of the known derivative. The method may be extended to include cellulose modifications other than the ones mentioned, merely by using the appropriate standards for comparative purposes.

Since infrared spectroscopy was first applied in textile research improved instrumentation and new techniques have resulted in an increased amount of available information on the structure of cellulose. Degree of crystallinity, hydrogen bonding, identification of natural or synthetic fibers alone and in blends, and identification and determination of the extent of cellulose modifications by infrared spectroscopy have been discussed. Much remains to be learned of the changes in molecular structure which occur during chemical and physical modification of cotton cellulose. Determination of the exact site of modification and the nature of the bonds formed with these modifications is the ultimate goal of the spectroscopist in the study of cellulose chemistry.

REFERENCES

1. R. T. O'Connor, E. F. DuPre , and E. R. McCall, *Anal. Chem.* **29**, 998 (1957).
2. R. T. O'Connor, E. F. DuPre , and E. R. McCall, *Textile Res. J.* **28**, 542 (1958).
3. R. T. O'Connor, E. R. McCall, and D. Micham, *Am. Dyestuff Reptr.* **49**, 214 (1960).
4. V. W. Tripp, E. R. McCall, and R. T. O'Connor, *Am. Dyestuff Reptr.* **52**, 598 (1963).
5. J. W. Rowen, C. M. Hunt, and E. K. Plyler, *J. Res. Natl. Bur. Std.* **39**, 133 (1947).
6. F. H. Forziati, W. K. Stone, J. W. Rowen, and W. D. Appel, *J. Res. Natl. Bur. Std.* **45**, 109 (1950).
7. H. J. Marrinan and J. Mann, *J. Appl. Chem. (London)* **4**, 204 (1954).
8. E. R. McCall, S. H. Miles, and R. T. O'Connor, *Am. Dyestuff Reptr.* **55**, 400 (1966).
9. J. A. Knight, M. P. Smoak, R. A. Porter, and W. E. Kirkland, *Textile Res. J.* **37**, 924 (1967).
10. J. Mann and H. J. Marrinan, *Trans. Faraday Soc.* **52**, 481, 487, and 492 (1956).
11. R. T. O'Connor, E. F. DuPré , and D. Mitcham, *Textile Res. J.* **28**, 382 (1958).
12. M. L. Nelson and R. T. O'Connor, *J. Appl. Polymer Sci.* **8**, 1325 (1964).
13. C. Y. Liang and R. H. Marchessault, *J. Polymer Sci.* **37**, 385 (1959).
14. R. H. Marchessault C. Y. Liang, *J. Polymer Sci.* **43**, 71 (1960).
15. E. R. McCall, S. H. Miles, and R. T. O'Connor, *Am. Dyestuff Reptr.* **56**, 35 (1967).
16. S. H. Miles, E. R. McCall, V. W. Tripp, and R. T. O'Connor, *Am. Dyestuff Reptr.* **53**, 440 (1964).

17. E. R. McCall, S. H. Miles, V. W. Tripp, and R. T. O'Connor *Appl. Spectry.* **18**, 81 (1964).
18. T. D. Miles, F. A. Hoffman, and A. Merola, *Am. Dyestuff Reptr.* **49**, 596 (1960).
19. J. P. Moreau and G L. Drake, *Am. Dyestuff Reptr.* **58** (3): 21-26 (Feb 10, 1969).
20. P. A. Wilks, Jr. and M. R. Iszard, Paper presented at the 15th Mid-America Spectroscopy Symposium, Chicago, Illinois, June 2–5, 1964.

Spectroscopy and Computer Dyeing

Braham Norwick

Beaunit Corp.
Textile Division
New York, New York

Although computer analysis for proportions and weights of dyestuffs required to match a shade in dyeing fabric dates back about 25 years, progress has been slow and results of debatable commercial value. A review of problems and recent progress, including current application in the Beaunit Textile Company dyehouses, is given. Rapid spectrophotometer techniques for quality control and standardization of dyestuffs purchased in wide variety and small amounts was the first step in this application. Efforts to achieve zero adds in dyeing of new shades are described. Use of infrared spectroscopy is noted, as well as a color numbering system based on *L*, *a*, and *b* values.

INTRODUCTION

A commercial field which seems a natural domain for spectroscopy is textile dyeing. However, in practice dyeing has stayed an art rather than become a science because problems of quantitative prediction are often solved less expensively by intuition than by computation based on clear theory. Dyehouses are now entering the transition phase between scientific control and control by rule of thumb, or art and intuition. The role of spectroscopy, reasons for slow progress, and continuing existence of blocks to further advance are reviewed here in the hope this will elicit suggestions for improvements.

SOME QUESTIONS IN DETERMINING A COLOR MATCH

Of many problems in the dyehouse, a large number resolve into the following: What is the fastest, least costly way to make a textile take the apparent color of something else and produce a color match? The material may be fiber, yarn, fabric or even a garment. It may be composed of one fiber or blends, and the latter range from two types of the same nature such as two cottons varying in dyeability and diameter, or may contain fibers chemically different, such as cotton and polyester. Colorants and the way they attach to fiber vary, so one can go from pigment adhesion through solid solution to the point of chemical reaction. The technique of application depends on these factors. Jigs pass a sheet of fabric back and forth through a bath. Pads take fabric into a bath and then through pressure rollers; in tubs fabric as a rope loops continuously in and out of dye liquor. There are a variety of systems where instead of fabric moving, dye liquor is forced through a stationary mass of fibers, yarn, or fabric.

What is a color match? A human being looks at an illuminated standard and sample to see how they differ. Some work has been concentrated on this difference, and books are available containing a multitude of shades, such as the Munsell system.[1] Comparable work established so-called uniform color spaces[2], based on the assumption that color is a combination of three factors, and that one can geometrically determine distance between two points in this space. The tangible Munsell system is incomplete for textiles, and old swatches are usually misleading in the textile industry. The system is impractical in a dyehouse. Machines for matching are an obvious possibility, but original equipment was not adequate. Devices based on tristimulus have in the past shown readings to be a function of their own optical geometry and the alignment and luster of the fabric. Slight variations in alignment give major differences in readings. One instrument makes it appear that a pink satin is green. If in one alignment pink reads green, and the eye doesn't see green, then rotation of the sample is not the answer. Improved optical geometry has been a partial solution. In addition, tristimulus matching does not solve the problem of metamerism. Colors which look alike in incandescent light will not look alike in broad daylight, at sunset, or under fluorescent illumination. What an eye sees is a function, among other things, of object reflectivity at different wavelengths and the energy of object illumination at these wavelengths. In order to make matches which stay a match if the textiles go from home to street, to office or plant with different lighting, a fair match is required at all wavelengths, otherwise one risks having some tristimulus values in one place and different values in another.

A further problem exists with most laboratory spectrophotometers. The eye sees things illuminated with a variety of energy distributions ranging into ultraviolet from the sun or other light sources, all considered as "white" light. Usual spectrophotometers illuminate with light split into bands essentially monochromatic, and read out on a uniform illuminant energy basis. Moreover, if the color or the fabric itself tends to fluoresce, the spectral distribution curve read out by sequential monochromatic illumination will look even less like the curve resulting from "white"- light illumination. Reversed optics is available on some equipment, and this problem is partially solved by the use of a filter instrument, such as the Color Eye.[3] This device is not automated; the operator must turn knobs and punch out values on the computer cards. While economical and stable, it is not a device one can put easily into a working dyehouse, since it is slow, requires continuous operator concentration, and yields poor results if the operator is plagued with interruptions or is rushed.

Many dyes used on acrylonitrile fibers, and for whites on all fibers, show fluorescence, and to match shades successfully, this must be taken into account. A problem not fully faced occurs with shades which flare violently as one goes from one light source to another, and is particularly noticed by men who buy a tan suit which turns out to be olive green most of the time. Automatic switching instrumentation for tristimulus values from A to C illuminant would be a help.

The preceding gives a picture of problems in determining a match. The material must appear the same shade as the standard, independent of illumination and of how much the standard may flare under different illuminants. How closely it does match can be stated in numerical terms, but there is a gray zone of how closely it must match. The latter is in the area of art rather than science. The smooth semimatte surfaces uniformly butted require smaller color differences than a textured or lustrous surface which joins in an uneven or gathered seam, especially if the adjacent is of different texture. A much greater difference is tolerable if the two materials almost never will be contiguous, such as a shoe and a hat or a blouse and a pair of stockings. Even in the uniform color spaces, the same distance resulting from different tristimulus values may give rise to different levels of consumer dissatisfaction. Carpet, velvet, and satin surfaces give rise to other problems.

FACTORS INVOLVED IN DYEING

In the practical dyehouse once a good match has been made it is not always repeated, even when one thinks all variables have been kept constant. Dyeing no less than experiments are expected to be repeatable in a scientifically-controlled area. It is helpful to make lists of things which may vary. For example, if one reaches a shade by making adds of dyestuff as the process continues, merely totalling the amount added and putting it in initially the next time is not necessarily going to give the same shade. Among factors which must be considered are:

1. Age of dyestuff, fabric, and finish.
2. Rate of wet out and dyeing.
3. Time of dyeing.

4. Fabric to bath ratio.
5. Temperature and rate of rise or fall.
6. Amount of water evaporated and added.
7. Relative movement of bath and fabric.
8. Nature of fabric structure, effects of swelling and squeeze.
9. Changes in temperature if fabric leaves bath and then re-enters.
10. Total amount of bath carried by fabric as it leaves.
11. Presence and nature of materials in the bath: buffers and ionization solubilizing agents and foaming oxidizing or reducing materials.
12. Temperature and conditions of drying.
13. Presence of air currents.
14. Nature of sizes and desizing procedure.

COLORANT NONUNIFORMITY CONTROL

If dyestuff varies, this is an evident problem. Dyestuffs may leave the manufacturer with controlled strength and still show variations. Just as wool in a heated area in winter may weigh 15% less than when exposed to humidity in

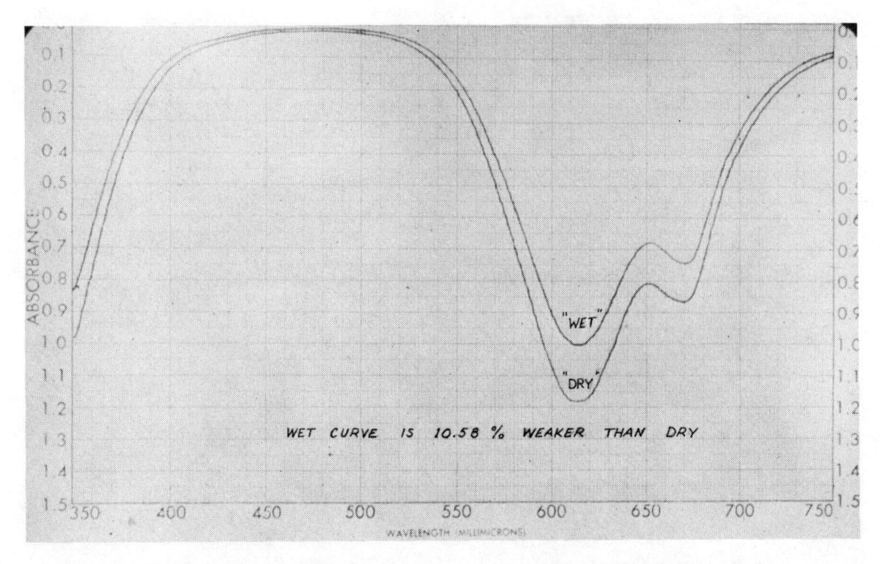

Fig. 1. Solophenyl Turquoise Blue; concentration 0.050 g/500 ml; cell path 1 cm.

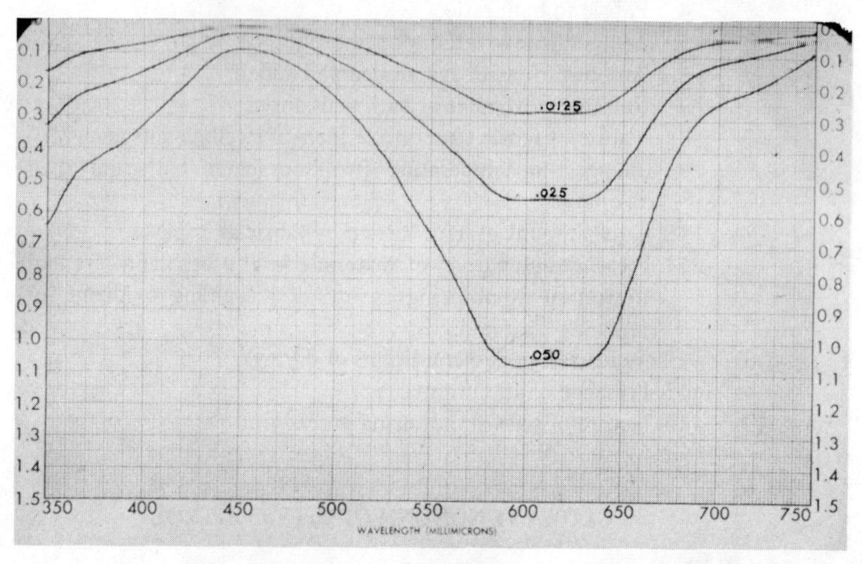

Fig. 2. Xylene Fast Blue P. concentration in g/500 ml; cell path 1 cm.

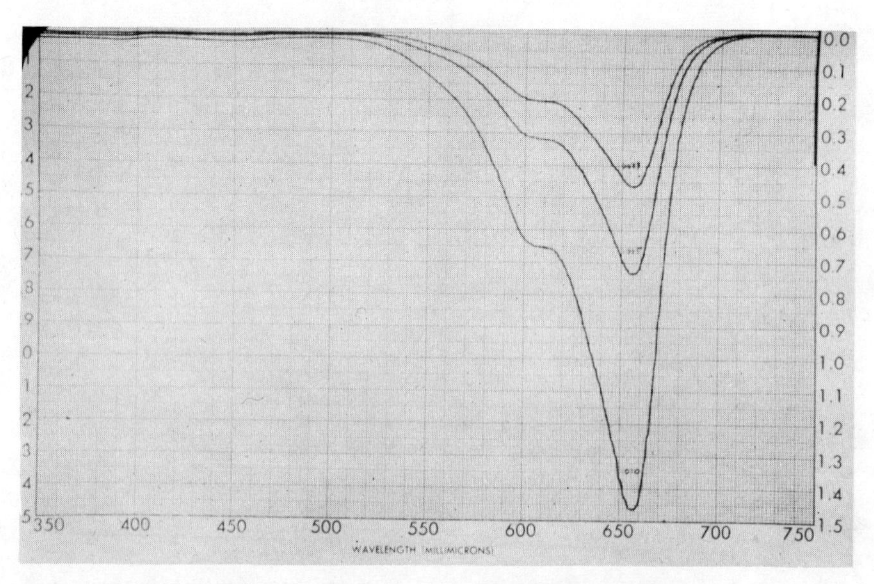

Fig. 3. Basic Blue 3; concentration 0.010 g/500 ml; cell path 1 cm.

summer, commercial dyestuffs pick up or lose moisture, and it is a misplaced effort to weigh to one part in a thousand an uncontrolled material which can change by several per cent in moisture content. An example of what is found in practice is illustrated in Fig. 1. Here the dyestuff samples appear to be of different strengths, because of moisture variations. If one thinks fabric weight is known when it is not, this is a source of trouble.

One source of poor reproducibility is variability of dyestuff received from the producer. Here transmission spectroscopy of known-concentration solutions is basic to control uniformity of shipments. Usually, Beer–Lambert relations hold, as in Figs. 2 and 3. Occasionally, distinct abnormalities are found (Fig. 4) as with Acid Red 99 and (Fig. 5) Acid Orange 63.

There are suppliers with inadequate control, mixing, and reproducibility, and some well-known manufacturers manage to send shipments which vary significantly more than current standards. This is based on initial shipment, which is called 100%, and over ± 3% variation is not accepted. Standardization is not easy. Many dyestuff companies still and probably must test by relatively costly dyeing techniques suitable for people working with many pounds of expensive products. They usually dye skeins. While this is satisfactory for a dyehouse handling yarns, in many dyehouses, despite their diverse activities, yarns are one of the things not usually dyed. In addition, they buy in relatively small quantity as a result of handling a variety of different fibers and fabrics in

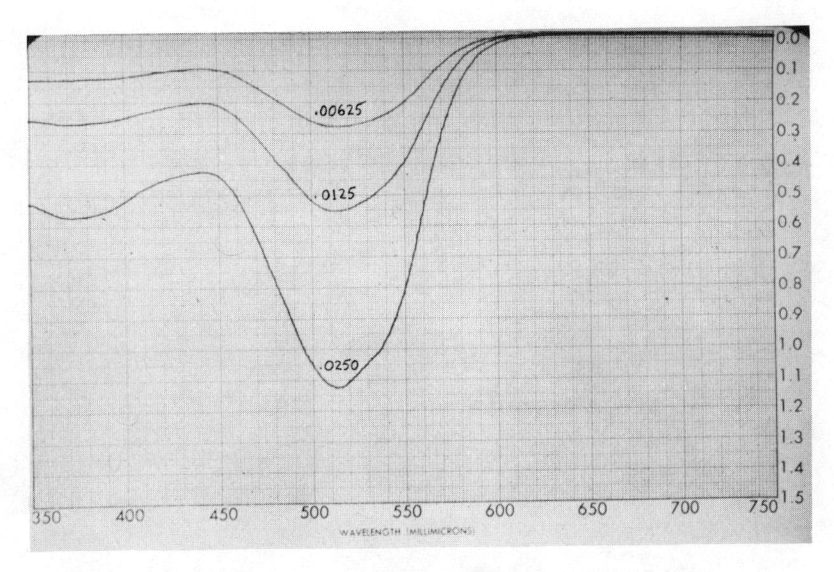

Fig. 4. Acid Red 99; concentration in g/500 ml; cell path 0.1 cm.

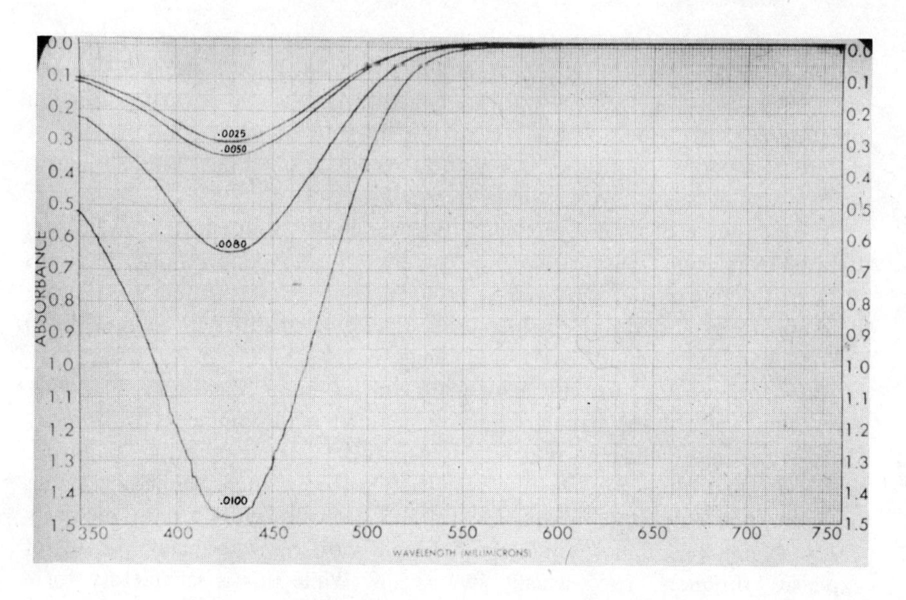

Fig. 5. Acid Orange 63; 0.010 g/500 ml; cell path 1 cm.

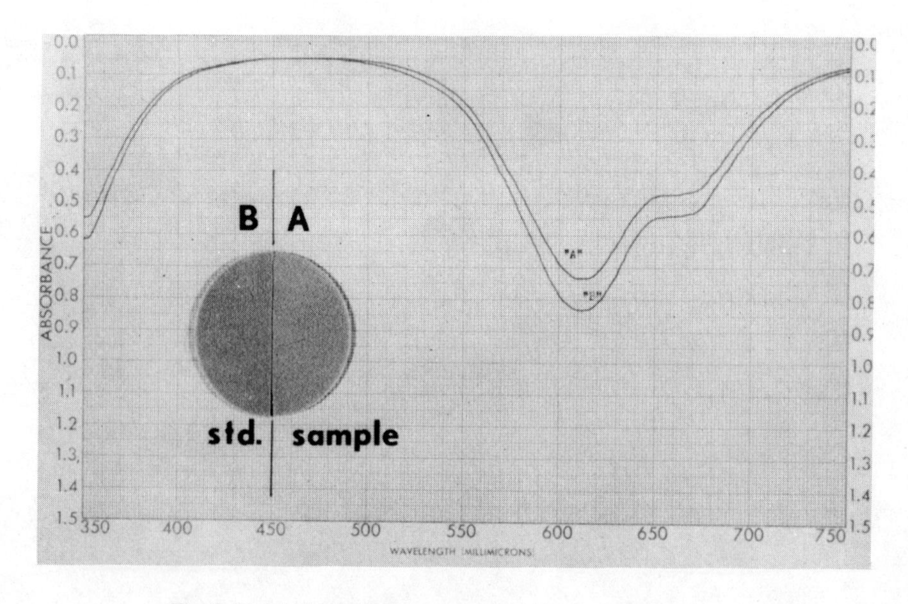

Fig. 6. Direct Turquoise; concentration 5/100; cell path 3 mm.

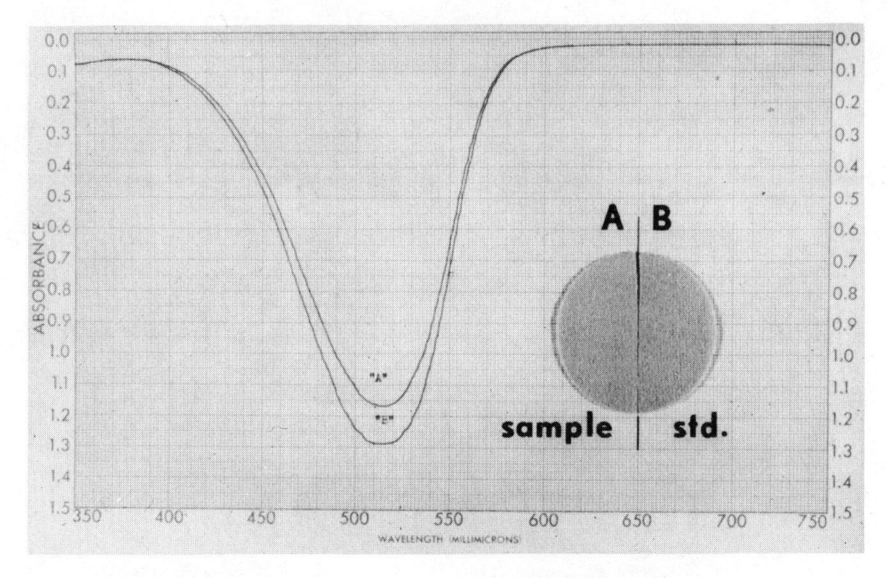

Fig. 7. Basic Red; concentration 10/50; cell path 3 mm.

small and scattered operations. A dyehouse usually can not afford, nor should it have to spend, as much on standardizing dyestuff as the manufacturer. In addition, reproducibility of dyeing is not as good as reproducibility of making a solution and checking light transmission. A suitable procedure is to employ a stable spectrophotometer (Perkin-Elmer 205) fitted with a digital read out and a flow-through cell. The problems met in the spectroscopy of dyestuff solutions have been studied by others.[4,5] Many dyestuff companies suggest helpful procedures. A typical problem is that acetate colors are usually insoluble and must be in stable solution if reproducible values are to be obtained. The accuracy of determination must be better than the accuracy with which one can produce and blend dyestuffs. In order to achieve reproducibility in the laboratory to better than 1/2 of 1%, one works in temperature- and humidity-controlled areas, must weigh and dye with reliable equipment, and recalibrate. The use of a flow-through cell makes for rapid and precise determinations, and is a partial solution to the problem of cleaning and zero recheck. The differences found by photometric inspection of dyestuffs are illustrated by the following: In Fig. 6 Direct Turquoise shows sample 'A' weaker than standard 'B'. The sample is 88.5% based on the standard of 100%. A similar case shows a Basic Red in which sample 'A' is 91% of the standard 'B' (Fig. 7). The next example is an Acid Orange which just got under the line at the time it was done. Here the 'A' sample is 96.4% of the 'B' standard (Fig. 8). Figure 9

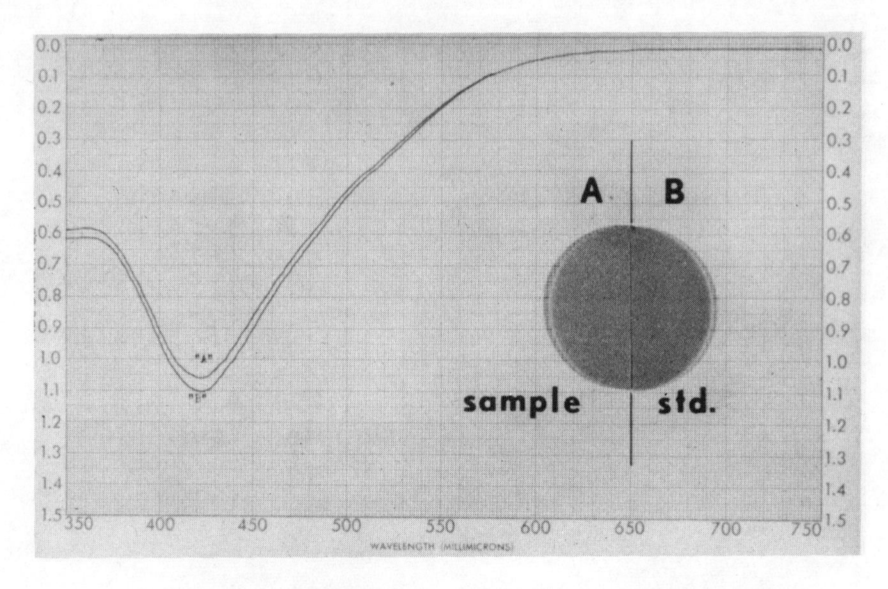

Fig. 8. Acid Orange; concentration 10/50; cell path 3 mm.

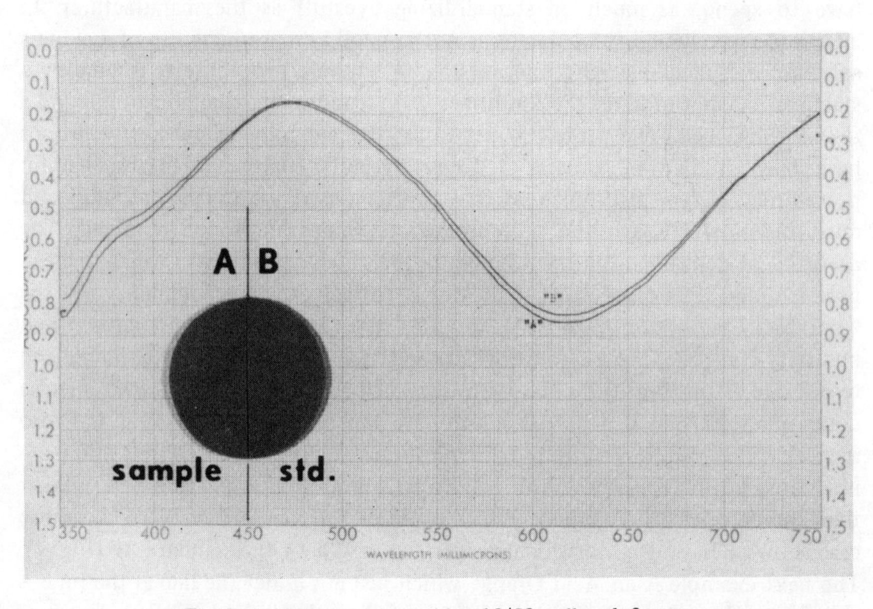

Fig. 9. Acid Blue; concentration 10/50; cell path 3 mm.

shows the standard 'B', or 100%, weaker than sample 'A' which is 103.2%. Some question why one objects to receiving dyestuff stronger than standard, since at the same price one should get more out of it. While true, and dyestuffs consciously and regularly sent stronger for the same price would be acceptable, if changes are frequent, as a matter of quality control in standardized procedures it is dangerous. It is usually easier to add dye than to bring a shade lighter. The time, labor, and material cost of making an add is less than that in bringing a shade back. For this reason dyehouses aim at lighter than the shades they hope to obtain, and in that way get shades by adds.

DYEHOUSE CONTROL IN PRACTICE

In many dyehouses there is no such thing as dyeing without adds. Since an add requires time for leveling out, it takes additional labor. In certain dyehouses three and four adds are normal, and the economics of the operation are predicated on it. If the dyer can be assured he will not overshoot, one cause for adds can be eliminated.

The Beer–Lambert law in absorption spectroscopy gives fair assurance about whether or not the dyestuff is in control. This does not always hold. For example, two sources of the same dye index number, Celanthrene Fast Pink 3B and Palacet Fast Pink FF 3B, show the same strength in the absorbance curves (Figs. 10 and 11), but do not act alike in the dye bath. One of the reasons can be seen in the infrared curves, which show differences (Fig. 12).

It would be desirable if one could as a practical matter monitor for real dyeing in real dye baths. We would know from the start if the dye bath was right, and assuming leveling, we would know when to take our patch. All problems have not yet been practically and economically solved. This is not to say one cannot monitor dye baths and dyeing processes. Figure 13 shows a dyeing as it goes to exhaustion, and indicates different constituents. Figure 14 shows uniform exhaustion rates.

Using standardized dyestuffs will not always give identical shades if used in identical proportions, and some reasons why variations occur were listed. Each has been implicated in actual problems at different times. Problems arise with good overall control but with varying traces of aluminum. Many dyestuffs such as Disperse Red 11 are labile with pH. Recently a brilliant red was popular, and this shade was sensitive to acid, turning dark. Stiffened cotton lace trim which had resin with acid catalyst could print its pattern onto the fabric. Washed garments have changed shade in different parts of the United States because traces of different metallic elements exist in the water. Since the whole garment changes, normally this does not result in complaints.

A problem not solved in all areas is that when a lot of many pieces is dyed,

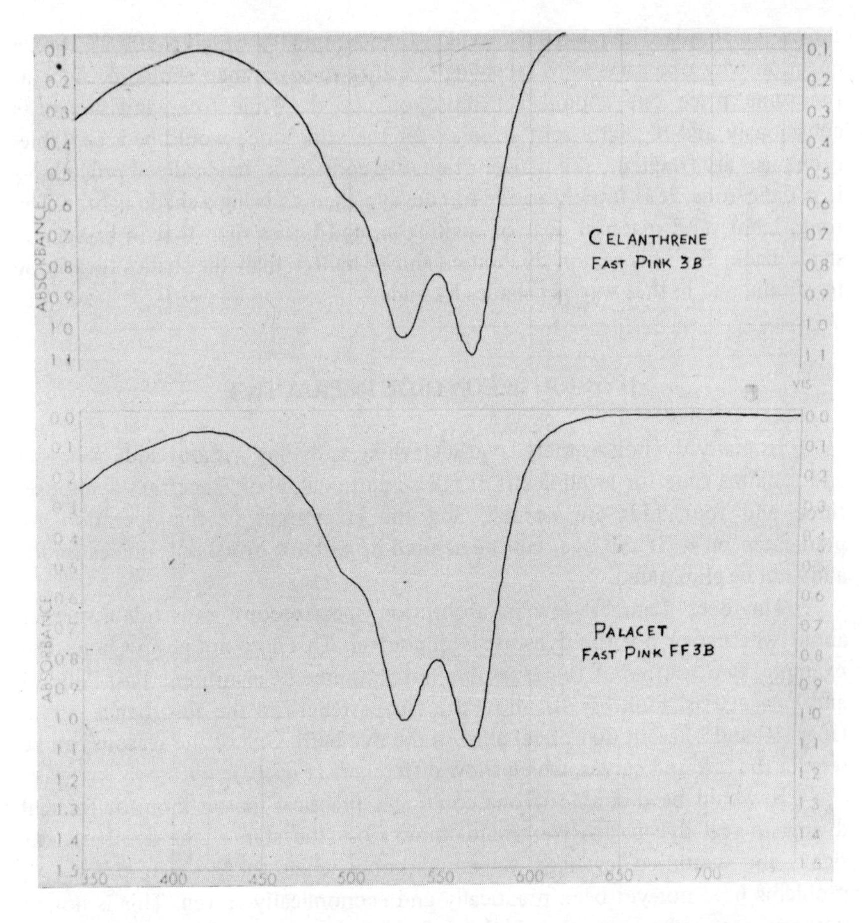

Fig. 10. Absorbance curves for celanthrene Fast Pink 3B and Palacet Fast Pink FF 3B.

more than one shade results. It is not easy to give a dyehouse an exact formula for a specific shade if that dyehouse is in the habit of producing, for reasons they do not control, more than one shade per dyelot. In the past there was much open jig dyeing where as many as five shades were segregated in a dyelot. In those days jigs were open to drafts, and as the fabric passed back and forth, speed and tension varied continuously. Today the leaders in the industry have enclosed jigs with uniform tension, and the laboratory has found a way of duplicating jig-dyeing results. In a jig a large roll of fabric is repeatedly unrolled through a small bath and up onto another roll. The bath-to-fabric weight ratio may be as low as 2½ to 1. Miniature jigs do not solve the problem. In the color-matching laboratory one would like to have a swatch of fabric no bigger than 25 cm on a side. It is easier to work with higher ratios, going up to 15–40 times fabric weight. If dyestuff is not added uniformly or temperatures vary in the system, one tends to get multiple shades. While this part of the operation is not the laboratory responsibility, it would be a foolhardy laboratory which would offer to supply formulas to a dyehouse not in itself well controlled. Every problem an uncontrolled dyehouse would have would be assigned to formulas issued by the laboratory. No matter how badly off a dyehouse may find itself originally, all improvements which result are considered normal operating level, and remaining problems due to laboratory interference. Prior to the introduction of laboratory control based on spectroscopic data it is advisable to make a study of:

1. Depth of shades normally run.
2. Average cost of dyeing.
3. Standards the dyeing must meet.
4. Times for carrying out dyeing at different levels and different standards.
5. Percentage of redyes: (a) caught early, (b) caught after fabric has left the dying area.
6. Time between shade submittal and shade delivery.
7. Amount of business lost because of inability to supply shades.
8. Cost of matching by existing procedures.

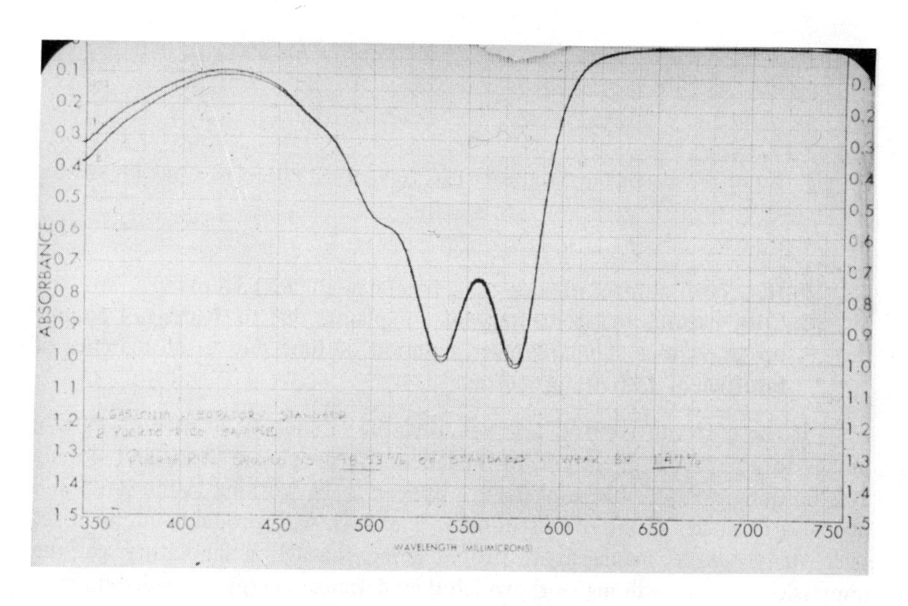

Fig. 11. Absorbance curves for Celanthrene Fast Pink 3b (1) and Palacet Fast Pink FF 3b (2).

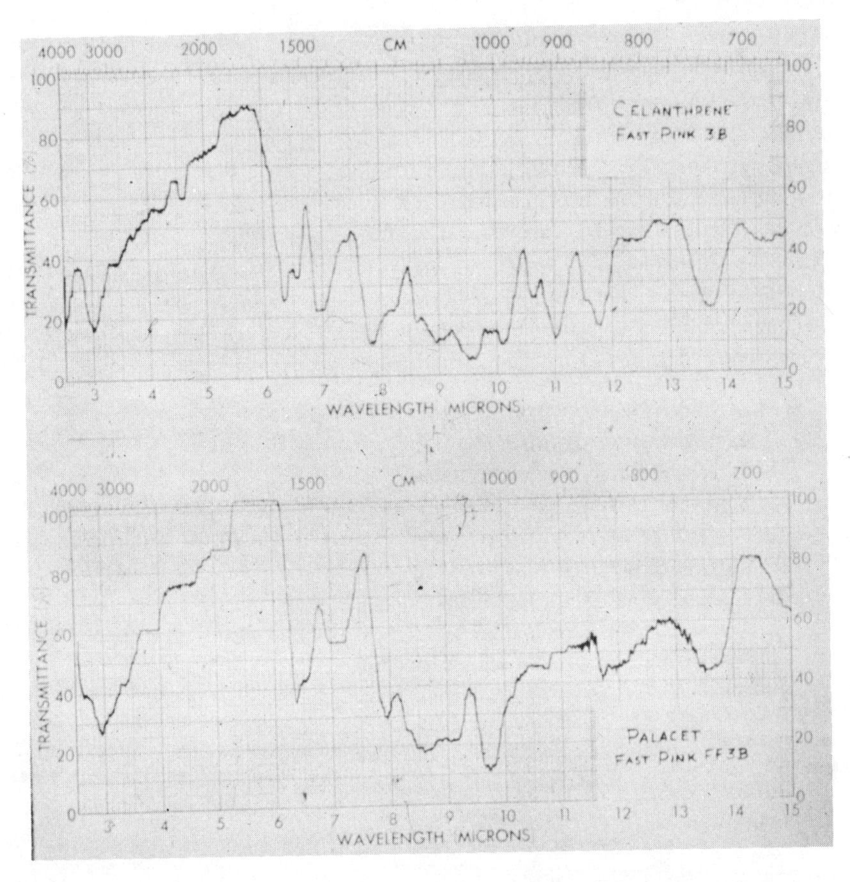

Fig. 12. Infrared spectra for Celanthrene Fast Pink 3B and Palacet Fast Pink FF 3B.

9. History of number of adds going from new shades to reruns.
10. Cost, nature, and percentage of complaints due to shades not coming up to various standards over a period of time due to mismatches of insufficient fastness properties.

If there are no records, one will find the laboratory accused of wasting money in a high-cost operation. In the laboratory people are sitting and looking at something not saleable, and do not appear to be working in the sense mill people are, since the latter must look at and work with products intended for sale. Many textile management people going through a laboratory get the impression no one is doing work, which they define as energy expended directly on products they will sell. This is especially true with automated

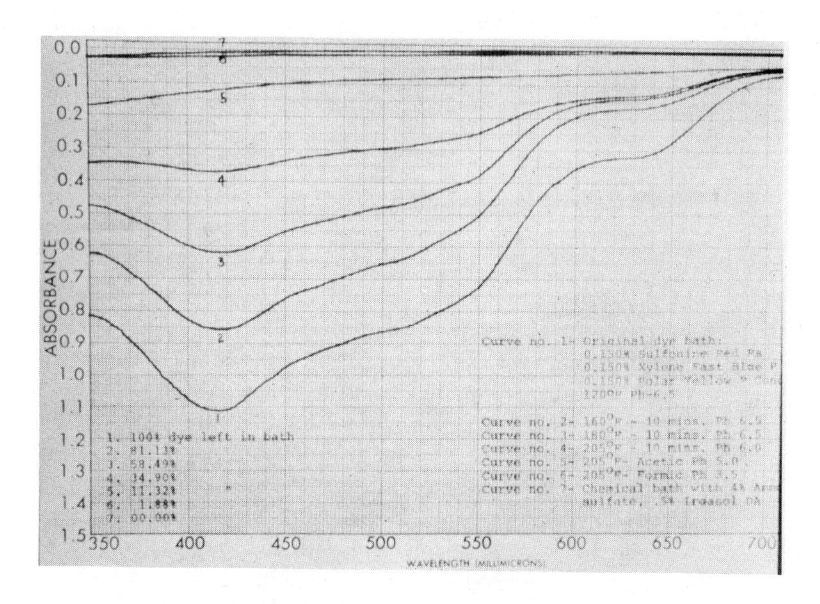

Fig. 13. Absorbance curves for a dye bath going to exhaustion.

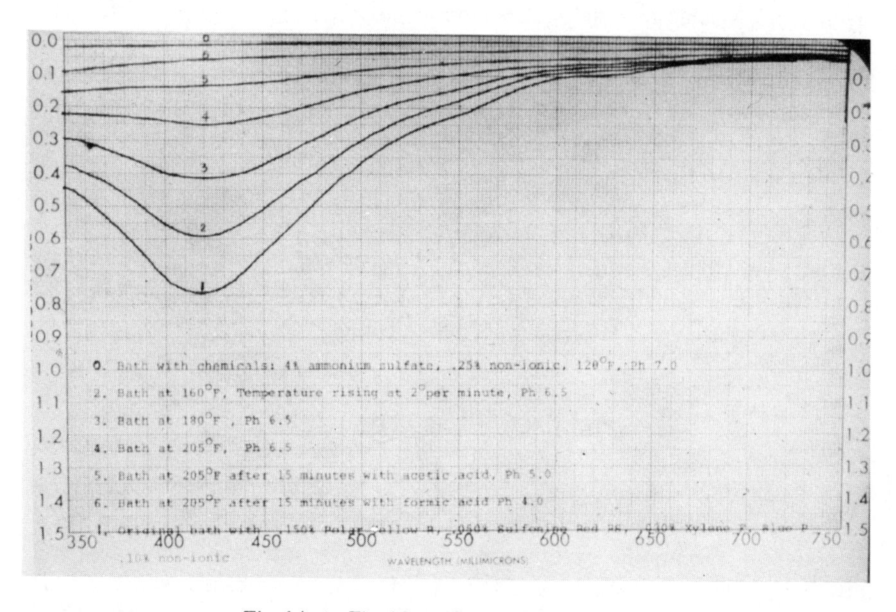

Fig. 14. As Fig. 13; uniform exhaustion rates.

spectrophotometers and laboratory dyeing equipment. The sooner the practical dyehouse begins to resemble the automated laboratories, the sooner they will begin to have the reproducibility of the laboratory.

SPECTROPHOTOMETER–CONTROLLED COLOR MATCHING

The discussion up to this point has touched on transmission matching and tristimulus appearance matching. When it comes to dyeing of fabrics, however, there is a somewhat different situation. The formulas are related to the Kubelka Monk equations going back to 1931,[6] which have been revamped and extended to the point where effective, computed color matching is being carried out in many places. Stearns at American Cyanamid did the first color matching purely on the basis of computation.[7,8] It does not help much if computations cannot be done quickly by people of average intelligence. Graphical and long-hand methods are usually satisfactory only in the laboratory. The only practical solution which begins to compete in time and labor cost with the judgment of a good dyer must make use of modern computer methods. In real dyehouses a desk-top calculator usually can't beat a good man in reaching correct conclusions. With the computer, either analog or digital – and with an advantage in using the digital in many but not all areas – one can run rings around any real dyer. It is probable that the best computer programs are yet to be written, and further sophistication is on its way. Nevertheless, some companies have reached the point where it is possible to take standardized dyestuffs, fabrics, and procedures and dye in existing equipment with zero adds. In one dyehouse recent figures indicated that 40 out of 42 new shades were dyed and approved without adds to the computed formula. It is possible to go from a significant per cent of rejects on new shades to the point where there are almost no rejects except where the customer says what he submitted wasn't the shade really wanted, only bluer, redder, or brighter, than the swatch actually sent. If a customer sends in a shade which is not exactly what is wanted, no computer or spectrophotometer will produce the required color.

THE USE OF THE COMPUTER

Once dyehouse procedures are standardized to the point where one obtains essentially a single shade per dyelot and there is evidence that a stable formula will repeat, the dyehouse is ready for computer matching. Computer matching is what it says. It is not intended to solve problems which lead to variations in shade. If these are not solved, adding the computer is like putting a Cadillac motor onto a wagon. One cannot introduce high-speed methods to only one aspect of an operation without seeing the whole system shake to pieces. One cannot with the computer eliminate knowledgable dyers, because even with the best formula there must come times when the questions are whether to run the fabric longer, change the temperature, add more dyestuff, and if so which. If one were at the stage where dye baths were continuously spectrally checked, and the color of the fabric as well, and had programs that told what to do each time

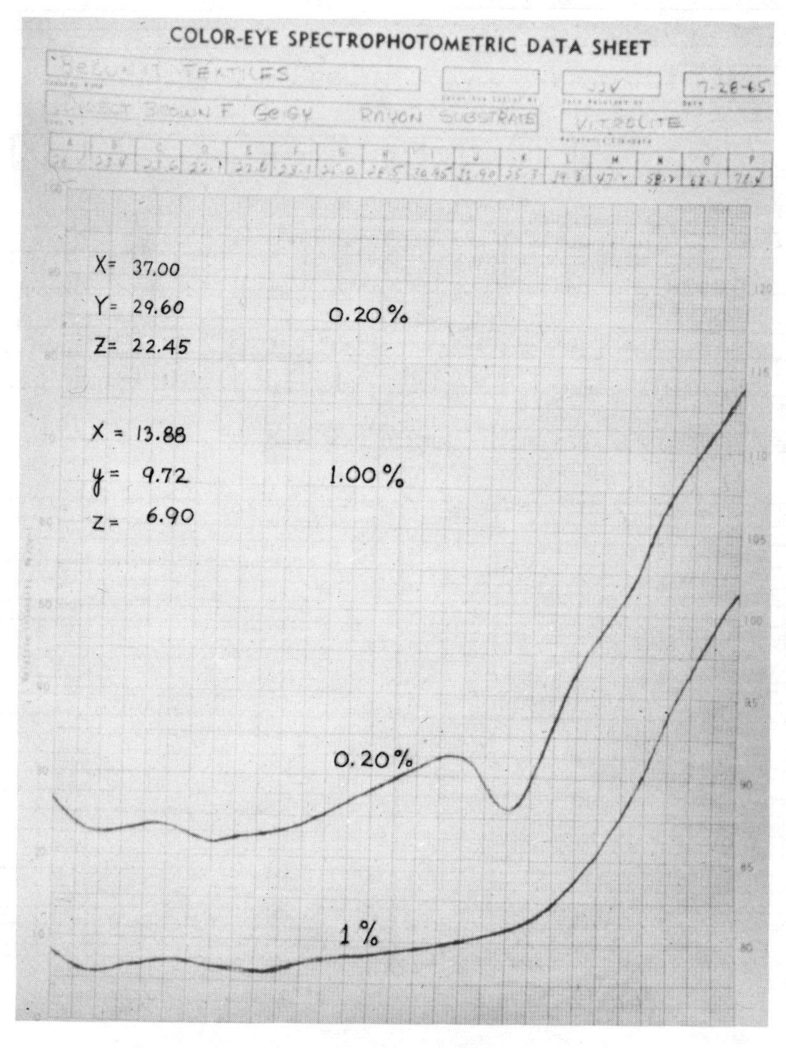

Fig. 15. Spectrogram of Dyeing of Direct Brown F on rayon.

B

BEAUCOLOR MATCHING SERVICE
NEW SHADE SUBMITTAL FOR COMPUTER MATCH

DYE CLASS: A C I D

METAMERIC INDEX CEILING:

FABRIC TYPE/STYLE: TRICOT S/

SUBSTRATE TO BE USED: MOCK DYEING
NYLON S/7168 ( )
NYLON S/CALENDER ( )
NYLON S/ 15 DENIER ( )
OTHER: ( ) ( ) ( )

DATE:

COLOR: NO.:

PLANT: FOUNTAIN INN

COST CEILING: /LB.

NEW SHADE: ( ) REMATCH: ( )

TARGET DESCRIPTION:

2 21

400 nm	420 nm	440 nm	460 nm	480 nm	500 nm	520 nm	540 nm

22 45

560 nm	580 nm	600 nm	620 nm	640 nm	660 nm	680 nm	700 nm

46 69

ACID DYE INVENTORY NYLON ONLY!!! CHECK IF TO BE PROGRAMMED

✓	DYESTUFF NAME	CODE		✓	DYESTUFF NAME	CODE	
	YELLOW 5GL	1					
	YELLOW R	2					
	RED G	3					
	RED RS	4					
	SCARLET B	5					
	BLUE S5R	6					
	BLUE GRL	7					
	BLUE 3G	8					
	CYANINE JR	10					
	BLUE B	9					
	RED B	11					
	ORANGE LSF	12					
	RUBINE 3GP	13					
	ORANGE G	14					
	GREEN G	15					

Fig. 16. Acid dye program on tricot.

S BEAUCOLOR MATCHING SERVICE
NEW SHADE SUBMITTAL FOR COMPUTER MATCH

DYE CLASS: ACID

METAMERIC INDEX CEILING: 78 79

FABRIC TYPE/STYLE: DOUBLE KNIT

SUBSTRATE TO BE USED: MOCK DYEING

DATE:

STYLE 413 ( )
STYLE 966 ( )
STYLE 982 ( )
STYLE 933 ( )
STYLE 923 ( )
STYLE 948 ( )
OTHER: SCOURED ( ) NOT SCOURED ( )

COLOR: NO.

PLANT: STATESVILLE

COST CEILING: /LB.

TARGET DESCRIPTION:

NEW SHADE ( ) REMATCH ( )

2 21

400 nm	420 nm	440 nm	460 nm	480 nm	500 nm	520 nm	540 nm

22 45

560 nm	580 nm	600 nm	620 nm	640 nm	660 nm	680 nm	700 nm

46 69

ACID DYE INVENTORY NYLON ONLY CHECK CODE IF TO BE PROGRAMMED!

DYESTUFF DESCRIPTION	CODE	✓
Navy 2BCS	216	
Rubine 3GP	217	
Red BG	218	
Blue 2GAN	219	
Brown B	220	
Blue DT	221	
Blue P	222	
Blue S3B	223	
Blue P2G	224	
Blue SWF	225	
Red P	226	
Yell O	227	
Yellow FLW	228	
Blue B Conc.	229	
Scarlet GWL	230	
Yellow 2GP	231	
Yellow GR	232	
Orange R Conc.	233	
Blue 5R	234	
Red SWB	235	
Red G	236	
Grey 2BLN	237	
Blue GLF	259	
Green GSN	238	
Red RS	239	
Red B	240	
Turq. 8GL	241	
Yellow R Conc.	242	
Grey P	243	
Orange P	244	
Cyanine G	245	
Violet FBL	246	
Rubine S5B	247	
Orange SFL	248	
Violet P3R	249	
Brown G	250	
Orange SLF	251	
Rubine 5BLF	252	
Navy LFWG	253	
Scarlet YLFW	254	
Scarlet GYL	255	

DYESTUFF DESCRIPTION	CODE	✓
Green 2GFL	256	
Blue 3BLF	257	
Red 4RL	258	
Yellow WGL	260	
Brown 2GLN	261	
Blue 5GLW	262	
Red FD, Conc. 150%	263	
Yellow 5GL	264	
Red SWG	265	

Fig. 17. Acid dye program on nylon double knits.

BEAUNIT TEXTILES
BEAUCOLOR MATCHING SERVICE
NEW SHADE SUBMITTAL FOR COMPUTER MATCH

TO: GASTONIA CENTRAL LABORATORY ATTENTION: [illegible]

DATE:

PLANT: STATESVILLE

CUSTOMER'S NAME:

ORDER (S):

FABRIC: DOUBLE KNIT NYLON

FABRIC STYLE:

END USE AND CODE:

SPECIAL FASTNESS REQUIREMENTS:

LAB DYEINGS REQUIRED FOR SHADE APPROVAL: YES ( ) NO ( )

LAB DYEINGS TO BE SUBMITTED TO: LOCATION:

SPECIAL INSTRUCTIONS:

DELIVERY DATE:

SUBMITTAL SWATCH

NOTE: SWATCH MUST BE AT LEAST 1¼ INCHES IN DIAMETER OR 1¼ INCHES SQUARE, TO BE HANDLED BY OUR EQUIPMENT. STANDARDS SUBMITTED NOT MEETING THESE REQUIREMENTS WILL NOT BE PROCESSED!

ONLY ONE SHADE PER SHEET, PLEASE!

MINIMUM SWATCH SIZE

PREFERRED SWATCH SIZE

COLOR NO. AND NAME

CC:

RETURN DUPLICATE COPY TO NEW YORK OFFICE, ATTENTIC

Fig. 18. Submittal form for computer match; nylon double knit.

BEAUNIT TEXTILES
BEAUCOLOR MATCHING SERVICE
NEW SHADE SUBMITTAL COMPUTER MATCH

TO: GASTONIA CENTRAL LABORATORY ATTENTION: JIM VERNON

DATE:

PLANT: STATESVILLE

CUSTOMER'S NAME:

ORDER (S):

FABRIC: WOVEN STRETCH

FABRIC STYLE:

END USE AND CODE:

SPECIAL FASTNESS REQUIREMENTS:

LAB DYEINGS REQUIRED FOR SHADE APPROVAL: YES ( ) NO ( )

LAB DYEINGS TO BE SUBMITTED TO: LOCATION:

SPECIAL INSTRUCTIONS:

DELIVERY DATE:

SUBMITTAL SWATCH

MINIMUM SWATCH SIZE

PREFERRED SWATCH SIZE

NOTE: SWATCH MUST BE AT LEAST 1¼ INCHES IN DIAMETER OR 1¼ INCHES SQUARE, TRIPLE THICKNESS, TO BE HANDLED BY OUR EQUIPMENT. STANDARDS SUBMITTED NOT MEETING THESE REQUIREMENTS WILL NOT BE PROCESSED!

ONLY ONE SHADE PER SHEET, PLEASE!

COLOR NO. AND NAME

CC:

RETURN DUPLICATE COPY TO NEW YORK OFFICE, ATTENTION:

Fig. 19. Submittal form for computer match; woven stretch.

anything went wrong, it is possible one could do without a good dyer. This is far off into the future. At the present level there are places and instances where the computer is not yet practical in use. There is no current good way to handle fluorescent shades. A new program as well as improved spectrophotometers are required. If the dyer does not know what to do when the shade comes a trifle off, and does not have the experience required, he can cause just as much trouble with the computer as without. If the dyer does not know the difference between something that is or is not a commercial match, the computer will not be able to tell him.

A primary aspect of computer matching is that one can use less- expensive dyestuffs to get a match. This is usually true. However, if one is dyeing light shades and many nonrepetitive lots of dark shades, the amount of money saved can be so slight as to make no difference in total costs. If time of dyeing and cost of chemicals other than dyestuff, plus labor, completely swamp cost for dyestuffs, savings in dyestuffs may be negligible. A scientific system requires expenditure of time and money, and must be paid for as much as anything else.

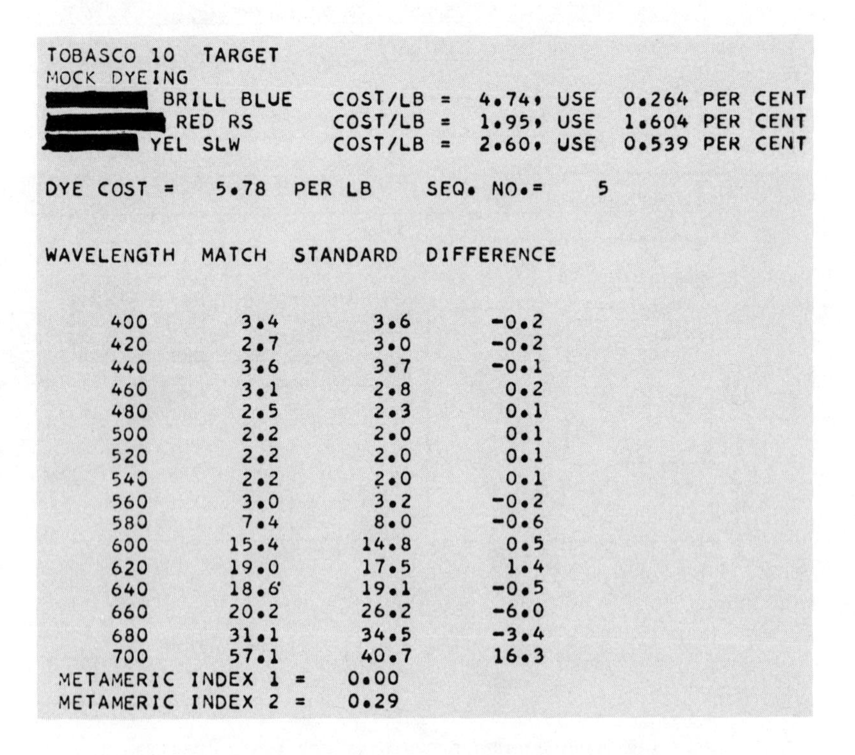

TOBASCO 10 TARGET
MOCK DYEING
BRILL BLUE COST/LB = 4.74, USE 0.264 PER CENT
RED RS COST/LB = 1.95, USE 1.604 PER CENT
YEL SLW COST/LB = 2.60, USE 0.539 PER CENT

DYE COST = 5.78 PER LB SEQ. NO.= 5

WAVELENGTH	MATCH	STANDARD	DIFFERENCE
400	3.4	3.6	-0.2
420	2.7	3.0	-0.2
440	3.6	3.7	-0.1
460	3.1	2.8	0.2
480	2.5	2.3	0.1
500	2.2	2.0	0.1
520	2.2	2.0	0.1
540	2.2	2.0	0.1
560	3.0	3.2	-0.2
580	7.4	8.0	-0.6
600	15.4	14.8	0.5
620	19.0	17.5	1.4
640	18.6	19.1	-0.5
660	20.2	26.2	-6.0
680	31.1	34.5	-3.4
700	57.1	40.7	16.3

METAMERIC INDEX 1 = 0.00
METAMERIC INDEX 2 = 0.29

Fig. 20. Typical cost and metameric index printout.

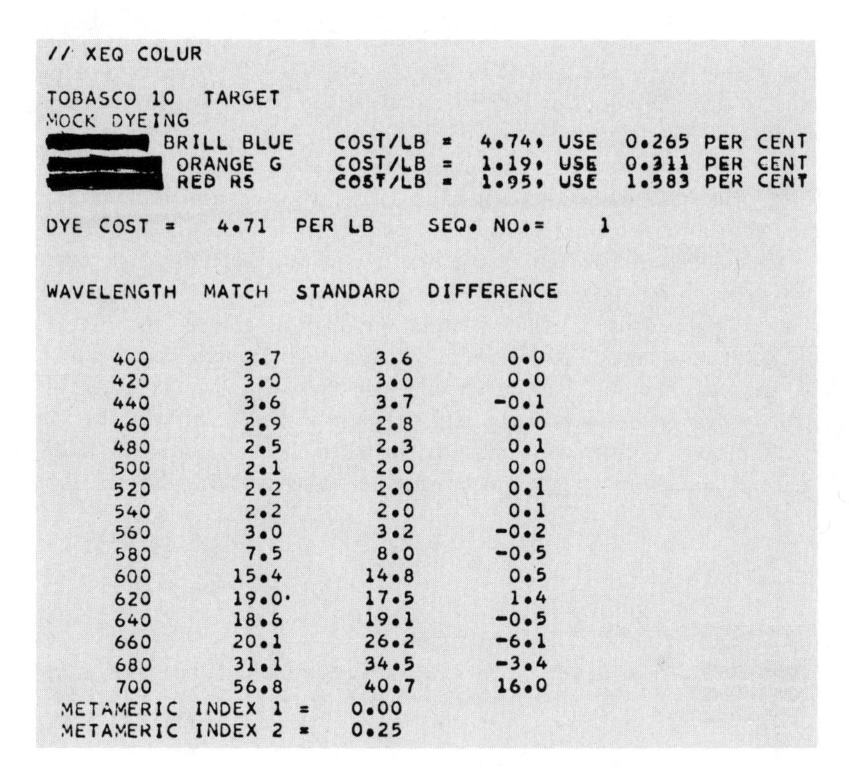

// XEQ COLUR

TOBASCO 10 TARGET
MOCK DYEING
BRILL BLUE COST/LB = 4.74, USE 0.265 PER CENT
ORANGE G COST/LB = 1.19, USE 0.311 PER CENT
RED RS COST/LB = 1.95, USE 1.583 PER CENT

DYE COST = 4.71 PER LB SEQ. NO.= 1

WAVELENGTH	MATCH	STANDARD	DIFFERENCE
400	3.7	3.6	0.0
420	3.0	3.0	0.0
440	3.6	3.7	-0.1
460	2.9	2.8	0.0
480	2.5	2.3	0.1
500	2.1	2.0	0.0
520	2.2	2.0	0.1
540	2.2	2.0	0.1
560	3.0	3.2	-0.2
580	7.5	8.0	-0.5
600	15.4	14.8	0.5
620	19.0	17.5	1.4
640	18.6	19.1	-0.5
660	20.1	26.2	-6.1
680	31.1	34.5	-3.4
700	56.8	40.7	16.0

METAMERIC INDEX 1 = 0.00
METAMERIC INDEX 2 = 0.25

Fig. 21. Typical cost and metameric index printout.

It is necessary to make six dyeings with every dyestuff brought into our program. While this may seem large, the reason is simple. Just as there are discrepancies from the Beer–Lambert law, one finds more discrepancies from the modified Kubelka Monk equations. These problems are illustrated by Fig. 15, showing a dyeing of Direct Brown F on rayon. As can be seen from the shape of the curves, there is a significant and startling difference between the 0.2% dyeing and the 1.0% dyeing. With only one or two or three dyeings one would be continually involved in corrections and modifications to make matches. Actually, this is what has happened with other systems based on a limited number of dyeings, where the goal has not been total elimination of adds, but only reduction in dyestuff costs.

We do dyeings on different substrates. Figure 16 shows an acid dye program on tricot with the names of the dyestuffs deleted. At the top it indicates 40 denier, 15 denier, and calendered styles. A program for 40 denier is incomplete, because one needs a program for bright and dull. Corrections are required if one goes from nylon 66 to 6. There are 15 dyestuffs on this program, but all are not computed all the time. Figure 17 shows an acid dye program on nylon double knits. The substrates include different yarn sources. Bulk nylon from different sources, even if the same luster and denier, needs modification. Fastness and leveling must be taken into account.

Laboratory submittal forms for new colors are required. Figure 18 is a form for a nylon double knit, and Fig. 19 for a woven stretch. Some typical computer print-out results are shown in Figs. 20, 21, and 22.

One does not need to match exactly the spectrophotometric curve of every sample. If one does, problems and costs may increase. If someone sends a dry leaf, a painted paper, or even fabric, the question arises as to what to do: Should one match under conditions at which the designer decided this was the shade desired, or should one match under all illuminations? One might think that underwear shades need only be matched in incandescent light. This alone involves problems. Swimwear must be matched in the equivalent of bright daylight. It is remarkable how much many colors seem to stay the same despite

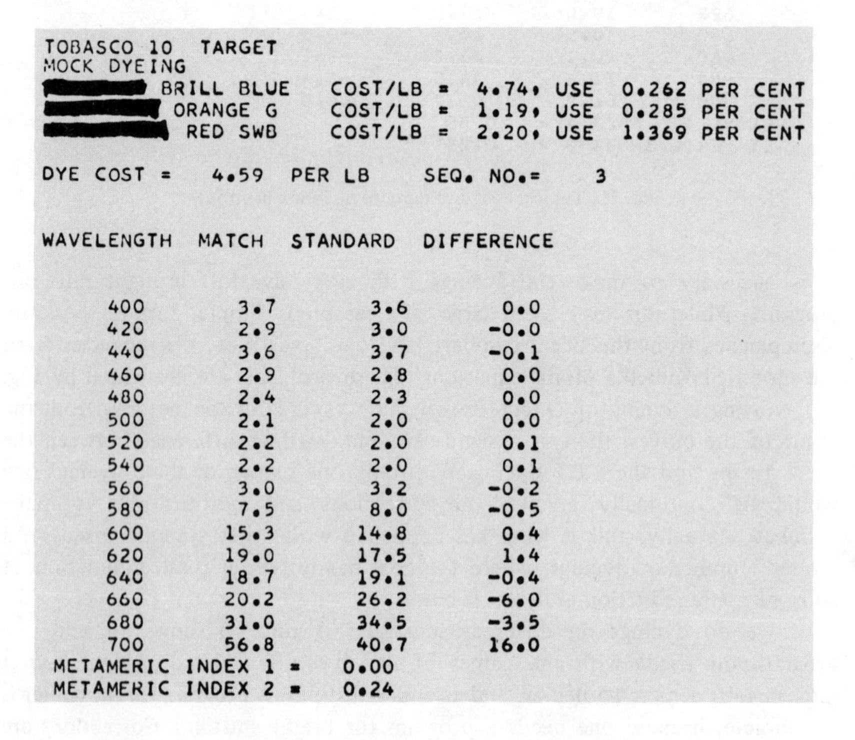

TOBASCO 10 TARGET
MOCK DYEING
[redacted] BRILL BLUE COST/LB = 4.74, USE 0.262 PER CENT
[redacted] ORANGE G COST/LB = 1.19, USE 0.285 PER CENT
[redacted] RED SWB COST/LB = 2.20, USE 1.369 PER CENT

DYE COST = 4.59 PER LB SEQ. NO.= 3

WAVELENGTH	MATCH	STANDARD	DIFFERENCE
400	3.7	3.6	0.0
420	2.9	3.0	-0.0
440	3.6	3.7	-0.1
460	2.9	2.8	0.0
480	2.4	2.3	0.0
500	2.1	2.0	0.0
520	2.2	2.0	0.1
540	2.2	2.0	0.1
560	3.0	3.2	-0.2
580	7.5	8.0	-0.5
600	15.3	14.8	0.4
620	19.0	17.5	1.4
640	18.7	19.1	-0.4
660	20.2	26.2	-6.0
680	31.0	34.5	-3.5
700	56.8	40.7	16.0

METAMERIC INDEX 1 = 0.00
METAMERIC INDEX 2 = 0.24

Fig. 22. Typical cost and metameric index printout.

B

BEAUCOLOR MATCHING SYSTEM
NEW SHADE SUBMITTAL FOR COMPUTER MATCH

DYE CLASS: D I S P E R S E

METAMERIC INDEX CEILING:

FABRIC TYPE/STYLE: TRICOT S/

SUBSTRATE: (MOCK DYEING)
- ACETATE S/08602 ( )
- ACETATE S/2085 ( )
- NYLON S/7168 ( )
- NYLON CALENDER ( )
- NYLON 15 DENIER ( )
- ACETATE/NYLON S/313 ( )
- OTHER:

DATE:

COLOR: NO:

PLANT: FI ( ) LOWELL ( )

COST CEILING:

TARGET DESCRIPTION: NEW: ( ) REMATCH: ( )

2 21

400 nm	420 nm	440 nm	460 nm	480 nm	500 nm	520 nm	540 nm

22 45

560 nm	580 nm	600 nm	620 nm	640 nm	660 nm	680 nm	700 nm

46 69

DISPERSE DYE INVENTORY ACETATE AND NYLON, CHECK IF TO BE PROGRAMMED

✓	DYESTUFF NAME DESCRIPTION	ACE CODE	NYL CODE	✓	DYESTUFF NAME DESCRIPTION	ACE CODE	NYL CODE
	YELLOW 8GGLF	78	50		SCARLET GN	90	----
	SCARLET RGLF	79	51		RED PGMF	92	----
	PINK BGLF	80	52		VIOLET P2B	97	----
	YELLOW 4RLF	81	53		YELLOW P2RLF	98	----
	RED 2BGLF	82	54		BLUE PBFL	99	----
	TURQUOISE 4G	83	55		ORANGE 3RGLF	102	----
	YELLOW 2RGLF	84	56		NAVY BLUE BJ	103	----
	BLUE BGLF	85	57		BLUE 5G	104	----
	BLUE AF	86	58		BLUE JBN	105	----
	YELLOW G	87	59		CERISE YRF	106	----
	BLUE BN NEW	88	60		BLUE GREEN BA	107	----
	TURQ BLUE GRL	---	61		YELLOW GH	108	----
	PINK PRF	94	62		SCARLET RNA	109	----
	TURQ BLUE G	95	63		ORANGE 2RN	110	----
	YELL RL PST	96	64		RED 3B	111	----
	BLUE BG	100	65				
	CERISE YL PAST	101	66				
	ORANGE 4RN	87	----				
	RED R CONC.	76	----				
	SCARLET GP	77	----				
	NAVY BLUE BR	88	----				
	RUBINE GF4BN	89	----				

Fig. 23. Program indiciating different codes for the same dyestuff on different fibers.

change of illuminant. If it is necessary to match at point of sale, or to work on the assumption of minimal differences at each wavelength, the program must be a sort of least-squares system for all wavelengths, and one gives away flexibility in matching. It is only when matching to replace a fabric which will be adjacent or contiguous in a garment or ensemble of the match shade, and which will be exposed to varying illumination, that one must make matches spectrally similar.

A problem in setting up the system has to do with paperwork and forms. A considerable amount of information is required on a continuing basis. Programs for dyeing tricot using acid and dispersed colors have been shown. The colors dye differently on nylon and on acetate. In Fig. 23 there are different code addresses for the same dyestuff if used on both fibers.

FADING AND SPECTROSCOPY

Spectroscopy has helped to explain some of the anomalies in lightfastness testing. The Ciba Company has shown (Fig. 24) strong fading in 4 hr of Fadeometer exposure, and relatively slight fading after 40 hr of actual sunshine

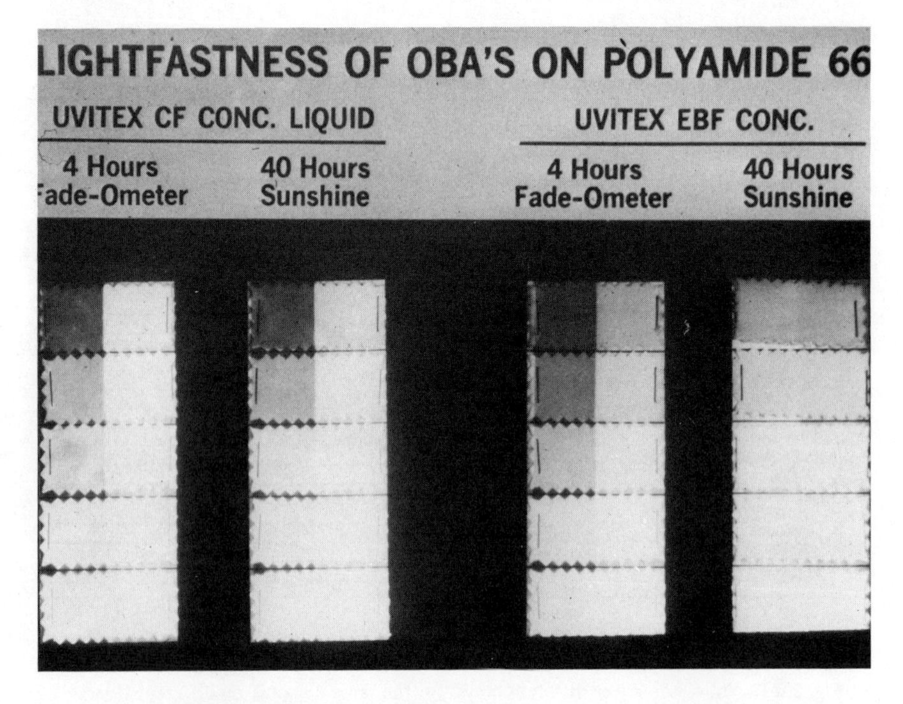

Fig. 24. Comparsion of effects of fadeometer and sunlight on identical swatches.

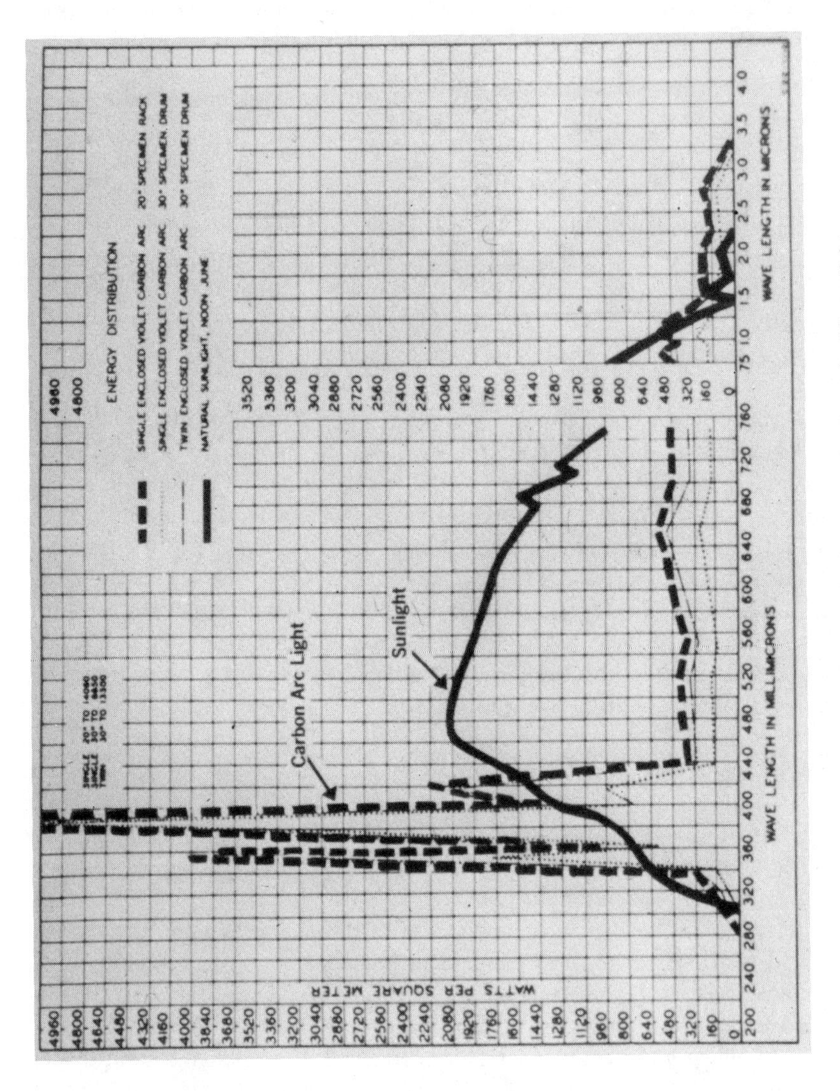

Fig. 25. Chart of energy output of a fadeometer compated with sunlight.

with Uvitex EBF conc. Figure 25 shows the energy output of a Fadeometer compared with daylight. The former concentrates energy in a region where the fluorescent color is most subject to breakdown. Under these circumstances the standard Fadeometer test can be misleading if one wishes to know what will happen to a fluorescent whitened fabric under conditions of actual wear.

COLOR NUMBERING SYSTEMS

A further approach to control in the dyehouse involves the possibility of easily applying meaningful identification numbers to fabric colors. Reporting the entire spectroscopic reflectance curve has not been helpful, since what seem minor variations in the curve lead to significant differences in dominant wavelengths. The *XYZ* system, and the uniform color space *UVW* system, while useful in the laboratory, are not readily visualized by the practical dyer. The use of *Y, x, y* values, while more readily visualized with the help of charts, suffers from the major change in dominant wavelength with slight changes in numerical values of *x* and *y*. For this reason a classification number must have many significant figures to be effective. The *L, a, b* system appears to be more appropriate for dyehouse color specification. Efforts have been made to try a seven-digit classification number where the first two numbers give the *L* value, the third number indicates the sign of the *a* and *b* values, the fourth and fifth values are the *a* and the sixth and seventh values are the *b* readings. Merely by inspection, one has a fairly good idea of the psychological color. While it is too early to decide whether this Beaunit color number will prove itself in practice, it has shown a number of attractive features.

PROBLEMS FOR THE FUTURE

The foregoing has reviewed in a limited way the application of spectroscopy and computer technology to this old industry. Part of the remaining problems are due to incomplete information about vision. Part are due to limited instrumentation; but many details lead to continuing problems. We have indicated why the dyer and the art of dyeing is still necessary. If a color is surrounded by other colors, it appears different in shade. The amount and direction it changes is a complex function. If called on to match a solid shade, which the designer sees surrounded in a print, we can almost guarantee missing the shade required if we match exactly our solid to the spectrophotometric curve of the small section in the surround. Similarly, a printer is called on to put a particular shade in a multicolored print so that one of the colors looks exactly like the solid shade on the other fabric. He is called upon to make a color judgment which at present cannot be furnished completely successfully on any mathematical program. Under these circumstances the match must be made by a human being. This problem will be solved with the computer in the future, but it is not solved at present.

REFERENCES

1. *Glossy Munsell Book of Color,* Munsell Color Company, Baltimore, Maryland (1958).
2. E. Q. Adams, *J. Opt. Soc. Am.* **32**, 168 (1922); D. L. MacAdams, *J. Opt. Soc. Am.* **32**, 247 (1942).
3. Wesley Coppock, Instrumental Assessment of Optically Brightened Paper, *Canadian Pulp and Paper,* 16 (1966).
4. Robert G. White, Allied Chemical Corporation, National Aniline Division, Publication Vol. 45 (1965).
5. E. I. Stearns, Measurement of Dye Strength, *Am. Dyestuff Rep.* **39**, 358 (1950).
6. P. Kubelka and F. Munk, Ein Beitrag zur Optik der Farbanstriche, *Z. Tech. Phys.* **12**, 593 (1931).
7. E. I. Stearns, *J. Opt. Soc. Am.* **34**, 112 (1944).
8. E. I. Stearns, *Am. Dyestuff Rep.* **1968** (February 29), 148.

Application of Internal Reflection Spectroscopy to the Quantitative Analysis of Mixed Fibers

Paul A. Wilks, Jr. and John W. Cassels

Wilks Scientific Corp.
South Norwalk, Connecticut

Internal-reflectance spectroscopy has proved to be an excellent method for the qualitative identification of fibers both in the raw state and in fabrics. In the present study the technique has been applied to the recording of infrared spectra of a number of fabrics woven from two or more types of fibers. Infrared spectra can be obtained directly from the fabric with no special treatment. A number of factors affect quantitative accuracy. With careful control of these factors spectra of quantitative value can be obtained on many mixed fabrics. Special techniques have been developed aimed at maximizing the quantitative accuracy.

INTRODUCTION

Internal-reflection spectroscopy[1] has been demonstrated to be a practical method for the qualitative identification of textile fibers.[2] Characteristic spectra can be obtained on raw fibers or woven fabrics by placing them in contact with a multiple-internal-reflection plate. Because fibers make inefficient contact with the reflector plate, a thin plate (1 mm thick or less) is generally required to provide a sufficient number of reflections to produce a strong spectrum. Such spectra when properly recorded are nearly identical to transmission curves on

*Presented before the 7th national meeting of the SAS Chicago, Illinois, May 1968.

the same material. Similar spectra can also be obtained on yarns or woven fabrics of mixed fibers, and under the proper circumstances can be made to yield quantitative information.

In transmission spectroscopy only two characteristics control the amount of absorption: the absorptivity of the sample and the path length of the radiation through the sample. In internal-reflection spectroscopy there are a number of variables inherent in the technique itself, plus some that are specifically associated with fibers; all of those must be controlled if a quantitative analysis with any degree of accuracy is to be obtained.

VARIABLES INHERENT IN THE INTERNAL-REFLECTION TECHNIQUE

Internal-Reflector Plate Material

Materials with different indices of refraction will produce different effective penetrations into the sample. For fiber studies KRS-5 and germanium are the most commonly used materials. Germanium has the highest index of refraction and is sometimes required for the higher-index fibers, but is quite fragile and subject to fracturing under the pressures required to obtain good fiber contact. KRS-5 provides good results under most circumstances. Because of its slightly plastic nature, it will not fracture under stress, although the surfaces can be permanently deformed when too much pressure is applied. Other materials which have limited use are silver chloride and silver bromide, both of lower index than the other materials and softer, but more economical.

Number of Reflections

Absorption is proportional to the number of reflections; thus the dimensions of the reflector plate are an important variable in internal-reflection spectroscopy. Most multiple reflection plates are approximately 50 mm long by 2 or 1 mm thick, giving 25 and 50 reflections, respectively, at 45°.

Angle of Incidence

The angle of incident radiation has a multiple effect on the spectrum: (1) The closer the angle of incidence is to the critical angle, the greater is the absorbance. (2) A change in the angle of incidence will change the number of reflections, and hence the absorbance. (3) Distorted spectra will be obtained when the angle of incidence is set too close to the critical angle.[3]

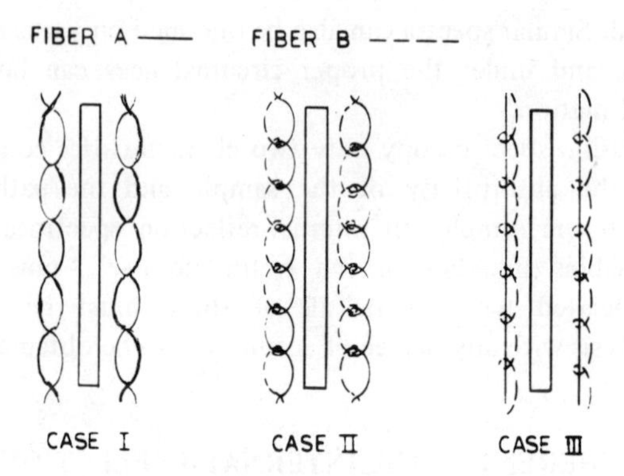

Fig. 1. Fabric woven of two different yarns.

Area of Plate Coverage

The area of the plate (like the angle of incidence) is analogous to the path length of a transmission cell; the amount covered by the sample can be varied to secure the proper absorbance range.

Efficiency of Contact

This is the most difficult of the variables to control accurately, particularly when dealing with fibers. A soft, plastic material makes perfect contact with the reflector plate, but a relatively stiff sample like a fiber makes only line contact at best. Differences in the amount of pressure applied can cause large variations in absorption. Hence for reproducible results a means for repeating pressures is required. Furthermore, quantitative analyses should be made by a band ratio technique rather than by direct measurement of a single band.

SPECIAL PROBLEMS ASSOCIATED WITH FABRICS

ATR is a surface-measuring technique. The spectrum recorded is of what is in actual contact with the plate. If a mixed fabric is formed of a yarn which is of itself made from the two fibers twisted together, then a good quantitative analysis should be possible. If, however, the fabric is woven of two different yarns (Fig. 1), then the chances of making accurate quantitative analyses are reduced.

In case I of Fig. 1, a symmetrical fiber, i.e., with both sides of the cloth containing the same amount of each fiber, poses no problem, but in case II, where the fabric differs from side to side in fiber content, the quantitative error can be minimized by contacting the "front" side of the fabric to one side of the reflector plate and the "back" side to the other. (Both pieces must be identical in size.) In case III of Fig. 1 the only solution is to chop up a piece of the fabric and then contact the resulting mixture to the plate.

THE QUANTITATIVE INTERNAL-REFLECTION METHOD

In Fig. 2 the method of mounting a fabric sample is shown. Pieces of fabric are cut to size and placed on either side of the reflector plate, and the sandwich is placed in the solid sample holder. Some sample should always be on both sides of the plate so that the metal parts of the holder do not come in contact with it. Smoothly-woven fabrics need no special backing, but rough or nubby fibers should be backed by a rubber pad to reduce surface damage to the ATR plate. A typical reflectance assembly is shown mounted in an infrared spectrophotometer in Fig. 3.

Calibration curves can be prepared from samples of known concentration. A strong band for each fiber is selected, preferably in an area of minimum absorbance from the other material. An alternate procedure is to make use of synthetic standards. Here a pure fabric containing one fiber is placed on one side of the reflector plate and a pure fabric containing the other is placed on the

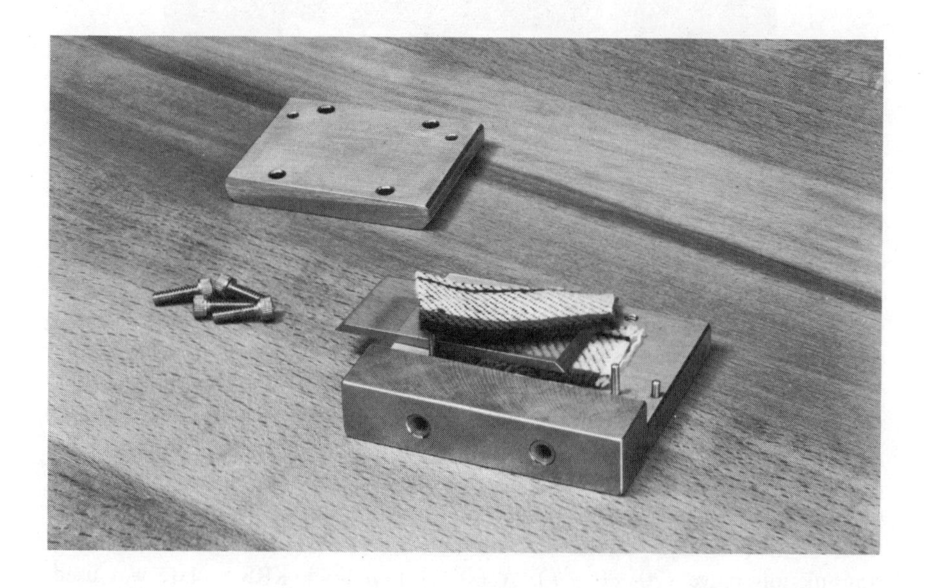

Fig. 2. Method for mounting a fabric sample.

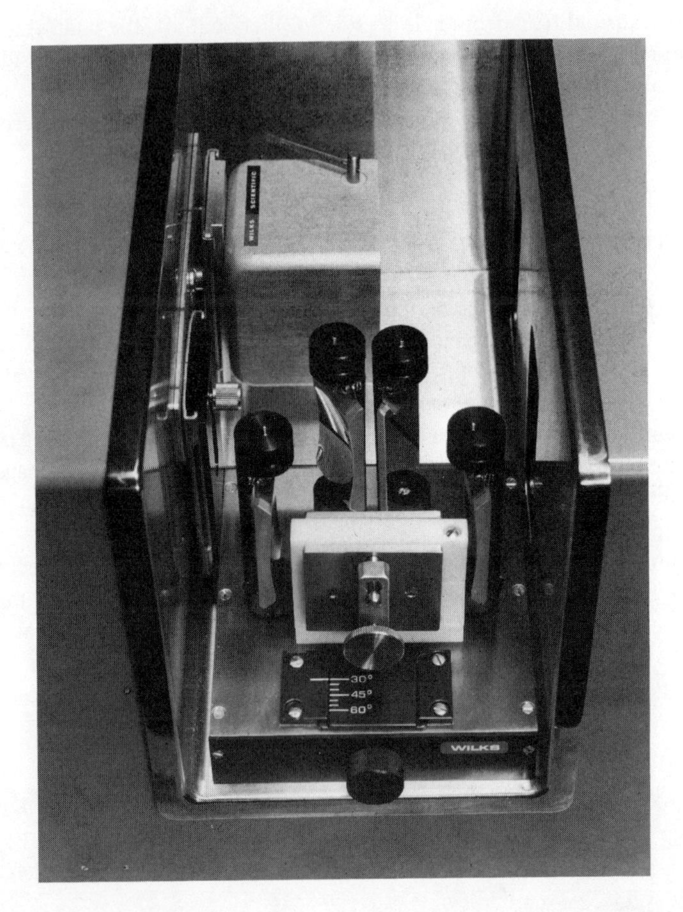

Fig. 3. Typical reflectance assembly mounted in an infrared spectrophotometer.

other side. The different calibration points are obtained by using fabric pieces of different sizes to provide the required ratios of plate coverage.

A typical calibration curve for a cotton–nylon mixture is shown in Fig. 4 together with two samples which were marked 36% and 70% nylon, respectively.

The measurements were made using a Wilks Model 50 variable-angle multiple-internal-reflection attachment mounted in a Beckman IR-5A spectrophotometer. A Jomax 10 Screwdriver was used to ensure reproducible pressure settings (although the band-ratio method makes absolute reproduction of pressures unnecessary). The angle of incidence was set at 60° where it was found that the major bands of both nylon and cotton fell within the ideal transmission range (20–80% *T*). A 60°, 1 mm thick KRS-5 plate was used.

Under these circumstances there appears to be practically no cotton absorbance in the region of the strong nylon bands between 6 and 7 μ, and at the strong cotton bands in the 9–10 μ region there is little nylon absorption. Hence this particular system lends itself very nicely to quantitative infrared analysis.

The sample labeled 36% nylon gave an analytical value of 35%, while the sample labeled 70% had an analytical value of 64% (Fig. 5). This latter sample had been suspected as having a low nylon value after failing physical tests, which may account for the discrepancy between the analysis and the label.

In spite of the many variables which must be controlled, internal-reflection spectroscopy appears capable of giving good quantitative results when used for fabric analysis. It has the following advantages over transmission methods:

1. No sample preparation is necessary on symmetrically-woven fabrics.
2. A combination of reflector plate dimensions, area coverage, and angles of incidence can be selected which will provide absorbance values in the range of best instrument accuracy.
3. Synthetic samples can be readily prepared for calibration purposes.
4. The large area of the reflector plate means that errors in the precise size of the sample will have relatively little effect when compared to errors introduced by sample thickness in transmission spectroscopy.

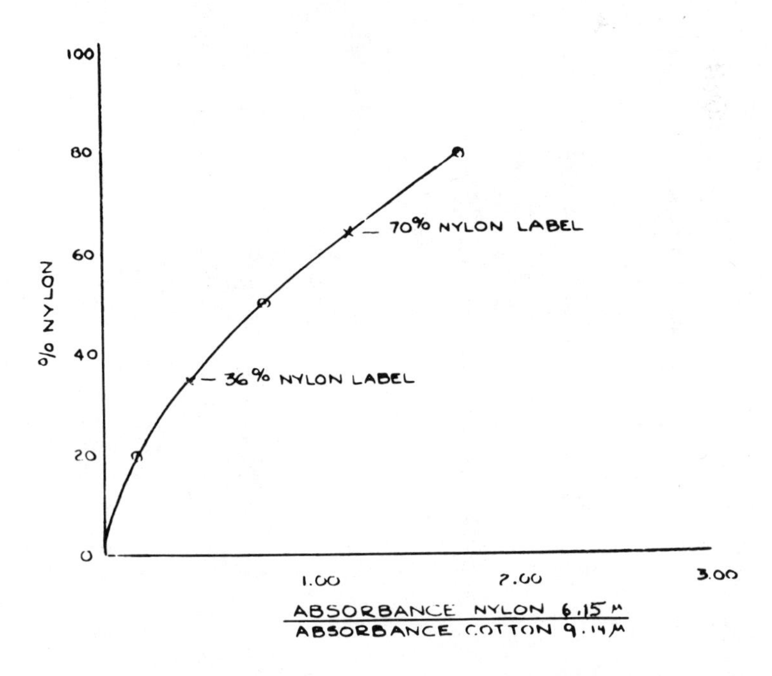

Fig. 4. Calibration curve for a cotton–nylon mixture.

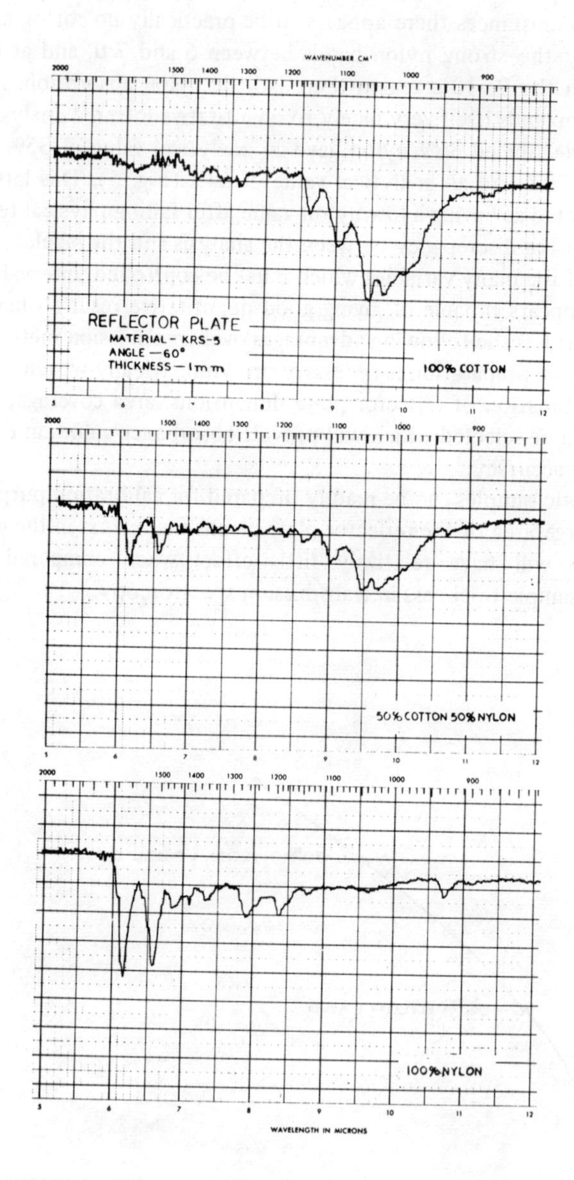

Fig. 5. Spectra of fibers at 60° angle of incidence.

The method should be applicable to other multiple-fiber systems than the one examined here.

REFERENCES

1. P. A. Wilks, Jr. and T. Hirschfeld, *App. Spectry. Rev.* **1** (1), 99–130 (1967).
2. P. A. Wilks, Jr. and M. R. Iszard, The Identification of Fibers and Fabrics by Internal Reflection Spectroscopy, Paper presented at 15th Mid-America Spectroscopy Symposium, Chicago, Illinois, June 2–5, 1964. (Copies available from Wilks Scientific Corporation.)
3. P. A. Wilks, Jr., *J. Appl. Spectry.* **22** (6), 782-784 (1968).

Index